An Introduction to Modern Experimental Organic Chemistry

Royston M. Roberts
John C. Gilbert
Lynn B. Rodewald
Alan S. Wingrove

The University of Texas at Austin

HOLT, RINEHART AND WINSTON, INC.

New York   Chicago   San Francisco   Atlanta
Dallas   Montreal   Toronto   London   Sydney

Preface

Since organic chemistry is fundamentally an experimental science, this book is written with the view that laboratory experience should be an integral part of the first course in organic chemistry. In recent years the trend in teaching the elementary course has been to incorporate increasingly more theory into the lecture material, particularly in the first semester, thereby increasing the difficulty in correlating laboratory experiments with the lecture part of the course. In some schools this problem has been "solved" by postponing organic laboratory until after one semester of lecture, or even later. However, this solution is unsatisfactory, not only because of the inherent scheduling difficulties, but also because it begs the question.

A better approach is to design an up-to-date laboratory course that can be correlated with the most modern lecture texts. This book, having been designed to provide the basis for such a laboratory course, emphasizes both the theoretical and practical aspects of a variety of organic reactions. The experiments on the halogenation of hydrocarbons, for example, are illustrative of the mechanistic and the potential synthetic utility of free-radical substitution. The experiments involving electrophilic aromatic substitution are also designed to promote an understanding of the mechanistic and synthetic aspects of this important class of reactions.

Another factor that must be considered in planning a modern laboratory text is the application of recently developed spectroscopic (nmr and ir) and chromatographic (tlc and glpc) techniques to experimental organic chemistry. Since it is important that the student become acquainted with these techniques as soon as possible, we have discussed the fundamentals of the methods in early chapters, and have applied them in experiments throughout the book.

It is possible to give students experience with the modern instrumental methods suggested in the experiments even though access to the necessary instruments may be limited. For example, the instructor can demonstrate the use of the instrument(s) to the students and then obtain typical analyses to post in the laboratories. Even if no instrumentation is available, students can still gain valuable experience by studying the illustrative spectra and chromatograms presented in the text.

A third important factor is to provide the student with useful experimental challenges. Toward this end, we have included procedures involving the *in situ* generation of volatile or highly reactive reagents, the handling of substances that are sensitive to moisture or air, and the safe removal of noxious gases formed in the course of a reaction. Several multi-step syntheses are described to afford the student the opportunity to experience the stimulation and satisfaction of successfully completing a series of reactions.

We have endeavored to maintain a good balance between the new and the old by retaining many classical procedures and synthetic experiments of proven value. For example, although we believe that an introduction to identification of organic compounds by spectroscopic methods must now be given in the first organic laboratory course, we have provided an adequate outline of the classical "wet" scheme of qualitative organic analysis, because of its pedagogical value and because students enjoy it— and thus learn a great deal of practical organic chemistry in the process.

The experimental procedures in this book are preceded by a rather complete discussion both of the theory upon which the experiment is based and of the important practical aspects—the how's and why's—involved in the experiment. In general, experimental directions are detailed and specific in the earlier chapters of the book but become less so as the student gains experience in the techniques of organic chemistry; in this way we hope to avoid the "cookbook" approach while at the same time providing sufficient information so that relatively little reference to other sources is required for a *complete* understanding of the experiments. In short, we have attempted to produce a laboratory *textbook* rather than a laboratory manual.

Acknowledgment is made to Sadtler Research Laboratories, Inc. for some of the spectra incorporated in this book. We are indebted to Professor R. Pettit and Dr. James C. Barborak for the procedure for tropylium iodide. We thank Professors Arthur Fry, Tom S. Ellis, James G. Traynham, and Frank Cartledge who read part or all of the text and tested some of the experiments. Special thanks are due to Professor Stanley H. Pine who provided us with a thorough critique and class testing of our book. Grateful acknowledgment is made to the Research and Development Center for College Instruction of Science and Mathematics of the University of Texas at Austin for financial support.

R.M.R./J.C.G./L.B.R./A.S.W.

Contents

INTRODUCTION   1

1   PHYSICAL CONSTANTS OF ORGANIC COMPOUNDS   7

1.1   Melting Point of a Pure Substance   7
1.2   Effect of Impurities on Melting Points; Mixtures   8
1.3   Other Kinds of Melting Behavior of Mixtures   11
1.4   Micro Melting Point Methods   11
1.5   Boiling Points   14
1.6   Index of Refraction   17
1.7   Density   18

2   SEPARATION AND PURIFICATION OF ORGANIC COMPOUNDS
I. DISTILLATION, RECRYSTALLIZATION, AND SUBLIMATION   20

2.1   Simple Distillation   20
2.2   Fractional Distillation   22
2.3   Fractional Distillation Columns   24
2.4   Fractional Distillation of Nonideal Solutions   26
2.5   Distillation under Reduced Pressure   28
2.6   Steam Distillation   29
2.7   Recrystallization   35
2.8   Sublimation   41

3   SEPARATION AND PURIFICATION OF ORGANIC COMPOUNDS
II. PHASE DISTRIBUTION:
EXTRACTION AND CHROMATOGRAPHY   44

3.1   Extraction   44
3.2   Technique of Simple Extraction   48
3.3   Chromatography   51
3.4   Gas Chromatography   52
3.5   Column Chromatography   57
3.6   Thin-Layer Chromatography   62
3.7   Paper Chromatography   65

4 SPECTROSCOPIC METHODS OF IDENTIFICATION
AND STRUCTURE PROOF   67

    4.1   Infrared Spectroscopy   67
    4.2   Nuclear Magnetic Resonance Spectroscopy   72

5 ALKANES. FREE RADICAL HALOGENATION   82

    5.1   Chlorination by Means of Sulfuryl Chloride   82
    5.2   Bromination. Relative Ease of Substitution of Hydrogens
        in Different Environments   90
    5.3   Triphenylmethyl Bromide, Triphenylmethyl Radical,
        Hexaphenylethane and Triphenylmethyl Peroxide   93

6 ALKENES. PREPARATION AND REACTIONS   98

    6.1   Generation of Alkenes by Elimination Reactions   98
    6.2   Addition Reactions of Alkenes   111

7 ELECTROPHILIC AROMATIC SUBSTITUTION   119

    7.1   Introduction to Electrophilic Aromatic Substitution   119
    7.2   Friedel-Crafts Alkylation of Benzene with Butyl Chlorides   120
    7.3   Relative Rates of Electrophilic Aromatic Substitution   128

8 ALKYNES. HYDRATION   136

9 THE DIELS-ALDER REACTION   144

10 KINETIC AND EQUILIBRIUM CONTROL OF A REACTION   152

11 ELIMINATION AND SUBSTITUTION REACTIONS   161

    11.1   Base-Induced Elimination Reactions   161
    11.2   Nucleophilic Substitution at Saturated Carbon   170
    11.3   Chemical Kinetics: Evidence for Nucleophilic Substitution
        Mechanisms   178

12 CARBENES AND ARYNES: HIGHLY REACTIVE INTERMEDIATES   188

    12.1   Carbenes   189
    12.2   Arynes   196

13 ORGANOMETALLIC CHEMISTRY   202

    13.1   The Grignard Reagent: Its Preparation and Reactions   202
    13.2   The Organolithium Reagent   211

14 OXIDATION REACTIONS OF ALCOHOLS
AND CARBONYL COMPOUNDS   214

    14.1   Preparation of Aldehydes and Ketones by Oxidation of
        Alcohols   214
    14.2   Classification of Alcohols and Carbonyl Compounds by
        Means of Chromic Acid   221
    14.3   Base-Catalyzed Oxidation-Reduction of Aldehydes; the
        Cannizzaro Reaction   223

15 SOME TYPICAL REACTIONS OF CARBONYL COMPOUNDS 226

15.1   Reactions Involving Nucleophilic Addition to the Carbonyl Group 226
15.2   Reactions of Carbonyl Compounds Involving the α- and β-Carbon Atoms 242
15.3   Identification of an Alcohol or Carbonyl Compound by Means of Three Reagents 249

16 CARBOXYLIC ACIDS AND THEIR DERIVATIVES 250

16.1   Carboxylic Acids 251
16.2   Carboxylic Acid Esters 252
16.3   Carboxylic Acid Halides (Acyl Halides) and Anhydrides 254
16.4   Carboxylic Acid Amides and Nitriles 256

17 MULTI-STEP ORGANIC SYNTHESES 265

17.1   Polystyrene 265
17.2   The Synthesis of Sulfathiazole 275
17.3   The Synthesis of 1-Bromo-3-chloro-5-iodobenzene 292

18 NONBENZENOID AROMATIC COMPOUNDS 304

18.1   Aromaticity 304
18.2   Ferrocene 307
18.3   Tropylium Iodile 308

19 OPTICAL ACTIVITY. CARBOHYDRATES 315

19.1   Introduction 315
19.2   Mutarotation of D-Glucose 319
19.3   The Hydrolysis of Sucrose 323

20 IDENTIFICATION OF ORGANIC COMPOUNDS 328

20.1   Separation of Mixtures of Organic Compounds 328
20.2   Classical Qualitative Organic Analysis 329
20.3   Modern Spectroscopic Methods of Analysis 353
20.4   Tables of Derivatives 355

APPENDIX I   Drying Agents. Desiccants 367

APPENDIX II   The Laboratory Notebook 371

APPENDIX III   List of NMR and IR Spectra 376

INDEX   379

Experimental Procedures

CHAPTER 1

1.4   Melting Points
A. Calibration of thermometer   12
B. Determination of capillary melting points   12
C. Unknown   14
1.5   Boiling Points
Micro boiling points   16

CHAPTER 2

2.1, 2.2   Distillation
Simple and fractional distillation   21
Alternate fractional distillation experiment   25
2.7   Recrystallization
A. General discussion—Selecting a solvent; solution; hot
filtration; crystallization; filtration; drying the
crystals   36
B. Experiments—Determination of solubilities; recrystallization
of an impure solid   39

CHAPTER 3

3.2   Simple Extraction
Selective liquid-liquid extraction   48
3.5   Column Chromatography
A. Preparation of column   61
B. Separation of *syn*- and *anti*-azobenzenes   61
3.6   Thin-Layer Chromatography
A. Separation of *syn*- and *anti*-azobenzenes   63
B. Separation of green leaf pigments   64
3.7   Paper Chromatography
A. Analysis of inks   65
B. Alternative experiment   66

CHAPTER 5

5.1 Chlorination of Hydrocarbons
   A. Cyclohexane   84
   B. *n*-Heptane   86
   C. 1-Chlorobutane   86
5.2 Bromination of Hydrocarbons
   Relative rates of bromination of various hydrocarbons   90
5.3 Triphenylmethyl Bromide
   A. Preparation of triphenylmethyl bromide   95
   B. Preparation and reactions of triphenylmethyl radicals   95

CHAPTER 6

6.1 Alkenes by Elimination Reactions
   A. Dehydration of 4-methyl-2-pentanol   103
   B. Dehydration of cyclohexanol   103
6.2 Addition Reactions of Alkenes
   A. Geometric isomerization of dimethyl maleate   114
   B. Hydrogenation of 4-cyclohexene-*cis*-1,2-dicarboxylic acid   114

CHAPTER 7

7.2 Friedel-Crafts Alkylation of Benzene
   A. *n*-Butyl chloride at room temperature   124
   B. *sec*-Butyl chloride at room temperature   125
   C. *sec*-Butyl chloride at 80°   125
   D. Isobutyl chloride at room temperature   125
7.3 Rates of Aromatic Substitution
   A. Qualitative rate comparisons   132
   B. Semiquantitative rate measurements   133
   C. Quantitative rate measurements   134

CHAPTER 8

Alkynes. Hydration
   A. Qualitative tests for the triple bond—Reaction with bromine
      in carbon tetrachloride; Baeyer test for unsaturation;
      reaction with silver ammonia complex; formation of
      silver acetylide and its decomposition   139
   B. Preparation of 3-methyl-3-hydroxy-2-butanone; the
      hydration of 2-methyl-3-butyn-2-ol   139
   C. Identification of the product; spectroscopic methods   140
   D. Identification of the product; semicarbazone formation   140

CHAPTER 9

Dienes. The Diels-Alder Reaction
   A. Preparation of *endo*-norbornene-*cis*-5,6-dicarboxylic
      anhydride   145
   B. Reaction of cyclopentadiene with *p*-benzquinone   146
   C. Reaction of 1,3-butadiene and maleic anhydride   147

CHAPTER 10

    Kinetic and Equilibrium Control
    Competitive reaction of semicarbazide with cyclohexanone and
      furfural. Effect of different pH's on the reaction by use
      of various buffers 157

CHAPTER 11

    11.1 Base-Induced Elimination Reactions
      A. Elimination with alcoholic potassium hydroxide 163
      B. Elimination with sodium methoxide 164
      C. Elimination with sodium ethoxide 166
      D. Elimination with sodium isopropoxide 166
      E. Elimination with potassium *t*-butoxide 166
    11.2 Nucleophilic Substitution
      A. Preparation *n*-butyl bromide 172
      B. Preparation of *t*-butyl chloride 176
      C. Preparation of *t*-pentyl chloride 176
    11.3 Chemical Kinetics
      A. General kinetic procedure 183
      B. Special kinetic procedures 185
      C. Treatment of data 185

CHAPTER 12

    12.1 Carbenes
    Reaction of dichlorocarbene with indene 192
    12.2 Arynes
    Preparation of benzenediazonium-2-carboxylate
      hydrochloride and its reaction with anthracene
      to produce tryptycene 198

CHAPTER 13

    13.1 The Grignard Reagent
    Phenylmagnesium bromide 206
    Triphenylmethanol 207
    Benzoic acid 208

CHAPTER 14

    14.1 Preparation of Aldehydes and Ketones
      A. Preparation of 2-methylpropanal 217
      B. Preparation of cyclohexanone 218
      C. Oxidation of cyclohexanone to adipic acid 218
    14.2 Classification by Means of Chromic Acid
    Use of chromic acid in acetone for qualitative analysis 222
    14.3 Base-Catalyzed Oxidation-Reduction
    The Cannizzaro reaction of benzaldehyde 223

CHAPTER 15

    15.1 Nucleophilic Addition to the Carbonyl Group
      A. Addition of derivatives of ammonia to carbonyl
        compounds 231
      B. Wittig syntheses 233

15.2 Reactions of Carbonyl Compounds Involving the α- and β-Carbon Atoms
   A. The haloform reaction   242
   B. The aldol condensation   243

CHAPTER 16
   16   Carboxylic Acids and their Derivatives
   A. Oxidation of an aromatic side chain; preparation of
      benzoic acid from ethylbenzene   257
   B. Reactions of benzoic acid   258
   C. Reactions of benzoyl chloride   259
   D. Preparation of benzonitrile; dehydration of an amide   260

CHAPTER 17
   17.1 Polystyrene
      Acetophenone   266
      1-Phenylethanol   267
      Styrene   268
      Polystyrene   272
   17.2 Sulfathiazole
      Nitrobenzene   284
      Aniline   284
      Acetanilide   285
      para-Acetamidobenzenesulfonyl chloride   285
      Sulfanilamide   286
      2-Aminothiazole   287
      Sulfathiazole   288
   17.3 1-Bromo-3-chloro-5-iodobenzene
      4-Bromoacetanilide   293
      2-Chloro-4-bromoacetanilide   294
      2-Chloro-4-bromoaniline   295
      2-Chloro-4-bromo-6-iodoaniline   295
      1-Bromo-3-chloro-5-iodobenzene   296

CHAPTER 18
   18   Nonbenzenoid Aromatic Compounds
      Acetylferrocene and other derivatives of ferrocene   309
      Tropylium fluoborate and iodide   311

CHAPTER 19
   19.2 Mutarotation of D-glucose
      Preparation of α-D-Glucose   321
      Rate of Mutarotation   321
   19.3 The Hydrolysis of Sucrose
      A. Hydrolysis   326
      B. Specific Rotation   326
      C. Formation of Osazones   326

CHAPTER 20
   20.2 Classical Qualitative Organic Analysis
      Preliminary Examination   329

Physical Constants 330
Elementary Analysis 330
Solubility Tests 331
Classification Tests 333
Preparation of Derivatives 333

Introduction

The laboratory part of an introductory course in organic chemistry is complementary to the lecture part; this is where you learn firsthand that the compounds and reactions described in lecture are not merely abstract notations. In addition, many of the theoretical concepts discussed in the lectures are amenable to experimentation at even an introductory level, and you will find that collecting and interpreting your own data will add reality to the theoretical framework of the subject.

Probably the most important purpose of the laboratory, however, is the introduction it provides to the experimental techniques of the practicing organic chemist. Learning to handle organic chemicals and manipulate apparatus is obviously a necessary part of a chemist's education. Less obvious, but equally important, is the learning of the scientific approach to laboratory work, which results in what is called good experimental technique. Some suggestions to aid in developing good technique are given in the following paragraphs.

Never begin any experiment unless you understand the overall purpose of the experiment and the reasons for each operation involved. This requires *studying* (not just reading) an experiment *before coming to the laboratory*. For this reason laboratory experiments are always assigned several days in advance, except under unusual circumstances. You will find that not only will your performance in laboratory be better if you are well-prepared, but also you will benefit much more from the experiments, in knowledge as well as in grades.

Neatness is an important part of good technique. Carelessness in handling chemicals may not only lead to poor results but is often unsafe. Similarly, careless assembly of apparatus is not only aesthetically displeasing but is also dangerous.

The procedures given in this book are meant to be followed closely. There is usually a reason why each operation is to be carried out exactly as described, although the reason may not, at first, be obvious to the beginning student. In the earlier chapters of the book, the operations required in the experimental procedures are described rather specifically, but in later chapters, after experimental techniques have become more familiar to you, the directions are less specific. In this way we hope to avoid your feeling that you are following "cookbook" directions. As you gain more practice in the techniques of the organic laboratory, you may even formulate alternative proce-

1

dures for performing a reaction, purifying a product, etc. However, for the sake of safety it will be wise for you to check any original plans with your instructor. In more advanced courses you may expect to use your chemical initiative more freely.

Stopping points. In some of the experimental procedures, stars (★) have been inserted to indicate points at which the experiment may be interrupted without affecting the results unfavorably. This has been done to help instructors and students in planning to make the best use of laboratory time. For example, some experiments may be started toward the end of a laboratory period and carried through to a starred point (★) and then discontinued until the next laboratory period, the reaction mixture being safely stored in the desk in the interim. In cases where it is already obvious from the text that the experiment may or should be interrupted, stars are not inserted.

One of the most valuable characteristics that you can develop as an experimentalist is that of being *observant* during all stages of an experiment. Was there a fleeting color change when a drop of reagent was added to the solution? Was a precipitate formed? Is the reaction exothermic? Such observations may seem insignificant at the time, but may later prove to be vital to the correct interpretation of an experimental result. All such observations should be *recorded*, which brings us to the subject of notebooks.

A permanent record of all laboratory work should be kept in a *bound notebook*, approximately 8 × 10 inches, rather than in a loose-leaf or spiral notebook. Pages of a bound book are less likely to tear out accidentally and be lost. Always write in ink. It will be easier to keep the notes neat and legible if the pages are lined horizontally. (Although notebooks with vertical as well as horizontal lines are sometimes recommended for laboratory notes, particularly in physical chemistry, they are less desirable for organic chemistry laboratory notes. In the special cases where graphs are necessary, they may be pasted in.) Use the first page as a title page and reserve two additional pages to be filled in as a table of contents. The pages should be numbered throughout the notebook.

Your instructor will probably have specific directions regarding the format of your notebook which you should follow. These may include a preliminary "write-up" giving the title of the experiment, complete equations for the main and side reactions, calculations of the theoretical yield in synthetic experiments, etc.[1] You may be allowed some flexibility in style, but proper recording of experimental results does not allow great literary license.

The overriding requirements of good experimental description are accuracy and completeness of observation and recording. The results should usually be summarized and conclusions should be drawn from each experiment. If the results are obviously not those expected, an explanation should be given.

Safety in the Laboratory

The chemical laboratory is potentially a dangerous place; it contains flammable liquids, fragile glassware, and corrosive and poisonous chemicals. However, if proper precautions are taken and safe procedure is followed, it is no more dangerous than an ordinary kitchen or bathroom. The accident rate in the chemical industry is actually lower than in many other types of industries. Unfortunately, however, there is usually less attention paid to safety measures in school laboratories than in industrial labora-

---

[1] See Appendix II for a suggested notebook format.

tories. Your own safety will largely depend on your knowledge of possible dangers and your adherence to proper procedures; some of these dangers and the ways to avoid them will be outlined here. Other specific precautions will be pointed out in the experimental directions at the appropriate places.

The most common types of danger in the organic laboratory may be divided into three categories: (1) fire and explosion, (2) chemicals, (3) glassware. Your eyes are particularly susceptible to injury from all of these dangers. Therefore, *Safety glasses must be worn at all times in the laboratory*.[2]

Fire and Explosion Precautions

1. Avoid using flames in the laboratory whenever possible.
2. If flames are used, observe the following precautions:

(a) Never heat a flammable liquid in an open container; *i.e.*, a flame should be used only when the container is protected by a condenser. Use a steam-bath or an electrical heating device if possible.

(b) When a flammable liquid is heated in a distillation or at reflux make certain that all connections are tight and strain-free. When a very volatile liquid is being heated, it is good practice to connect a rubber tube to the atmospheric opening of the assembly and run the tube over the edge of the bench, away from your burner (and your neighbor's).

(c) Never pour flammable liquids from one container into another within several feet of a flame. Do not pour flammable liquids into the center trough of a laboratory bench, as the vapors may be carried near a flame farther down the bench.

(d) Never allow the flammable distillate from a condenser to drip free into a receiver several inches below it, particularly if it is near the flame of your neighbor; use an adapter to conduct the distillate into the receiver.

3. Never heat a closed, pressure-tight assembly of apparatus, even if it includes a condenser. The increase in pressure resulting from the heating may cause the assembly to blow apart or explode and, if a flame is being used, any flammable liquid present in the assembly may be ignited.

4. Whenever an exothermic reaction is being carried out, prepare a cold- or ice-water bath and have the reaction vessel positioned so that the bath may be placed around it if the reaction shows signs of getting out of control.

5. Familiarize yourself with the location of the nearest fire extinguisher and be sure you know how to use it. Your instructor will explain its operation. Also ascertain the location of any fire blankets, safety showers, or eye-wash fountains in the laboratory.

Chemical Precautions

1. Do not allow any chemicals to come in contact with your skin unnecessarily. Transfer solid materials by means of spatulas or other appropriate utensils. Should any chemical be spilled on the skin, wash the area immediately with copious amounts of soap and water. Inform your instructor if in doubt as to the toxicity of the chemical. Be particularly careful to avoid contact of open cuts or abrasions with poisons. Never

---

[2] Opinion is divided as to whether ordinary eye glasses afford satisfactory protection. Consult your laboratory instructor if you wish to wear only your own corrective glasses.

use organic solvents such as acetone or alcohol to wash an organic chemical from the skin; such solvents may actually increase the rate of absorption of the chemical into the skin. Always wash your hands at the end of the laboratory period.

2. Never taste any chemical unless specifically instructed to do so.

3. Avoid inhalation of fumes and vapors of chemicals and solvents, as much as possible. The laboratory should be well ventilated. When working with volatile materials in open systems, use a fume hood if one is available. If a noxious gas or vapor (*e.g.*, hydrogen chloride) is produced in a reaction, attach an efficient gas trap.

Although the odor of an organic compound is often a useful criterion of its type or even of its identity, discretion should be used in smelling any substance. Some compounds are extremely irritating, if not toxic. Hold any unknown substance about six inches from the nose and, using your hand, gently waft the vapors toward you.

4. Clean up any spilled chemical immediately; if you wait, it may cause damage and the cleanup may be more difficult.

Glassware: Precautions and Cleaning

The cardinal rule in handling laboratory glassware is *never apply undue pressure or strain to any piece of glassware.*

This rule applies to insertion of thermometers or glass tubes into rubber stoppers, rubber tubes, or corks. "If you have to force it, don't do it!" Make the hole larger or use a smaller piece of glass. Lubricate the rubber or cork with a little water or glycerol. Always grasp the glass part very close to the rubber or cork part when pushing the glass into it.

Hopefully, these directions will become less necessary, as more and more laboratories are making the transition to all ground glass-jointed apparatus. However, the cardinal rule stated above applies to this glassware as well; take care that strain does not develop because of carelessly positioned components. Strained glassware will break when heated, or simply upon standing, at times.

If ground glass-jointed glassware is available, it is important that the joints be properly lubricated so that they do not "freeze" and become difficult, if not impossible, to separate. Proper lubrication can be accomplished by putting a thin layer of grease on opposite sides of the male joint, mating the two joints and then rotating them together in order to cover the surfaces of the joints with a thin coating of grease. The quantity of lubricant used is important since too much grease may cause ultimate contamination of the reaction whereas too little will permit the joints to freeze.

Glassware is most easily cleaned *immediately* after use. Most chemical residues can be removed by washing the glassware with detergent and water, or with common organic solvents like benzene and acetone. More stubborn residues may require the use of powerful chemical cleaning solutions such as chromic acid (*concentrated* sulfuric acid and chromic anhydride) or potassium hydroxide in ethanol. Gentle warming of the organic solvent or cleaning solution on a steam bath will generally hasten the cleaning process. Before you resort to the use of any cleaning agents other than detergent and water, consult your instructor for permission and directions concerning their safe handling.

Brown stains of manganese dioxide on glassware may sometimes be encountered; these can generally be removed by rinsing the apparatus with a 30% aqueous solution of sodium bisulfite.

It is good practice to wipe off any lubricant from ground glass-jointed glassware with a towel or tissue wetted with a solvent such as acetone or chloroform before washing with detergent and water. Otherwise the lubricant will stick to the brush used to scrub the glassware and be carried by the brush onto all surfaces it touches.

In Case of Accident

Fire

Your first consideration is to remove yourself from any danger, *not* to extinguish the fire. Notify your instructor immediately. To help prevent spread of the fire, remove any containers of flammable solvents from the immediate area and turn off any burners, if this is possible. For the most effective use of a fire extinguisher, direct its nozzle toward the base of the flames.

If your clothing is on fire, DO NOT RUN; rapid movement will only fan the flames. Roll on the floor to smother the fire and to help keep the flames away from your head. Your neighbors can help to extinguish the flames by using fire blankets, laboratory coats or other items that are immediately available. Do not hesitate to aid your neighbor if he is involved in such an emergency, since a few seconds delay may result in serious injury.

A laboratory safety shower, *if close by*, can be used to extinguish burning clothing, as can a carbon dioxide extinguisher if the flames are not near the head. A carbon tetrachloride extinguisher, however, should NOT be employed.

If burns are minor, apply a burn ointment. In case of serious burns, do not apply any ointment but seek professional medical treatment at once.

Chemical Burns

Areas of the skin with which corrosive chemicals have come in contact should be immediately washed thoroughly with soap and water. If the burns are minor, apply burn ointment; for treatment of more serious burns, see a physician.

Bromine burns can be particularly serious. These burns should first be washed with soap and water and then thoroughly soaked with 10% sodium thiosulfate solution for three hours. Apply cod liver oil ointment and a dressing; see a physician.

If chemicals, particularly corrosive or hot reagents, come in contact with the eyes, immediately flood the eyes with water *from the nearest outlet*. A specially designed eyewash fountain is useful if it is available in the laboratory. *Do not touch the eye.* The eyelid as well as the eyeball should be washed with water for several minutes. If the injury appears to be serious, see a physician as soon as possible.

Cuts

Minor cuts may be treated by ordinary first-aid procedures; seek professional medical attention for serious cuts. If severe bleeding indicates that an artery has been severed, apply a tourniquet just above the injury.

A person who is injured severely enough to require a doctor's treatment *should be accompanied* to the doctor's office or infirmary, even if he protests that he is all right. Persons in shock, particularly after suffering burns, are often more seriously injured than they appear to be.

Physical constants of organic compounds

Physical properties such as melting point (mp), boiling point (bp), index of refraction (*n*), and density (*d*) have long been utilized in identification and characterization of organic compounds. More than one compound may exhibit the same constant for one or two of these properties; however, it would be highly fortuitous if more than one compound exhibited the same constant for each of the properties. It should be seen then that a list of physical constants is a highly useful characterization of a substance. In addition, the observed melting point or boiling point may give information about the purity of the substance under consideration. Other properties such as color, odor, and crystal form are also useful.

In Chapter 4 we shall see how more recently developed spectroscopic techniques, based on molecular-physical properties, are extremely useful in identification of organic compounds.

1.1   Melting Point of a Pure Substance

The melting point of a substance is defined as the temperature at which the liquid and solid phases exist in equilibrium with one another without change in temperature. If heat is added to a mixture of the solid and liquid phases of a pure substance at the melting point, ideally no rise in temperature will occur until all the solid has been converted to liquid (melted); if heat is removed from such a mixture, the temperature will not drop until all the liquid has been converted to solid (frozen). Thus the melting point and freezing point of a pure substance are identical. The relationship between phase composition, total supplied heat, and temperature for a pure compound is shown in Figure 1.1. It should be understood that *heat* is being supplied to the compound at a constant rate, and thus that the elapsed time of heating is a cumulative measure of the supplied heat. At the lower temperature (below the melting point) the compound exists in the solid phase and the addition of heat causes the temperature of the solid to rise. As the melting point is reached, the first small amount of liquid appears; equilibrium is established between the solid and liquid phases. As heat continues to be added, the temperature does not change, the additional heat causing conversion of solid to liquid with, however, both phases remaining in equilibrium. As the last of the solid melts, the

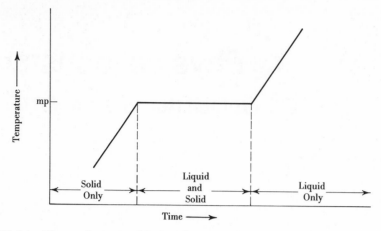

FIGURE 1.1 Phase changes with time and temperature.

subsequently supplied heat causes the temperature to rise linearly at a rate which de-pends on the rate of heat supply.

We may describe the interconversion of the liquid and solid phases in terms of their respective vapor pressures, which are directly related to the rates at which the molecules pass from one phase to the other. For simplicity we may consider only transfer of molecules between the solid and liquid phases. The term vapor pressure may then seem to be a misnomer, but the same "escaping tendency" which produces the measurable equilibrium vapor pressure is responsible for the direct transfer of molecules between the solid and liquid phases. Figure 1.2 represents these transfers between the three

FIGURE 1.2 Phase equilibria.

phases by means of reversible arrows. At the melting point, the vapor pressures of the solid and liquid phases are equal, and no *net* transfer of molecules from one phase to the other occurs unless there is a change in the heat content of the system.

1.2 Effect of Impurities on Melting Points; Mixtures

Consider a mixture of solid and liquid A at its melting temperature. If a small amount of a second pure material, B, is dissolved in the liquid A, solid A will begin to melt. This is because the addition of B lowers the vapor pressure of liquid A (Raoult's Law), and the vapor pressure of solid A remains the same; thus, the rate at which mole-cules of solid A pass into the liquid phase is greater than the rate of the reverse process.

If the temperature is kept constant by supplying the heat required for the melting process, all of solid A will melt; if no heat is added, the temperature will drop because heat energy is consumed by the melting solid. Since the vapor pressure of a solid decreases more rapidly than the vapor pressure of its solution as the temperature drops, they will become equal at a lower temperature, and equilibrium will be established again at that temperature. This lower temperature then represents the freezing and melting point of the mixture of A and B, that is, "impure A." The more of the impurity B that is added, the lower will be the melting point of the mixture.

Binary mixtures of most organic substances exhibit this kind of behavior; that is, the addition of a second substance progressively lowers the melting point of the first, as may be represented by the melting point-composition diagram of Figure 1.3. In this

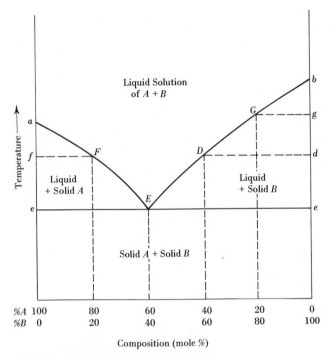

FIGURE 1.3   Melting point-composition diagram.

diagram, a represents the melting point of pure substance A and b the melting point of pure substance B. The point F represents the temperature (f) at which crystals of pure A are in equilibrium with a molten solution of B (20%) in A (80%). If a mixture of this composition is prepared and melted, and the temperature is then lowered to f, pure crystals of A will form in the presence of the melt. As additional heat is removed, A will continue to crystallize, reducing the percentage of A in the melt. The equilibrium temperature will correspondingly decrease because of the reduced vapor pressure of A in the molten solution. When the equilibrium temperature reaches e, at which time the melt has the composition indicated at E, then and *only then* does B, the impurity, also begin to crystallize. A and B will crystallize in the constant ratio 60% A to 40% B, and the melt composition will no longer change. During this stage, the melt of the specified composition is in equilibrium with both solid A and solid B. The temperature remains constant until the entire sample has crystallized and then again begins to fall as the solid cools. Point *E* is called the *eutectic point*, and it defines the composition at which *A* and *B* can cocrystallize (eutectic mixture) and the temperature (*e*) at which the eutectic mixture freezes (and melts).

Now consider the reverse process, the melting of a solid mixture of 80% *A* and 20% *B*, which is of more interest to the organic chemist. Heat is applied and the temperature rises. When it reaches *e*, *A* and *B* will melt together at a constant ratio (eutectic composition), and the temperature will remain constant (remember the eutectic melt is in equilibrium with both solid *A* and solid *B*). Eventually *B* will be entirely melted (as it is the minor component), leaving only solid *A* in equilibrium with a melt of eutectic composition. As more heat is applied, the remaining *A* will continue to melt, raising the percentage of *A* in the molten solution above the eutectic composition. Since the vapor pressure of *A* in the solution is thus increased, the temperature at which solid *A* is in equilibrium with the molten solution will rise, and the relationship between the equilibrium temperature and the composition of the molten solution will be represented by the curve *EF*. When the temperature reaches *f* the last of *A* melts. Thus this impure sample *A* exhibits a depressed melting "point" occurring over a relatively broad range (*e–f*). Notice that we have been considering the case of substance *A* with impurity *B*. If the composition of the solid had been to the right of point *E* in Figure 1.3 we would conversely speak of substance *B* with impurity *A*, and the rising temperature during the melting process would follow curve *ED* or *EG*, with melting ranges of *e–d* or *e–g*.

It may be noted that the actual melting *range* of a mixture containing 20% impurity may be broader than that of a mixture containing 40% impurity; for example, in Figure 1.3, compare the range *e–g* with *e–d*. However, in practice, particularly in the capillary tube method (see below), it is very difficult to observe the initial melting point. If the amount of impurity is small, the amount of liquid produced in the early stages (near the eutectic temperature) is very small. However, the temperatures at which the last crystals disappear (*d*, *g*) can be determined quite accurately, so that the mixture containing the smaller amount of impurity will have both a higher final melting point and a narrower (*observed*) range.

It should be noted that a sample whose composition is exactly that of the eutectic mixture will exhibit a *sharp* melting point at the eutectic temperature. Thus a eutectic mixture is sometimes mistaken for a pure compound. In such a case the sample may be identified by adding a small amount of either component (assuming one knows what the components are) and observing that the melting point rises.

The melting point depression produced by the introduction of an impurity into a pure compound may be used to advantage in identification of that compound. This important procedure is known as a *mixture-melting point*[1] determination. It is perhaps best presented in terms of an example. Assume that a compound melts at 133°, and you suspect it is either urea or *trans*-cinnamic acid (both melt at 133°). If the compound is mixed intimately with urea and the melting point of this mixture is found to be lower than either the pure compound or pure urea, then urea is acting as an impurity, and the compound cannot be urea. If the mixture-melting point is identical to that of the pure compound and of pure urea, then the compound is tentatively identified as urea. It should be noted that one can discard candidates by this method with far more certainty than one can elect them. Such a procedure is often used by the chemist when he encounters an identification problem and has on hand samples of likely candidates.

[1] Frequently referred to (ungrammatically) as a "mixed-melting point" determination.

1.3 Other Kinds of Melting Behavior of Mixtures

Although the melting·temperature behavior (eutectic formation) described above is typical of most binary mixtures, there are many exceptions to this general pattern. A second component does not always lower the melting temperature of an organic compound. Figure 1.4(a)–(c) illustrate some unusual melting point-composition diagrams.

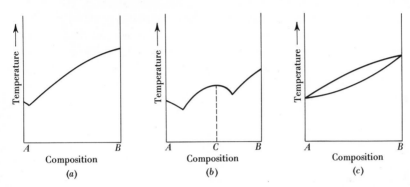

FIGURE 1.4 Unusual melting point-composition diagrams.

In Figure 1.4(a) the eutectic composition contains so little of component *B* that even a fairly small amount of *B* as impurity may actually raise the observed melting point of *A*. Figure 1.4(b) illustrates the formation of a compound (*C*) from *A* and *B*, and Figure 1.4(c) is representative of one type of solid solution formation. For further information consult Weissberger [2] or a physical chemistry text.

1.4 Micro Melting Point Methods

The determination of an exact melting point of an organic compound requires enough material so that equilibrium can be established between the liquid and solid phases and the temperature at the equilibrium state measured, usually by means of repeated cooling and heating cycles which show a plateau in a temperature-time plot. The amount of material required for such procedures is often more than the chemist has available, so that micro methods have been developed which are not so exact but which are convenient and require negligible amounts of sample.[3] The most commonly used method is the capillary tube melting point procedure which is described in detail below. The *microscopic hot-stage* [4] and simpler variations of it are also popular; the equipment is somewhat more expensive, but even smaller amounts of crystals are required. In all micro methods the melting point is actually determined as a melting

[2] E. L. Skau, J. C. Arthur, Jr., and H. Wakeham, in *Techniques of Organic Chemistry*, A. Weissberger, editor, 3rd Ed., Interscience Publishers, Inc., New York, 1959, Vol. 1, Part 1, Chapter 7, Pp. 287 *f*.

[3] It must be admitted that there is less emphasis on extremely precise melting point determinations since the development of spectroscopic methods for characterization of organic compounds (see Chapter 4).

[4] The Kofler hot-stage technique is described by E. E. Jelley, in *Techniques of Organic Chemistry*, A. Weissberger, editor, 3rd Ed., Interscience Publishers, Inc., New York, N. Y., 1959, Vol. 1, Part 2, Chapter 21, p. 1347.

range, encompassed by the temperature at which the sample is first observed to melt and that at which the melting process is complete. When properly carried out with a *pure* crystalline compound, a range of no more than 0.5 to 1.0° should be observed. However, few commercially obtained compounds exhibit such a narrow melting range.

A wide variety of types of heating devices have been used in connection with capillary tube melting point determinations, ranging from a simple beaker containing a high-boiling liquid heated by a burner and stirred manually to elaborate electrically heated and mechanically stirred devices. Heated metal blocks are also used in connection with capillary tubes. Descriptions of these devices may be found in Weissberger. In our procedure we describe the use of a Thiele tube (Figure 1.5), but much of the description applies equally well to the use of other types of heating devices for capillary tubes. The Thiele tube is shaped so that heat applied by a burner to the sidearm is distributed to all parts of the vessel by convection currents in the heating fluid and no additional stirring is required.

EXPERIMENTAL PROCEDURE

A. CALIBRATION OF THERMOMETER

Determine the 0 and 100°[5] readings on your thermometer in the following ways.

Place about 50 ml of a mixture of finely chopped ice and water in a 100- or 150-ml beaker and stir it briskly with your thermometer (*Caution:* the bulb is made of thin glass; do not hit it against the beaker), taking care that the mercury bulb is entirely immersed in the ice-water mixture. Record the temperature reading when it becomes constant.

Set up a simple distillation apparatus (see Figure 2.2) and place about 50 ml of distilled water in a 100-ml flask. Distil until the temperature becomes constant and record the reading. Be sure that the bulb of the thermometer is positioned correctly; the top of the mercury bulb should be about even with the sidearm of the flask or still head. The rate of distillation should be vigorous enough to insure equilibrium between liquid and vapor but not so fast as to produce superheating. Consult your instructor if in doubt.

Make a permanent record in your notebook of any corrections at the 0 and 100° readings and apply these in all future temperature measurements.

B. DETERMINATION OF CAPILLARY MELTING POINTS

Place the sample in a closed-end capillary tube of the commercial variety or one prepared from a piece of 10-mm soft glass tubing by heating and drawing to an outside diameter of about 1–1.5 mm. (If you are to make your own capillaries, your instructor will demonstrate the procedure.) The easiest way to fill the capillary is to place a bit of the powdered sample on a small watch glass or the bottom of an inverted beaker and tap the open end of the capillary into the solid a few times. The solid may be made to "filter" down to the closed end of the tube by inverting the tube and scratching it gently with a file, or by tapping the closed end of the tube briskly on a solid surface. The solid should be tightly packed in the tube, and this can best be

---

[5] In laboratories at altitudes of 1000 ft or more above sea level, a correction for the variation of boiling point with pressure (*Cf.* p. 15) must be made. The boiling point of water is lowered approximately 1° by each 1000-ft increment in altitude above sea level, owing to the lowering of the atmospheric pressure.

accomplished by finally dropping the capillary through a larger piece of glass tubing about two feet in length onto a hard surface. The size of the sample should be such as to fill the capillary to a depth of 2–3 mm after the compacting process. It should not be larger. Attach the capillary to a thermometer by means of a small rubber band (conveniently obtained as a slice of ordinary rubber tubing). The sample itself should be directly adjacent to the bulb of the thermometer. The rubber band should be positioned such that even at 200° it will remain above the level of the heating fluid (see Figure 1.5). This accomplished, the thermometer is placed into the heating vessel and supported by

FIGURE 1.5   Thiele melting point apparatus.

means of a bored cork cut away on one side so as to make visible the thermometer markings in that vicinity. This cut also serves the purpose of making the apparatus an open system. *One should never heat a closed system.* By application of heat from a small Bunsen burner (microburner) raise the temperature of the heating fluid slowly (about 2° per minute). Note the temperature at which melting is first observed and the temperature at which the last of the solid melts and record these as the melting range of the solid. As one can obviously spend a great deal of time approaching the melting point of a high-melting solid this way, it is usually convenient to prepare two samples of the solid under consideration and determine the approximate melting point of the first by heating rapidly. Then let the heating fluid cool to 10–15° below this approximate point and insert the second tube and reheat slowly to the melting point.

The observed melting point is dependent on a number of factors: sample size, state of subdivision of the sample, and heating rate, as well as purity and identity of the sample. The first three cause the observed melting point to differ from the actual melting point due to the time lag in heat transfer from heating fluid to sample and conduction within the sample. Also, if heating is too fast, the thermometer reading will lag behind the actual temperature of the heating fluid.

Following the prescribed procedure, determine the melting point (or range) of the following:

1. Pure urea;
2. Pure *trans*-cinnamic acid;
3. Mixtures of the above two compounds in the proportions given below:

	urea	*trans*-cinnamic acid
mixture (a)	4	1
mixture (b)	2	1
mixture (c)	1	1
mixture (d)	1	2
mixture (e)	1	4

It is not necessary to weigh out each sample individually. Place a small amount (about 0.5 to 1.0 g) of each pure compound on watch glasses or smooth paper. Subdivide them with a spatula into small piles, estimating the proper ratios, mix them intimately with the spatula, and introduce them into capillary tubes.

Make a plot of melting temperatures (vertical) *versus* composition (horizontal); in the case of the mixtures, record the temperature of disappearance of the last bit of solid. Estimate the eutectic temperature and composition.

C. Unknown

Determine the melting point of an unknown supplied by your instructor and identify it by consulting a list of possibilities and their melting points which will be provided.

1.5   Boiling Points

The molecules of a liquid are in constant motion. When at the surface, some of these molecules are able to escape into the space above the liquid. Consider a liquid contained in a closed *evacuated* vessel. The number of molecules in the gas phase can increase until the rate at which molecules reenter the liquid becomes equal to the rate of their escape; the rate of reentry is proportional to the number of molecules in the gas phase. At this point no further *net* change is observed in the system, and it is said to be in a state of dynamic equilibrium.

The molecules in the gas phase are in rapid motion and are continually colliding with the walls of the vessel, resulting in the exertion of pressure against the walls. The magnitude of this pressure at a given temperature is called the *equilibrium vapor pressure* of the particular liquid substance at that temperature. This vapor pressure is temperature dependent as shown in Figure 1.6. This dependence is easily understood in terms of the escaping tendency of molecules from the liquid. As the temperature is increased, the average kinetic energy of the molecules increases, facilitating their escape into the gas phase. The rate of reentry also increases, and equilibrium is soon established at the higher temperature. However, the number of molecules in the gas phase is now larger than at the lower temperature, so the vapor pressure is greater.

Now consider a liquid sample at a given temperature, placed in an open container so that molecules of the vapor phase above the liquid may escape its confines. The vapor above such a sample is composed of molecules of air as well as of the sample. The *total* (external) pressure above the liquid is, according to Dalton's Law of Partial

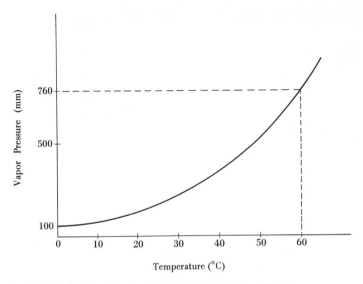

FIGURE 1.6 Dependence of vapor pressure on temperature.

Pressures, equal to the sum of the partial pressures of the sample and of the air [see equation (1)]. The partial pressure of the sample is equal to its *equilibrium vapor pres-*

$$P_{\text{Total}} = P_{\text{Sample}} + P_{\text{Air}} \tag{1}$$

sure at the given temperature. If the temperature is raised (thus increasing the equilibrium vapor pressure of the sample), the number of molecules from the sample in the space immediately above the liquid will increase, in effect displacing some of the air. At the higher temperature the partial pressure of the sample will be a larger percentage of the total pressure. The same trend will continue as the temperature is further increased, until the equilibrium vapor pressure becomes equal to the external pressure, at which point all of the air will have been effectively displaced from the vessel. Further evaporation will have the effect of displacing gaseous molecules of the sample. Consideration of these facts leads to the conclusion that the equilibrium vapor pressure of the sample has an upper limit dictated by the external pressure. At this point the rate of evaporation increases dramatically (as evidenced by the formation of bubbles in the liquid), and the process commonly known as boiling occurs. *The temperature at which the vapor pressure of a liquid just equals the external pressure* is defined as the *boiling point* of the liquid. As the observed boiling point is obviously directly dependent on the external pressure, it is necessary in reporting boiling points to state the external pressure, for example, "bp 152° (752 mm)." The *standard boiling point* is typically measured at atmospheric pressure, 760 mm of Hg. The standard boiling point of the substance in Figure 1.6 is shown by the dotted lines to be 60°.

Boiling points are useful for identification of liquids and some low-melting solids. Higher melting solids usually boil at temperatures too high to be measured conveniently. Many compounds, notably the higher boiling ones, decompose at their standard boiling points.

Impurities, both volatile and nonvolatile, affect boiling points. These effects will be discussed in Chapter 2.

EXPERIMENTAL PROCEDURE

Micro boiling points. A simple micro boiling-point apparatus may be constructed from two 1-mm capillary melting-point tubes and 4-mm soft glass tubing in the following way. Seal the ends of two capillary tubes in a flame and join the tubes at the seal. Make a clean cut about 3–4 mm from the joint as shown in Figure 1.7(a). Seal a piece of 4-mm soft glass tubing at one end and cut it to a length about 1 cm shorter than the prepared capillary ebullition tube.

FIGURE 1.7   Micro boiling-point apparatus.

Attach the 4-mm tube to a thermometer with a rubber ring, with the rubber ring near the top of the tube and the bottom of the tube even with the mercury bulb of the thermometer. Place about two drops of the liquid whose boiling point is desired in the bottom of the tube by means of a capillary pipet. Introduce the capillary ebullition tube as shown in Figure 1.7(b). If the liquid level of the sample is below the joint-seal of the capillary, add enough more sample to bring it above the seal.

Immerse the thermometer and attached tubes in a heating bath (Thiele tube or other melting-point apparatus), *taking care that the rubber ring is above the liquid level.* Raise the temperature rather quickly until a rapid and continuous stream of bubbles comes out of the capillary ebullition tube. (A decided change from the slow evolution of bubbles caused by thermal expansion of the trapped air will be seen when the boiling temperature of the liquid is reached.) Discontinue heating at this point. As the bath is allowed to cool down slowly, the rate of bubbling will decrease. At the moment the bubbling ceases entirely and the liquid begins to rise into the capillary, note the temperature of the thermometer. This is the boiling point of the liquid sample.

Remove the capillary ebullition tube and expel the liquid from the small end by gentle shaking. Replace it in the sample tube and repeat the heating and cooling procedure. With a little practice and care the observed boiling points may be reproduced to within 1 or 2°.

Your instructor will provide the liquids for which you should determine the boiling points.

1.6   Index of Refraction

The velocity of light propagation is not constant in all media. One direct and observable consequence of this variation is the bending of a beam of light as it passes through the interface of two different media. The light bends because its "speed" changes. (The effect is commonly observed; most people are aware that an object seen under water, a fish, for example, is not where it seems to be because of the refraction at the interface between the water and the air.) The angle of refraction (angle of bending) is dependent on the density of the media, the types of molecules present, the temperature of the media, and the wavelength of the light. If one medium such as air is chosen as a standard and all angles of refraction are measured at the same temperature with the same wavelength of light, then the measured angle is a property of the second medium at the interface and may be used as an identifying characteristic of the substance composing that medium.

The value that is actually reported as a physical constant of a liquid is not the angle of refraction, but the *index* of refraction (n), which is defined by the equation

$$n = \sin i / \sin p \tag{2}$$

where i is the angle which incident light (in air) makes with a perpendicular to the interface and p is the angle which the refracted light (in the liquid) makes with the perpendicular (see Figure 1.8). Since monochromatic light gives more precise values than white light, most indices of refraction are reported for the D line of sodium ($\lambda = 589.3$ mμ.). This is indicated by a subscript, and the temperature by a superscript, so that a typical notation is $n_D^{20} = 1.3330$. In the instrument most frequently used by organic chemists, the Abbe refractometer, white light is used as a source, but compensation by a prism system gives indices for the sodium D line.

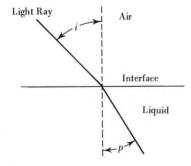

FIGURE 1.8   Refraction of light.

Impurities affect the index of refraction, and comparison of the observed index with the known index for the pure compound can be used as an indication of purity as well as for identification.

1.7   Density

The density of a substance is used less often than the physical constants discussed above for identification. Density is effectively a measure of the concentration of matter, and is generally measured and reported in units of grams per milliliter at 20°. The density of a liquid is rather easily measured, by making use of a small container, called a *pycnometer*, whose volume is accurately known. The container is weighed accurately and filled with the liquid whose density is to be determined. After bringing the pycnometer and its contents to 20° or some other desired temperature and readjusting the volume of the contents, if necessary, a second weighing will then yield the accurate weight of the substance whose volume is known, and the density may be calculated by dividing the weight in grams by the volume in milliliters.

Densities of solids are more difficult to determine. One way in which this measurement is often obtained is known as the *flotation method*. The density of the solid is determined as equal to the density of a liquid (measured as above) *in which* a single crystal of the solid will remain suspended, neither rising to the top nor settling to the bottom (Figure 1.9). Such a liquid may be obtained by first finding a suitable liquid *A*,

FIGURE 1.9   Density of solids by the flotation method.

in which the crystal rises to the top (liquid density is greater than the crystal density). A second liquid *B* is found in which the crystal sinks to the bottom. Of course only liquids in which the solid is insoluble are suitable. A mixture of *A* and *B* is then prepared in which the crystal remains suspended. The final measurement is made by determining the density of this liquid mixture, which is equal to the density of the solid.

EXERCISES

1. Describe errors in procedure which may cause an observed capillary melting point of a pure compound (a) to be *lower* than the correct mp; (b) *higher* than the correct mp; (c) *broad* in range (over several degrees).
2. Why is filter paper usually a poor material on which to powder a solid sample before introducing it into a capillary mp tube?
3. Criticize the following statements: (a) An impurity always lowers the mp of an organic compound. (b) A sharp mp for a crystalline organic substance always indicates a pure single compound. (c) If the addition of a sample of compound *A* to compound *X* does not lower the mp of *X*, *X* must be identical with *A*. (d) If the addition of a sample of compound *A* lowers the mp of compound *X*, *X* and *A* cannot be identical.

4. Give an explanation for the observed marked increase in rate of bubble evolution as the boiling point is reached in a micro bp determination.

5. Why is the micro bp taken as the temperature at the time the sample liquid begins to rise into the capillary tube? Justify this procedure in terms of the definition of boiling point.

REFERENCES

MELTING POINT

E. L. Skau, J. C. Arthur, Jr., and H. Wakeham, in *Techniques of Organic Chemistry*, A. Weissberger, editor, 3rd Ed., Interscience Publishers, Inc., New York, 1959, Vol. I, Part I, Chapter 7.

INDEX OF REFRACTION

N. Bauer, K. Fajans, and S. Z. Lewin, in *Techniques of Organic Chemistry*, A. Weissberger, editor, 3rd Ed., Interscience Publishers, Inc., New York, 1960, Vol. 1, Part II, Chapter 18.

DENSITY

N. Bauer and S. Z. Lewin, in *Techniques of Organic Chemistry*, A. Weissberger, editor, 3rd Ed., Interscience Publishers, Inc., New York, 1959, Vol. 1, Part I, Chapter 4.

2

Separation and purification of organic compounds

I. Distillation, Recrystallization and Sublimation

The practicing organic chemist seldom finds reactions that yield only the product (or products) he set out to obtain. This is because products of side reactions that proceed concurrently with the main reaction, varying amounts of unchanged starting materials, and solvents accompany the desired product in the reaction mixture. The chemist devotes a significant amount of effort to the separation of the desired products from such impurities. The purpose of this and the following chapter is to present an introduction to the theory and practice of the most important methods used by the modern chemist to separate and purify organic compounds.

The most commonly used method of purifying *liquids* is distillation. This process consists of vaporizing the liquid by heating and condensing the vapor in a separate vessel to yield a *distillate*. If the impurities present in the initial liquid are nonvolatile they will be left behind in the distillation residue, and a *simple distillation* will effect purification. If the impurities are volatile, a *fractional distillation* will be required.

2.1 Simple Distillation

When a nonvolatile impurity such as sugar, for example, is added to a pure liquid, it has the effect of reducing the vapor pressure of the liquid. This is because the presence of the nonvolatile component effectively lowers the concentration of the volatile constituent, *i.e.*, no longer are all of the molecules at the surface those of the volatile substance, and thus its ability to vaporize is lowered. The effect of a nonvolatile constituent on the vapor pressure of a mixture is shown in Figure 2.1. In this figure, curve 1 corresponds to the vapor pressure-temperature dependence for a pure liquid. The curve intersects the 760-mm line at 60°. Curve 2 is for the same liquid which now, however, contains a nonvolatile impurity. Note that the vapor pressure at any temperature is reduced by a constant amount by the presence of the impurity (Raoult's Law, see below). The temperature at which this curve intersects the 760-mm line is higher due to the lower vapor pressure, and the boiling point is higher, 65°.

A simple distillation may be carried out in an apparatus such as that in Figure 2.2. The thermometer is placed in the position shown for the purpose of determining the boiling temperature of the distillate (the material being collected in the receiver). In

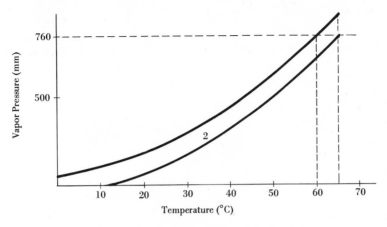

FIGURE 2.1 Dependence of vapor pressure on temperature.

the case of a pure liquid, the temperature shown on the thermometer, the *head temperature*, will be identical to the temperature of the boiling liquid in the distilling vessel, the *pot temperature*, if the liquid is not superheated. The head temperature, which thus corresponds to the boiling point of the liquid, will remain constant throughout the process of distillation.

When a nonvolatile impurity is present in a liquid being distilled, the head temperature will be the same as for the pure liquid, since the material condensing on the thermometer bulb is uncontaminated by the impurity. However, the pot temperature will be elevated owing to the decreased vapor pressure of the solution. The pot temperature will also rise during the course of the distillation since the concentration of impurity will increase steadily as the volatile component is distilled away, further lowering

FIGURE 2.2 Typical apparatus for simple distillation, atmospheric pressure or vacuum.

the vapor pressure of the liquid. However, the head temperature will remain constant, as in the case of a pure liquid.

The quantitative relation between vapor pressure and composition of homogeneous liquid mixtures (solutions) is known as Raoult's Law and may be expressed as in equation (1). P_R represents the partial pressure of component R and, at a given tem-

$$P_R = P_R^\circ N_R \tag{1}$$

perature, is equal to the vapor pressure of pure R at that temperature (P_R°) times the mole fraction of R in the mixture (N_R). The mole fraction of R is defined as the fraction of *all* molecules present that are molecules of R; it is obtained by dividing the number of moles of R in a mixture by the sum of the number of moles of all components [equation (2)].

$$N_R = \frac{n_R}{n_R + n_S + n_T + \ldots} \tag{2}$$

It should be noted that the partial vapor pressure of R above an ideal [1] solution containing R depends only on its mole fraction and is in no way dependent on the vapor pressures of the other components. If all components other than R are nonvolatile, the total vapor pressure of the mixture will be equal to the partial pressure of R, since the vapor pressure of nonvolatile compounds may be taken as zero. Thus the distillate from such a mixture will always be pure R. If, however, two or more of the components are volatile, then the total vapor pressure will equal the sum of the partial vapor pressures of each of these volatile components [Dalton's Law, equation (3), where R, S, and

$$P_{TOTAL} = P_R + P_S + P_T + \cdots \tag{3}$$

T refer to the volatile components only]. The process of distilling such a liquid mixture is significantly different from that of simple distillation discussed above, for here the distillate may contain each of the volatile components. Separation in this case will require fractional distillation.

2.2   Fractional Distillation

For simplicity we shall consider here only binary, ideal solutions, which contain two volatile components, R and S.[2] Ideal solutions are defined as those in which the interactions between *like* molecules are the same as those between *unlike* molecules. Only ideal solutions obey Raoult's Law strictly, but many organic solutions approximate the behavior of ideal solutions. Some important examples exhibiting significant deviation from ideality are discussed in Section 2.4.

As vapor pressure is a measure of the ease with which molecules escape the surface of a liquid, the number of molecules of component R in a given volume of the vapor above a liquid mixture of R and S is proportional to the partial vapor pressure of component R. The same is true for component S, and this leads to equation (4), where N_R' and N_S' are the mole fractions of R and S in the *vapor* phase. The partial vapor

---

[1] The definition of an ideal solution is given in the next paragraph.
[2] Solutions containing more than two volatile components are often encountered, but their behavior on distillation may be understood by extension of the principles given for binary systems.

$$\frac{N'_R}{N'_S} = \frac{P_R}{P_S} = \frac{P^\circ_R \, N_R}{P^\circ_S \, N_S} \qquad (4)$$

pressures, P_R and P_S, are determined by the composition of the liquid solution (Raoult's Law). Since the solution will boil when the sum of the partial vapor pressures of R and S is equal to the external pressure, the boiling temperature of the solution is determined by its composition.

The relationship between temperature and the composition of the liquid and vapor phases of ideal binary solutions may be best illustrated by a diagram such as that of Figure 2.3 for mixtures of benzene (bp 80°) and toluene (bp 111°). The lower curve

FIGURE 2.3 Boiling point-composition diagram for the system benzene-toluene.

gives the boiling points of all mixtures of these two compounds, and the upper curve, calculated using Raoult's Law, gives the composition of the vapor phase in equilibrium with the boiling liquid phase. For example, a liquid mixture with the composition 58 mole % benzene, 42 mole % toluene (A, Figure 2.3) boils at 90°, and the vapor phase in equilibrium with it has the composition 78 mole % benzene, 22 mole % toluene (B, Figure 2.3). Note that the *vapor phase is much richer in the more volatile component than the boiling liquid with which it is in equilibrium*, at any given temperature. For example, if component R is more volatile than component S, so that

$$\frac{P^\circ_R}{P^\circ_S} > 1$$

then from equation (4)

$$\frac{N'_R}{N'_S} > \frac{N_R}{N_S}$$

This is the basis of fractional distillation, as will be seen below.

If a mixture of composition *A* (Figure 2.3) is distilled, the first few drops of distillate (produced by *condensation* of some of the vapor phase) will have composition *B*, much richer in benzene than the original mixture from which it was distilled. Conversely, the liquid remaining in the distilling flask will be depleted in benzene and enriched in toluene, since more benzene than toluene was removed, and the boiling point will rise (from *A* to *A'*, for example). If distillation is continued, the boiling point of the mixture will continue to rise (from *A'* to *A''*, etc.) until eventually it approaches or reaches the boiling point of toluene. Meanwhile the composition of the distillate will change from *B* to *B'* to *B''*, etc., until at the end it will be nearly pure toluene.

Now let us return our attention to the first few drops of distillate which had the composition *B*. If these are collected separately and then redistilled, the boiling point will be the temperature at *C*, 85°. If only the first small fraction of distillate from *C* were collected, it would have the composition *D*, 90 mole % benzene, 10 mole % toluene. Repetition of this process could theoretically (but impractically) give a *very small amount* of pure benzene. Similarly, by collecting the *last* small fraction of each distillation and redistilling in the same stepwise manner, one could obtain a very small amount of pure toluene. By collecting larger amounts of initial and final distillation fractions, practical amounts of materials can be separated, but a proportionately larger number of individual distillations is required. Such a procedure is extremely tedious and time-consuming. Fortunately the repeated distillations can be accomplished virtually automatically in a single step by means of a *fractional distillation column*, the theory and use of which is described below.

2.3   Fractional Distillation Columns

There are a large number of types of fractional distillation columns, but all can be described in terms of a few fundamental characteristics. The column provides a vertical path, through which the vapor must pass from distilling flask to condenser; this path is significantly longer than in a simple distilling apparatus (Figure 2.4). As vapor from the distilling flask passes up the column, some of it condenses. If the lower part of the column is maintained at a higher temperature than the upper part of the column, as the condensate drains down the column it will be partially revaporized. The uncondensed vapor, together with that produced by revaporization of condensate, rises higher in the column and goes through a series of condensations and vaporizations. These amount to repeated distillations [3] and, as described above with reference to Figure 2.3, the vapor phase produced in each step becomes richer in the more volatile component. The condensates, which drain down the column, are at each level richer in the less-volatile component than the vapor with which they are in contact. Under ideal conditions, equilibrium becomes established throughout the column between the liquid and vapor phases, with the vapor phase at the top consisting almost entirely of the more volatile component and the liquid phase at the bottom being very rich in the less volatile component. The most important requirements for producing this state are (1) intimate and extensive contact between the liquid and vapor phases

---

[3] Each step along the route *A-B-C-D-E-F* of Figure 2.3 represents a single ideal distillation. Because one type of fractional distillation column, a "bubble-plate" column, was designed to bring about one such step for each "plate" it contains, the efficiency of any fractional distillation column is often described in terms of "theoretical plates."

FIGURE 2.4 Apparatus for fractional distillation, atmospheric pressure or vacuum.

in the column, (2) maintenance of the proper temperature gradient along the column, (3) sufficient length of the column, and (4) sufficient difference in the boiling points of the components of the liquid mixture.

If the first two requirements are well met, compounds having small differences in boiling points can be separated satisfactorily with a long column, since the length of the column required and the difference in boiling points of the components are reciprocally related. The most common way of providing the necessary contact between liquid and vapor phases is to fill the column with some inert material providing a large surface area, for example, glass, ceramic or metal pieces in a variety of shapes (helices, "saddles," woven mesh, etc.). A particularly effective way of achieving this liquid-vapor contact is to use a *spinning band* of metal or Teflon rotating at high speed within the column. This has the advantage of a low "hold-up" of material in the column, compared to "packed" columns with the same efficiency. (Hold-up refers to the amount of liquid and vapor required to maintain equilibrium conditions in the column.)

The necessary temperature gradient from pot to still head will, in most distillations, be established automatically by the condensing vapors if the rate of distillation is properly adjusted. Frequently, the temperature gradient can be maintained only by insulating the column with asbestos, glass wool, or, most effectively, with a silver-coated vacuum jacket. The insulation helps to reduce heat losses from the column to the atmosphere. For longer columns, particularly when poorer insulation is used, it is often necessary to supply additional heat by means of electrical resistance wire wrapped about the column.

A factor intimately related to the temperature gradient in the column is the rate of heating of the pot and the rate at which vapor is removed at the still head. If the heating is vigorous and the vapor is removed too rapidly, the whole column will heat up almost uniformly and there will be no fractionation (separation of components). On the other hand, if the pot is heated too vigorously and if vapor is removed too slowly at the top, the column will "flood" with returning condensate. Proper operation of a fractional distillation column requires judicious control of heating and "reflux ratio" —the ratio of the amount of vapor condensed and returned down the column to the amount taken off as distillate at the still head in the same time period. In general, the higher the reflux ratio, the more efficient the fractionation.

The most convenient way to measure and control the reflux ratio is to use a still head equipped for total reflux and variable "take-off." For example, with a still head of this type (see Figure 2.5), the number of drops of condensate falling from the tip of the

FIGURE 2.5   Total reflux-partial take-off still head.

condenser (P) in a given length of time gives a measure of the total reflux rate and the number of drops allowed to pass through the stopcock (S) and fall into the receiver at R in the same time period is a measure of the take-off rate. If one drop of distillate is taken off for every ten drops of total reflux, the reflux ratio is 9 to 1. For very precise fractionation, reflux ratios of 100 to 1 may sometimes be used with an efficient column.

2.4   Fractional Distillation of Nonideal Solutions

Although most homogeneous liquid mixtures behave as ideal solutions, there are many examples known in which the behavior is nonideal; in these solutions the unlike molecules are not indifferent to one another's presence. The resultant deviations from Raoult's Law occur in either of two directions: Some solutions display greater vapor pressures than expected and are said to exhibit *positive deviation;* others display lower vapor pressures than expected and are said to exhibit *negative deviation.*

In the case of positive deviation, the forces of attraction between the different molecules of the two components are weaker than those between the identical molecules of each component, with the result that, in a certain composition range, the combined vapor pressure of the two components is greater than the vapor pressure of the pure, more volatile component. Hence, mixtures in this composition range (between X and Y in Figure 2.6) have boiling temperatures *lower* than the boiling temperature

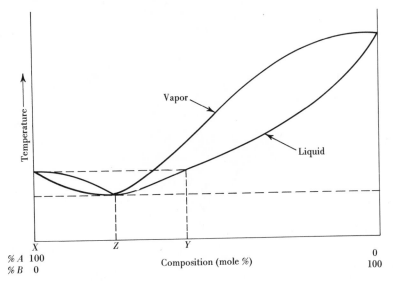

FIGURE 2.6   Minimum-boiling azeotrope.

of either pure component. The *minimum-boiling* mixture in this range (the composition of Z, Figure 2.6) must be considered as if it were a third component. It has a constant boiling point because the vapor in equilibrium with the liquid has the same composition as the liquid itself. Such a mixture is called a *minimum-boiling azeotropic mixture*, or *azeotrope.*[4] Fractional distillation of such mixtures will not yield both of the components in pure form but only the azeotrope and the component present in excess of the azeotropic composition. It is for this reason that pure ethanol cannot be obtained by fractional distillation of aqueous solutions containing less than 95.57% ethanol (the azeotropic composition), even though the boiling point of this azeotrope is only 0.15° lower than the boiling point of pure ethanol.[5]

In the case of negative deviation from Raoult's Law, the forces of attraction between

[4] This discussion is limited to binary mixtures. However, ternary mixtures are known that exhibit the same property, for example, water (7.4%), ethanol (18.5%) and benzene (74.1%).

[5] As optimum fractional distillations of aqueous solutions of ethanol containing less than 95.7% yield this azeotropic mixture, "95% ethyl alcohol" is the common form of commercial solvent ethanol. Pure or "absolute" ethanol can be obtained by chemical removal of the water or by a distillation procedure involving the use of a ternary mixture of ethanol, water and benzene.[4] This procedure is described in the references at the end of the chapter.

the different molecules of the two components are stronger than those between the identical molecules of each component, with the result that, in a certain composition range, the combined vapor pressure of the two components is less than the vapor pressure of the pure, less volatile component (see Figure 2.7). Hence, mixtures in this

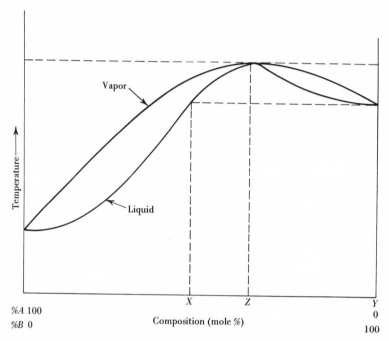

FIGURE 2.7   Maximum-boiling azeotrope.

composition range (between X and Y in Figure 2.7) boil at even *higher* temperatures than the boiling temperature of the pure, higher-boiling component. There is one particular composition corresponding to a *maximum-boiling azeotrope* (Z, Figure 2.7). Fractional distillation of mixtures of any composition other than that of the azeotrope will result in elimination from the mixture of whichever of the two components is in excess of the azeotropic composition (Z), so that the composition of the pot residue will approach that of Z. As an illustration, the maximum-boiling azeotrope of formic acid (bp 100.7°) and water boils at 107.3°; any mixture having other than the azeotropic composition (77.5% formic acid, 22.5% water) will yield this mixture as residue upon fractional distillation.

Data on other azeotropes, of both types, may be found in the references at the end of the chapter; extensive tables are given in the handbooks listed there.

2.5   Distillation under Reduced Pressure

As has been pointed out before, the boiling point of a liquid is that temperature at which the total vapor pressure is equal to the external pressure. It is most convenient to distil liquids under conditions such that the external pressure is the atmospheric pressure. However, in many instances, boiling temperatures at this pressure are too high to be desirable, because the compound being distilled may decompose, oxidize, or undergo molecular rearrangement at temperatures below its normal boiling point. Sometimes impurities present may catalyze such reactions at higher temperatures. These problems may often be alleviated by carrying out the distillation at pressures less than atmospheric, for under these conditions the boiling temperature is, of course, lower.

Reduced pressures may be obtained by connecting an aspirator ("water pump") or a mechanical oil pump to the distillation apparatus. An aspirator will commonly reduce the pressure to about 25 mm[6] and an oil pump to below 1 mm; the exact pressure obtained in the lower range is highly dependent on the condition of the pump and its oil and on the tightness of the connections of the distillation apparatus.

Two useful *approximations* of the effect of lowered pressure on boiling points are the following:

1. Reducing from atmospheric pressure to 25 mm lowers the boiling point of a high-boiling compound (250–300°) by about 100–125°.

2. Below 25 mm, each time the pressure is halved, the boiling point is lowered about 10°. More accurate estimates of boiling points at various pressures may be made by the use of charts and nomographs found in references at the end of this chapter. Calculations may also be made using integrated forms of the Clausius-Clapeyron equation, as described in physical chemistry textbooks.

2.6   Steam Distillation

The separation and purification of *volatile* organic compounds which are immiscible with water or nearly so is often accomplished by steam distillation, a technique which involves the codistillation of a mixture of water and organic substances. The virtues and limitations of this technique can best be illustrated by consideration of the principles which underlie steam distillation.

The partial pressure, P_i, at a given temperature of each component, i, of a mixture of immiscible, volatile substances is equal to the vapor pressure, P_i°, of the pure compound at the same temperature [equation (5)] and does not depend on the mole frac-

$$P_i = P_i^\circ \tag{5}$$

tion of the compound in the mixture; that is, each component of the mixture vaporizes independently of the others. This behavior is in sharp contrast to that exhibited by solutions of miscible liquids, for which the partial pressure of each constituent depends on its mole fraction in the solution [Raoult's Law, equation (1)]. Now, the total pressure, P_T, of a solution (mixture) of gases, according to Dalton's Law [equation (3)] is equal to the sum of the partial pressures of the constituent gases so that the total vapor pressure of a mixture of immiscible, volatile compounds is given by equation (6).

$$P_T = P_a^\circ + P_b^\circ + \cdots P_i^\circ \tag{6}$$

Note from this expression that the total vapor pressure of the mixture at any tem-

---

[6] The vacuum produced by a water aspirator is limited by the vapor pressure, and hence by the temperature of the water used; in cold climates, pressures as low as 8–10 mm may be obtained.

perature is always greater than the vapor pressure of even the most volatile component at that temperature, due to the contribution of the vapor pressures of the other constituents of the mixture. The boiling point of a mixture of immiscible compounds must then be *lower* than that of the lowest boiling component.

Demonstration of the principles just outlined is available from discussion of the steam distillation of a mixture of water (bp 100°) and bromobenzene (bp 156°), substances which are insoluble in one another. The vapor pressure versus temperature plot for a mixture of these substances, Figure 2.8, along with the corresponding plots

FIGURE 2.8   Vapor pressure *vs.* temperature plots for bromobenzene, water, and a mixture of bromobenzene and water.

for the pure liquids, shows that the mixture should boil at about 95°, the temperature at which the total vapor pressure equals atmospheric pressure. As would be predicted from theory, this temperature is below the boiling point of water, the lowest boiling component in this example. The ability to distil a compound at the relatively low temperature of 100° or less by means of a steam distillation is often of great use, particularly in the purification of substances which are heat-sensitive and which would decompose at higher temperatures. It is useful also in the separation of compounds from reaction mixtures which contain large amounts of nonvolatile residues such as the notorious "tars" which so often are formed during the course of an organic reaction.

The composition of the condensate from a steam distillation depends upon the molecular weights of the compounds being distilled and upon their respective vapor pressures at the temperature at which the mixture distils. Consider a mixture of the two immiscible components, A and B. If the vapors of A and B behave as ideal gases, the ideal gas law can be applied, and the following two expressions obtained:

$$P_A^\circ V_A = (g_A/M_A)(RT) \quad \text{and} \quad P_B^\circ V_B = (g_B/M_B)(RT) \tag{7}$$

where $P°$ is the vapor pressure of the pure liquid, V is the volume in which the gas is contained, g is the weight in grams of the component in the gas phase, M is its molecular weight, R is the universal gas constant and T is the absolute temperature ($°K$). Dividing the first equation by the second, one obtains

$$\frac{P_A° V_A}{P_B° V_B} = \frac{g_A M_B (RT)}{g_B M_A (RT)} \tag{8}$$

As the (RT) factors in the numerator and the denominator are identical and because the volume in which the gases are contained is the same for both ($V_A = V_B$), the above expression becomes

$$\frac{\text{grams of } A}{\text{grams of } B} = \frac{(P_A°)(\text{Molecular weight of } A)}{(P_B°)(\text{Molecular weight of } B)} \tag{9}$$

For a mixture of bromobenzene and water, which have vapor pressures of 120 and 640 mm, respectively, at 95° (see Figure 2.8), the composition of the distillate would be calculated from equation (9) as follows:

$$\frac{g_{\text{bromobenzene}}}{g_{\text{water}}} = \frac{(120)(157)}{(640)(18)} = \frac{1.64}{1}$$

Consequently, on the basis of weight, more bromobenzene than water is contained in the steam distillate, even though the vapor pressure of the bromobenzene is much lower at the temperature of the distillation. Because organic compounds generally have molecular weights much higher than that of water, it is possible to steam-distil compounds having vapor pressures of only about 5 mm at 100° with a fair efficiency on a weight to weight basis. Even solids can often be purified by steam distillation.

A steam distillation is generally accomplished in one of two ways. The first, and usually most efficient method, involves placing the organic compounds to be distilled in a round-bottomed flask fitted with a Claisen head which is equipped with a still head connected to a water-cooled condenser (Figure 2.9). The Claisen head helps to prevent splattering of the mixture into the condenser during distillation. Steam can be produced externally in a generator such as that shown in Figure 2.10 and then introduced into the bottom of the distillation vessel via a tube (Figure 2.9), or it can be obtained from a laboratory steam line. If the latter source is used, a trap is usually placed between the line and the distillation flask to allow removal of any water present in the steam (Figure 2.11). When an external source of steam is used, water may condense in the distillation flask, filling it to undesirable levels. This problem can generally be circumvented by gently heating the flask with a Bunsen burner.[7]

If only a small quantity of steam is required to distil the mixture completely, a second method of steam distillation may be employed. This method involves com-

---

[7] Better, the flask may be heated by immersion in an oil bath which is in turn heated either by an electric hot plate or by electric resistance coils. Alternatively, the flask may be heated with a "Glas-Col heating mantle," a heating device constructed of woven glass fiber and spun glass insulation in which electric resistance coils are embedded. Each mantle is form-fitted to accept a flask of given size. For a steam distillation, a steam bath may not be used for heating purposes as it will not supply enough heat.

FIGURE 2.9 Apparatus for steam distillation. The long tube is replaced by a stopper if steam is generated *in situ*.

FIGURE 2.10 A steam generator. The round bottom flask is initially half-filled with water and boiling chips are added before heating. The safety tube serves to relieve internal pressure if steam is generated at too rapid a rate.

FIGURE 2.11 Water trap.

bining water and the organic compounds to be distilled in the distillation flask. The flask is then heated directly with a Bunsen burner[7] to generate steam *in situ*. This method is, in general, not applicable for distillations in which large amounts of steam are needed; one would have to replenish the supply of water in the flask or use an inconveniently large flask.

In summary, steam distillation provides a method for separation of volatile liquid and solid organic compounds which are insoluble in water, or nearly so, from nonvolatile compounds under comparatively mild conditions. The technique obviously is not applicable to substances which decompose upon prolonged contact with steam or hot water, react with water, or have a vapor pressure of 5 mm or less at 100°.

EXPERIMENTAL PROCEDURE

Simple and fractional distillation. The purpose of this experiment is to demonstrate the operation of a fractional distillation column. Detailed directions are given for distillation of benzene-toluene mixtures, but similar results may be obtained with other mixtures such as carbon tetrachloride-toluene and methanol-water. The last mixture has the advantage of being less flammable than the other two.

If time permits, it will be instructive to compare results of distillation of the same mixture from a simple distillation setup (Figure 2.2), from an unpacked column (either a Vigreux column or a column designed to hold packing, but without the packing), and from a packed column.[8] Your instructor will inform you as to how many types of equipment you are to use, following the directions given below in each case.

Prepare three bottles or Erlenmeyer flasks of 50-ml capacity or larger and label them *A*, *B*, and *C*. In a 100-ml round-bottomed flask, place 30 ml of benzene, 30 ml of toluene,[9] and a few boiling chips to assure smooth boiling. Arrange the apparatus for distillation and have your instructor check it to be sure that all connections are tight. When a column is used, be sure that it is aligned in a vertical position. Place a wire gauze having an asbestos pad in the center *directly* beneath the pot (distilling flask).[10]

Heat the flask [11] with a burner [7] placed so that the tip of the flame touches or is just below the wire gauze, and protect the flame (also still head and column, if you are using one) from wind currents so that you can adjust the heating as precisely as possible. As soon as the mixture starts to boil and the refluxing vapors have reached the thermometer, regulate the flame so that distillation continues steadily at a rate of *only* about *1 drop of distillate every 1 to 2 seconds*.

Collect the first distillate in receiver *A*. When the head temperature reaches 86°, change to receiver *B* and, at 106°, change to receiver *C*. Continue distillation until about 1 or 2 ml of liquid remain in the pot, then turn off the flame. Measure the volumes of the distillation fractions in receivers *A*, *B*, and *C* in a graduated cylinder and record them in your notebook. Allow the liquid in the still head or column to drain into the pot; measure and record the volume of this residue.

[8] The packing may consist of glass beads, broken pieces of glass tubing (Hempel column), or glass helices, which are listed in order of increasing efficiency. A very satisfactory and inexpensive substitute for glass helices may be made from a copper (or, better, but more expensive, stainless-steel) cleaning sponge. The sponge is pulled apart and stuffed into the glass column rather compactly with the aid of a pencil. *Note:* The rapid corrosion of copper by organic halides makes this type of packing unsuitable for a mixture containing carbon tetrachloride.

[9] The benzene and toluene should be dry. If you are not sure that they are, each may be dried by placing a 50-ml portion in a simple distillation apparatus and distilling until the distillate shows no cloudiness. Discontinue distillation and use the *residue* (not the distillate) in the fractional distillation experiment. This is a method of drying moist solvents that form minimum-boiling azeotropes with water.

[10] A square of asbestos board with a small (*ca.* 1-cm) hole in the center may be used in place of the wire-asbestos gauze.

[11] The apparatus being used is open to the atmosphere at the receiving end of the condenser. This allows for pressure equilibration. At no time in the laboratory should a *closed system* be heated. If pressure equilibration is not allowed for, material expansion within the system, resulting in elevated pressures, may cause the apparatus to *explode*.

If you are to do another distillation using different equipment, transfer about 1 ml of each distillation fraction to labelled small test tubes, then recombine the remainder of the distillation fractions with the residue. Repeat the distillation, taking the same distillation "cuts" and compare the volumes of fractions *A*, *B*, and *C*, and the residue with those obtained with the other equipment.

Submit samples of fractions *A*, *B*, and *C* from each distillation for gas chromatographic analysis.[12] Compare the compositions of fractions *A* and *C* from the distillations using different equipment (see Figure 3.8).

Alternate fractional distillation experiment. The experimental procedure above may be modified to give the fractionation data in different form. Carry out the above experiment with one or more (as assigned by your instructor) types of equipment with the modification that 50-ml graduated cylinders are used to collect the three suggested distillation cuts. Prepare a table in your laboratory notebook for the purpose of recording approximately 25 successive readings of temperature and total accumulated volume of distillate. Begin the distillation and as soon as the mixture starts to boil and the refluxing vapors have reached the thermometer, regulate the flame so that distillation continues at a rate of about one drop of distillate every 1–2 seconds. When the distillation has been adjusted to a steady rate, note and record the head temperature and the total accumulated distillate in the graduated cylinder. Continue the distillation, recording the temperature and total volume for each additional increment of 3 ml of distillate. Although three distillation cuts will be taken for the ranges 79–85°, 86–105° and 106–111°, each in a different graduated cylinder, continue to record the temperature *vs.* the *total* accumulated volume data. During the collection of the second cut (86–105°) record measurements for increments of 1 ml of distillate, returning to 3 ml increments for the third and final cut. Continue the distillation until about 1 or 2 ml of liquid remains in the pot, then turn off the flame. Record the volume of each of the three distillation cuts taken. Allow the liquid in the still head or column to drain into the pot and measure and record the volume of this residue. Continue the experiment, following the directions given in the last two paragraphs of the above procedure.

Plot, on graph paper, the head temperature (vertical) *vs.* the *total* accumulated volume of distillate for each distillation carried out above. It will be useful to use a single piece of graph paper (and the same coordinates) so that the curves obtained for each distillation involving different equipment will overlay one another. Compare the results for each type of distillation equipment, drawing any conclusions suggested by the data concerning the relative efficiencies of the various columns.

EXERCISES

1. Explain why a packed fractional distillation column is more efficient at separating two closely boiling liquids than an unpacked column.
2. If heat is supplied to the distillation flask too rapidly, the ability to separate two liquids by fractional distillation may be drastically reduced. In terms of the general theory of distillation presented in the discussion, explain why this is so.

---

[12] Columns containing as the fixed phase either silicone gum rubber or a polyglycol such as Carbowax give good separation of benzene and toluene (see Chapter 3.4 for discussion of gas chromatography).

3. Explain why the column of a fractional distillation apparatus should be aligned as close to a vertical configuration as possible.

4. Explain the role of the boiling stones which are normally added to a liquid which is to be heated to boiling.

5. The bulb of the thermometer placed at the head of a distillation apparatus should be adjacent to the exit to the condenser. Explain the effect on the temperature reading of placement of the thermometer bulb (1) below the exit to the condenser and (2) above the exit.

6. (a) A mixture of 80 mole percent *n*-propylcyclohexane and 20 mole percent *n*-propylbenzene is distilled through a *simple distillation apparatus* (assume that no fractionation occurs during the distillation). The head temperature is found to be 157.3° as the first *small* amount of distillate is collected. The standard vapor pressures of *n*-propylcyclohexane and *n*-propylbenzene are known to be 769 mm and 725 mm, respectively, at 157.3°. Calculate the percentage of each of the two components in the first few drops of distillate.

(b) A mixture of 80 mole percent benzene and 20 mole percent toluene is distilled under exactly the same conditions as in part (a). Using Figure 2.3, determine the distillation temperature and the percentage composition of the first few drops of distillate.

(c) The standard boiling points of *n*-propylcyclohexane and *n*-propylbenzene are 156.9° and 159°, respectively. Compare the distillation results in parts (a) and (b). Which of the two mixtures would require the more efficient fractional distillation column for separation of the components? Why?

2.7   *Recrystallization*

One of the most necessary and useful techniques to be mastered by an organic chemist is that of *recrystallization*. Many organic compounds are solids, and thus solids are frequently encountered as reaction products. As they are likely to contain impurities, they need to be purified. Recrystallization is an important method of doing so. Essentially, the process is one in which the crystal structure is completely disrupted, either by fusion (melting) or by dissolution of the solid, and then crystals are allowed to regrow so that the impurities are left in either the melt or the solution.

Purification occurs because foreign molecules can seldom fit into the crystal structure of another compound. The fusion method is seldom used because the crystals are usually formed in the presence of a rather viscous oil (containing the impurities) from which they are difficult to separate. In solution recrystallization, a solvent is generally chosen in which the impurities are more soluble than the substance being purified. Crystals grown from such a solution will be of much greater purity than the original solid if the technique is performed properly and if the impurities are present only to the extent of a few percent. If impurities comprise more than a few percent of the original solid, two or more successive recrystallizations may be necessary.

Solution recrystallization as a technique involves several steps. These are (1) selection of the appropriate solvent, (2) dissolution of the solid to be purified in the solvent at or near its boiling point, (3) filtration of the hot solution to remove insoluble impurities, (4) crystallization from the solution as it cools, (5) separating the crystals from the supernatant solution, (6) washing the crystals to remove the adhering solution, and (7) drying the crystals.

EXPERIMENTAL PROCEDURE

A. GENERAL DISCUSSION

Selecting a solvent. A solvent must satisfy certain criteria in order to be used as a recrystallization solvent. (1) Its temperature coefficients for the solute and impurities should be favorable; that is, the compound being purified should ideally be quite soluble in the hot solvent but somewhat insoluble in the cold (this will minimize losses), and the impurities should remain at least moderately soluble in the cold solvent. Another possibility here is that the impurities be insoluble in the hot solution, from which they may be filtered. (2) The boiling point of the solvent should be low enough so that it can be easily removed from the crystals in the final drying step. (3) It is generally preferable that the boiling point of the solvent be lower than the melting point of the solute. (4) The solvent should not react chemically with the compound being purified.

If the compound has been previously studied, consultation of the chemical literature will generally give information concerning a suitable solvent. If not, it will be necessary to resort to trial and error methods using *small* amounts of material. Some general solubility principles should be kept in mind if this needs to be done. Normally, polar compounds are insoluble in nonpolar solvents and soluble in polar solvents. Conversely, nonpolar compounds are more soluble in nonpolar solvents. These solubility relationships are frequently summarized with the phrase "like dissolves like." Therefore, a highly polar compound is unlikely to be soluble in a hot nonpolar solvent but may well be too soluble in a cold very polar solvent, so that a solvent of intermediate polarity will be optimum. Occasionally, *mixed solvents* are found to work well.

Solution. The following operations are best carried out within a ventilation hood to avoid inhalation of solvent vapors. The solid to be purified is placed in an appropriately sized Erlenmeyer flask,[13] along with a few milliliters of the desired solvent. With *constant* stirring,[14] the mixture is then heated to boiling using either a hot plate or steam bath. A Bunsen burner may be used, with caution against too rapid heating, *if* the solvent is *not* flammable. More solvent is added *slowly* to the boiling mixture until just enough boiling solvent is present to dissolve the solid. When it appears that additional solvent is not dissolving any more of the solid, *particularly when only a relatively small quantity of solid remains*, enough solvent has probably been added. The remaining solid is likely to consist of insoluble impurities and may be removed in the hot filtration step (below). Many beginning laboratory students obtain poor results in their recrystallizations, because, when trying to dissolve the last traces of solid, quantities of solvent are added which are far in excess of that needed. During the solution of the impure solid, time should be allowed between each small addition of fresh solvent, for some solids dissolve only slowly.

When mixed solvents are used, they of course must be miscible. The crystals being purified should be soluble in one of the solvents, but insoluble or only slightly soluble in the second. The crystals are first dissolved in that pure, boiling solvent in which they are soluble. The second solvent is then added to the boiling solution until the solution turns *cloudy*. The second solvent decreases the dissolving ability of the solvent medium; when the solubility limit is reached, the solute begins to come out of solution, resulting

---

[13] It is good laboratory technique to save a few crystals of the impure solid; they may be needed later to induce crystallization if problems are encountered at that step.

[14] Stirring will help to prevent bumping of the boiling mixture. When relatively large quantities of undissolved solids are present, boiling stones are not very effective.

in a cloudy appearance. Finally, more of the first solvent is added *dropwise* until the solution just becomes clear again. Occasionally it is advantageous to carry out the hot filtration step before adding the second solvent to prevent crystallization during filtration.

If colored impurities are present, these may often be removed by adding a small amount of decolorizing carbon to the hot (*not boiling*) solution. Seldom is more decolorizing carbon needed than that which can be held on the tip of a small spatula. The impurities, especially colored ones, adsorb on the surface of the carbon particles and are removed during filtration. If too much carbon is used, some of the substance being purified will be adsorbed and subsequently lost in this step. After adding the carbon, the solution should be heated to boiling for a few minutes, while being continuously stirred or swirled to prevent bumping of the boiling mixture.

Hot filtration. To remove insoluble impurities (including dust and decolorizing carbon, if used), the hot solution is best filtered by gravity filtration. Suction filtration is not desirable, for under reduced pressure at least some of the hot solvent will, through rapid evaporation or boiling, be lost, resulting in a too concentrated solution.

Small Stick
(to break contact
between funnel
and flask)

Fluted
Filter
Paper

Short
Stemmed
Funnel

Steam

FIGURE 2.12   Apparatus for effecting hot filtration.

A *short-stemmed* glass funnel and a fluted filter paper should be used for filtration into a second Erlenmeyer flask (Figure 2.12). The fluted paper is used to allow rapid filtration. Crystallization from the solution occasionally occurs in the filter paper or on the glass surface of the funnel. This may be avoided by adding a further small amount of *hot* solvent to the solution (dilution); however, before the crystallization step (see below) this additional solvent would need to be evaporated to avoid unnecessary loss of purified crystals. Alternatively, and better, a few milliliters of the recrystallization solvent may be placed in the Erlenmeyer flask and this liquid heated to boiling. The condensing vapors will heat the glass funnel, preventing crystallization (Figure 2.12). For low-boiling solvents, a steam bath may be used; for solvents boiling higher than about 80°, an electric hot plate or heated oil bath must be used. A Bunsen burner can be used *only* if the solvent is nonflammable.

Decolorizing carbon, being very finely divided, is not always completely removed by this filtration. If not, a small amount of "filter-aid" (a finely divided silica that is easily filterable) may be added to the filtrate. The solution is heated again just to boiling, and refiltered using a fresh fluted filter paper. The carbon is trapped in the filter-aid and thus removed.

Crystallization. The hot filtrate is allowed to cool slowly by standing at room temperature. Crystallization should then ensue. Rapid cooling by immersion in water, etc., is undesirable because the crystals formed will tend to be quite small. Their large surface area may then facilitate *adsorption* of impurities from the solution. Generally the solution should not be agitated while cooling, since this will also lead to formation of small crystals. However, formation of very large crystals (larger than approximately 2 mm) may cause *occlusion* (trapping) of solution within the crystals. Such crystals are difficult to dry and will, when dried, have deposits of impurities in them. If large crystals seem to be forming, agitation may be used to lower the average crystal size. Judgment of proper crystal size will become easier with experience.

If crystallization does not occur after cooling, the super-saturated solution can usually be made to yield crystals by *seeding*. A tiny crystal of the original solid is added to the cooled solution. Crystals will usually form quite rapidly. Another method that may be used to induce crystal formation is to scratch the inside surface of the flask at or just above the surface of the solution with a glass rod.

Occasionally, the solute will separate as an "*oil*" rather than as crystals from the solution. This occurs when the limit of solubility of the solute is reached on cooling at a temperature that is above the melting point of the substance. This should be avoided and usually can be by diluting the solution with more solvent, reheating, and again allowing to cool. Sometimes choice of a different solvent is necessary.

After filtering the crystals (see below), a second "crop" of crystals can usually be obtained by further cooling with ice-water, or by concentrating the solution by boiling away some of the solvent and cooling. The crystals obtained as a second (or even third) crop may not be as pure as the first. Melting points of the crystals obtained in various crops can be used as a criterion of their purity.

Filtration. The cool mixture of crystals and solution is now filtered by using a Büchner funnel and a vacuum filter flask attached to an aspirator (Figure 2.13). Before filtration, the filter paper should be wetted with the solvent in order to "seal" it to the funnel.

A stirring rod or spatula may be used as an aid to transferring the crystals to the funnel. The last quantity of crystals may be transferred by washing them out of the flask with some of the filtrate ("mother liquor"). *Before removing the funnel from the filter flask, release the pressure by opening the screw clamp or stopcock on the trap.*

The crystals should be pressed as dry as possible with a cork. They should then be washed by releasing the pressure and just covering the crystals with *cold* fresh solvent. After a few seconds, this solvent is removed by suction. The purpose of this washing step is to remove the solution with which the crystals are wet. This must be done for successful purification; the solution, of course, contains the impurities. If the solvent was properly chosen, the crystals should now be pure, except for solvent.

FIGURE 2.13 Apparatus for vacuum filtration.

Drying the crystals. Most of the solvent may be evaporated by allowing the aspirator to pull air through the mass of crystals on the funnel for a few minutes. By means of a spatula, the crystals are then transferred to a clean watch glass. Complete drying is accomplished by allowing them to air-dry for a few hours, or (faster) by placing the watch glass in an oven. (*Caution:* the temperature of the oven should be about 20° below the melting point of the crystals.) Heat drying may not be used for crystals that sublime readily at atmospheric pressure (see Section 2.8).

B. EXPERIMENTS

1. Determination of solubilities. Place about 20 mg (a small spatula-tip full) of finely crushed resorcinol in a 10 x 75-mm test tube and add about 0.5 ml of water to the tube. Stir with a glass rod and determine whether resorcinol is soluble in water at room temperature. If it is not completely soluble, gently heat the test tube with a micro-burner to determine whether resorcinol is soluble in hot water. Record your observations of the solubility of resorcinol in water on the basis of the following definitions: soluble—20 mg of solute will dissolve in 0.5 ml of solvent; slightly soluble—some but not all of the 20 mg of solute will dissolve in 0.5 ml of solvent; insoluble—none of the solute appears to dissolve.

Repeat the solubility test using benzene and then petroleum ether (bp 60–80°).[15] *Do not heat these solvents with an open flame* as they are highly flammable! Rather, use a steam cone. Record the observed solubility of resorcinol in each of the solvents.

In like fashion determine the solubilities of napthalene, benzoic acid and anthranilic acid in water, in benzene and in petroleum ether. If any of these solutes is soluble in

---

[15] Petroleum ether is a mixture of aliphatic hydrocarbons, obtained from petroleum refining. Such a mixture may vary in composition and boiling range, depending on the distillation "cut" taken. In order to define the liquid being used, the boiling range is usually given, *i.e.*, petroleum ether (bp 60–80°). The name *ligroin* is occasionally encountered in place of petroleum ether.

the hot solvent, but only slightly soluble or insoluble in the cold solvent, allow the hot solution to cool slowly to room temperature and compare the size, color, and crystal form of the resulting crystals with the original solid material. Note which solvents you would consider best suited for recrystallization of each of the solutes.

2. Recrystallization of an impure solid. Place 1 g of crude phthalic acid and 5 ml of water in a 50-ml Erlenmeyer flask and swirl the flask while heating it with the flame from a microburner. Gently boil the solution until no more of the solid appears to dissolve and then add additional water, in 1-ml portions, to dissolve the remaining solid. You may find that some solid particles do not dissolve even when more water is added. These are probably highly insoluble impurities, and no more water should be added. Pure phthalic acid is colorless; if the hot solution is colored, the colored impurities may be removed by adding a small spatula tip-full of decolorizing carbon and heating the suspension to the boiling point with stirring. (*Caution:* Do not add decolorizing carbon to a boiling solution.) To remove insoluble impurities and decolorizing carbon, if used, perform a hot filtration according to the following directions. To a 50-ml Erlenmeyer flask add 1–2 ml of water. Place a short-stemmed glass funnel, containing a fluted filter paper, in the flask with a small stick (such as a toothpick, for example) between the funnel and the flask to create an air space. Heat this flask with a Bunsen burner until the water has nearly boiled away. Remove the burner and quickly heat the phthalic acid solution to boiling. Filter the hot solution by gravity through the fluted filter paper into the second flask. Be sure to pour the hot liquid on the upper portion of the filter paper in order to maximize the efficiency of the filtration. In this way, the solution will come in contact with a larger area of the filter paper, and thus will be filtered more rapidly. The more rapidly the solution can be filtered, the less likely the possibility of inducing crystallization in the funnel. Be careful, however, not to allow any solution to pass between the edge of the filter paper and the funnel, for this solution would not be filtered. Rinse the empty flask with 1–2 ml of hot water and filter the rinse water into the flask containing the phthalic acid solution. The filtrate should now be colorless and free of decolorizing carbon. If it is not colorless, repeat the treatment with decolorizing carbon. If carbon is still present in the filtrate, add some filter-aid, reheat to the boiling point, and filter the hot solution through a *new* filter paper using the procedure described above.* Allow the filtrate to stand undisturbed until it has cooled to room temperature and no more crystals form. Then place the flask in ice water to complete the process of crystallization. Collect the white crystals by suction filtration, using the mother liquor (filtrate) to rinse any crystals remaining in the flask into the Büchner funnel. Wash the filter cake free of mother liquor with two small portions of *cold* water and press the crystals as dry as possible on the funnel with a clean spatula or cork. Spread the crystals onto a piece of filter paper and allow them to dry completely. Determine the weight and melting point of the purified product. If the melting point is not within a range of 2–3°, recrystallize the phthalic acid from water. Another treatment with decolorizing carbon should not be necessary unless your crystals are colored.

EXERCISES

1. Assume that after the dissolution, hot filtration and rinsing steps in the above experiment, 1 g of phthalic acid remains dissolved in 12 ml of hot water. Also

assume the solubility of phthalic acid at 0° to be 0.45 g per 100 ml of water.

(a) Calculate the weight of phthalic acid which would be *unrecovered* by following the above procedure.

(b) Calculate the weight of a second crop of crystalline phthalic acid which might be obtained by evaporating the mother liquor to 5 ml and cooling to 0°.

(c) If, during the solution step, water in excess of the minimum amount necessary to dissolve the phthalic was added, so that 1 g of acid was dissolved in 25 ml of hot water, how much phthalic acid would be *unrecovered* by following the above procedure?

2. If a second crop of phthalic acid was obtained by the procedure suggested in exercise 1(b), would you expect it to be as pure as the first crop? Explain why or why not.

3. Explain why the rate of dissolution of a crystalline substance may depend on the *size* of its crystals.

2.8 Sublimation

An alternative to crystallization for the purification of some solids is the process of *sublimation*. This method uses to advantage the differing vapor pressures of solids in a way analogous to simple distillation. The impure sample is vaporized directly from the solid state by heating it at a temperature below the melting point and the vapor is then condensed (crystallized) directly to the solid state on a cold surface; both processes occur *without* the intermediacy of the liquid state. Figure 2.14 shows a typical phase diagram, relating the solid, liquid, and vapor states of a substance with pressure and

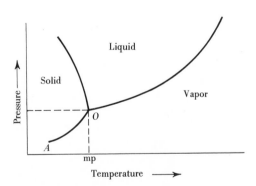

FIGURE 2.14 Single-component phase diagram.

temperature. Note that with conditions of temperature and pressure below those depicted at point O, the liquid state cannot exist (*i.e.*, is thermodynamically unstable). The vapor pressure of the solid at any temperature below the melting point is given by the curve OA. It is the equilibrium between solid and vapor, represented by this curve which is of importance for sublimation.

FIGURE 2.15 Sublimation apparatus.

Two types of sublimation apparatus commonly used are pictured in Figure 2.15. They have two characteristics in common: (1) a chamber that may be evacuated by attachment to a vacuum pump (the impure sample is placed at the bottom of this chamber) and (2) a finger-like projection in the center of this chamber that may be cooled to provide a surface upon which the sublimed crystals may form. The "cold-finger" in (a) is cooled by circulating water, and that in (b) by a medium such as Dry-Ice in acetone, or ice-water. Figure 2.16 shows a simple type of sublimation apparatus that may be assembled inexpensively from two test tubes (one with a sidearm), rubber stoppers, and glass tubing.

For effective purification using this technique, two criteria must be met: (1) the solid must have a relatively high vapor pressure, and (2) the impurities must exhibit

FIGURE 2.16 Test tube sublimator.

vapor pressures substantially different from the primary material (preferably lower). Should the first criterion *not* be met, the time necessary to pass any significant quantity of material through the vaporization-crystallization sequence would usually be prohibitively long. In this case, either recrystallization (Section 2.7) or chromatographic techniques (Chapter 3) must be used.

The "mechanics" of purification by sublimation may be simply described as follows. The impure solid is heated to a temperature higher than that of the surface (cold-finger) upon which the pure material is to be condensed, but lower than the melting point. Since vapor pressure and thermodynamic instability of the solid vary directly with the temperature (see curve *OA*, Figure 2.14), material will be transferred via the vapor phase to the colder surface. Crystals developing on the cold surface will tend to be very pure, for the molecules of the impurities will usually not be incorporated into the growing crystal structure and therefore will not condense on the cold surface. Should the vapor pressure of an impurity be similar to that of the material being purified, sublimation will not be an effective method of purification, for crystals of both substances will tend to form on the cold surface (criterion 2 above).

Very few organic compounds exhibit vapor pressures adequate for sublimation at atmospheric pressure. Generally, reduced pressure (vacuum) is needed to increase the rate of evaporation of the solid. This process is analogous to the use of vacuum distillation for high-boiling liquids.

Sublimation is generally restricted to relatively nonpolar substances having fairly symmetrical structures. In these cases, crystal forces are lower and vapor pressures tend to be higher. The ease with which a molecule may escape from the solid to the vapor phase is determined by the strength of the intermolecular attractive forces between the molecules of the solid. The most important attractive force is of an electrostatic nature. Symmetrical structures will have relatively symmetrical distributions of electron density, and as such will have smaller dipole moments than less symmetrical structures whose electron distribution will be more polarized. A smaller dipole moment infers a higher vapor pressure. Van der Waals forces are also important, but are generally less important than electrostatic attractions. In general, van der Waals forces increase in magnitude with increasing molecular weight, and thus large molecules, even if symmetrical, are less adaptable to purification by sublimation.

REFERENCES

A. DISTILLATION
1. *Techniques of Organic Chemistry*, A. Weissberger, editor, Interscience Publishers, New York, 1965, Vol. IV.
2. *Azeotropic Data*, American Chemical Society, Washington, D. C., 1952.
3. T. Earl Jordan, *Vapor Pressure of Organic Compounds*, Interscience Publishers, New York, 1954.

B. RECRYSTALLIZATION
R. Stuart Tipson, in *Techniques of Organic Chemistry*, A. Weissberger, editor, 2nd Ed., Interscience Publishers, New York, 1956, Vol. III, Part I, Chapter 3.

C. SUBLIMATION
R. Stuart Tipson, in *Techniques of Organic Chemistry*, A. Weissberger, editor, Interscience Publishers, New York, 1965, Vol. IV, Chapter 8.

3

Separation and purification of organic compounds

II. Phase Distribution: Extraction and Chromatography

Two widely applicable and relatively simple methods of separating a desired compound from its impurities, or of isolating each of the individual components of a mixture, are extraction and chromatography, both of which are based upon the principle of *phase distribution*. A substance may establish an equilibrium distribution between two *insoluble* phases with which it is in contact, in a ratio dependent upon its relative stability in each of those phases. The techniques being discussed involve the *selective* removal of one or more of the components of a gaseous, liquid or solid mixture by contact of the mixture with a second phase.

The process upon which such a distribution depends may be one of two varieties: (1) *partitioning*, based upon differing relative solubilities of the components in immiscible solvents (selective dissolution); and (2) *adsorption*, based upon the selective adherence of the components of a liquid or gaseous mixture to the surface of a solid phase. The various techniques of chromatography involve both of these processes, whereas the techniques of extraction involve only the former.

3.1 Extraction

A most widely employed method of separating organic compounds from mixtures in which they are found or produced is that of liquid-liquid extraction. In fact, virtually every organic reaction requires extraction at some stage in the purification of its products.

In its simplest form, extraction involves the distribution of a solute between two immiscible solvents. The distribution is expressed quantitatively in terms of the *distribution* (or *partition*) *coefficient*, K [equation (1)]. This expression indicates that a solute, A, in contact with a mixture of two immiscible liquids, S and S', will be distributed (partitioned) between the liquids so that at equilibrium the ratio of concentrations of A in each phase will be constant, at constant temperature.

$$K = \frac{\text{concentration of } A \text{ in } S}{\text{concentration of } A \text{ in } S'} \tag{1}$$

Ideally, the distribution coefficient of *A* is equal to the ratio of the individual solubilities of *A* in pure *S* and in pure *S'*. In practice, however, this correspondence is generally only approximate since no two liquids are completely immiscible. The extent to which they dissolve in one another alters their solvent characteristics and thus slightly affects the value of *K*.

It is evident that for *A* to dissolve completely in one or the other of two immiscible liquids, the value of *K* must be infinity or zero.[1] Neither of these limiting values is actually attained. However, so long as *K* is larger than 1.0, and the volume of solvent *S* is equal to or larger than the volume of solvent *S'* (see below), the solute will be found in *greater amounts* in solvent *S*. The amount of solute that will remain in the other solvent, *S'*, will depend on the value of *K*.

Equation (1) may be rewritten as shown in equations (2) and (3).

$$K = \frac{\text{grams of } A/\text{ml of } S}{\text{grams of } A/\text{ml of } S'} \tag{2}$$

$$K = \frac{\text{grams of } A \text{ in } S}{\text{grams of } A \text{ in } S'} \times \frac{\text{ml of } S'}{\text{ml of } S} \tag{3}$$

Note that when the volumes of *S* and *S'* are equal,[2] the ratio of the grams of *A* in *S* and in *S'* will equal the value of *K*. If the volume of *S* is doubled, and the volume of *S'* kept the same, the ratio of the grams of *A* in *S* to the grams of *A* in *S'* will be *increased* by a factor of two. This must necessarily follow because *K* is a *constant*. Therefore, if *A* is to be recovered from solvent *S*, the amount of *A* recovered will be increased by using larger quantities of solvent *S*.

A further consequence of the distribution law [equation (1)] is of practical importance in performing an extraction. If a given total volume of solvent *S* is to be used to separate a solute from its solution in *S'*, it can be shown to be more efficient to effect several successive extractions with portions of that volume than one extraction with the full volume of solvent. Thus, *more* butyric acid will be removed from water solution by two successive extractions with 50-ml portions of ether than will be removed in a single extraction with 100 ml of ether. Three successive extractions with 33-ml portions would be still more efficient (see exercises). There is, however, a point beyond which the further effort of additional extractions no longer yields a commensurate return. The larger the distribution coefficient, the fewer the number of repetitive extractions that are necessary to separate the solute effectively. This is an important consideration because it is desirable to keep the total volume of extracting solvent to a minimum, not only for reasons of expense, but also because of the time involved in eventual removal of the solvent by distillation.

Consider now a solution of two compounds in solvent *S'*. It should be evident that for effective separation of these two compounds by extraction with solvent *S*, the distribution coefficient of one should be significantly greater than 1.0, whereas the distribution coefficient of the other should be significantly smaller than 1.0. If these conditions are met, one compound will be mainly distributed in solvent *S* and the other in

[1] For simplicity, we shall define *S* as that solvent in which the solute is more soluble. Therefore, in the remaining discussion the value of *K* will by definition always be greater than 1.0.

[2] To be strictly correct, the volumes of *solution* should be used in this expression. However, when the solutions are reasonably dilute, volumes of *solvent* may be used without appreciable error.

solvent S' at equilibrium. Physical separation of the two liquid layers would then result in at least a partial separation of the two compounds.

When the coefficients are of similar magnitude, separation by the extraction technique may be quite ineffective since the relative concentrations of the compounds in each of the two liquid phases may be little changed from those of the original mixture. In this case, other methods of separation should be used, such as adsorption chromatography (Section 3.5), fractional distillation (Section 2.2), fractional crystallization, etc. The specialized techniques of *fractional extraction* and *countercurrent distribution* can also be used to separate compounds with similar but not identical distribution coefficients. Details of these techniques are supplied for the interested student in the references at the end of the chapter.

Another type of experimental problem often encountered is that involving separation of one component that is only slightly soluble in the extracting solvent from a mixture whose other components are essentially insoluble. Large quantities of solvent would have to be used in order to effect the separation in only one or two extractions, and the handling of such quantities may be extremely unwieldy. Alternatively, it would be tedious to do a very large number of extractions with smaller quantities of solvent. The method of *continuous extraction*, in which a relatively small volume of extracting

FIGURE 3.1 Light-solvent extractor.   **FIGURE 3.2** Heavy-solvent extractor.

Porous Thimble
(to Hold Solid)

Vapor

Siphon
Arm

Extraction
Solvent

Boiling
Flask

FIGURE 3.3 Soxhlet extractor.

solvent is used, is a possible solution to this problem. By means of specialized apparatus, the solution of extracting solvent and solute is continuously separated into a boiling flask from the mixture being extracted. The solution is subjected to continuous
distillation and the condensed distillate returned as fresh extracting solvent to the extraction vessel and reused. In the process, the extracted material builds up in an increasingly concentrated solution in the boiling flask. This is because more dilute solution is continuously draining into the flask, while, at the same time, the solvent is being
distilled away.

For continuous liquid-liquid extraction, the solvent (as condensate from the condenser) is made to pass either up or down, depending on relative densities, through the
solution containing the desired compound. If the solvent is less dense than the solution
being extracted, an apparatus such as that shown in Figure 3.1 may be used. If it is
more dense, apparatus such as that shown in Figure 3.2 may be used.

For separation of the components of a solid mixture by continuous solid-liquid
extraction, a Soxhlet extraction apparatus (Figure 3.3) is convenient. The solid is
placed in a *porous* thimble in the chamber as shown, and the extracting solvent in the
boiling flask below. The solvent is heated to reflux, and the distillate, as it drops from
the condenser, collects in the chamber. By coming in contact with the solid in the
thimble, the liquid effects the extraction. After the chamber fills to the level of the
upper reach of the siphon arm, the solution empties from this chamber into the boiling
flask by a siphoning action. This process may be continued automatically and without

attendance for as long as is necessary for effective removal of the desired component, which will then be contained with the solvent in the boiling flask.

When choosing an extracting solvent for the isolation of a component from a solution there are some general principles to be kept in mind. (1) The solvent must, of course, be immiscible with the solvent of the solution. (2) It must be chosen to have the most favorable distribution coefficient for the component in question, and to have unfavorable coefficients for the impurities or other components. (Since distribution coefficients for most compounds in various solvents are not available, the solvents for the various experiments will be suggested to the student. As he gains experience in, and understanding about, the solubility principles of organic compounds, the student should be able to suggest appropriate solvents.) (3) A solvent must be chosen, just as in recrystallization, that does not react chemically in an undesirable manner with the components of the mixture. (4) After extraction, the solvent should be readily separable from the solute. Usually the solvent is removed by distillation.

The distribution coefficients of organic acids and bases may be quite dramatically affected by pH when one of the solvents is water. An organic acid that is insoluble in water at pH 7 may be quite soluble in dilute aqueous sodium hydroxide or sodium bicarbonate. In such a case, the acid is converted by the base into a salt which is very water soluble. In similar fashion, an organic base that is insoluble in water at pH 7 may be quite soluble in dilute hydrochloric acid. Such a solubility change depends on a reaction between the organic base and hydrochloric acid to produce a water-soluble salt. Therefore, organic acids and bases may be separated selectively from organic solutions by extraction with aqueous solutions of appropriate pH. Since most organic acids and bases are insoluble or only slightly soluble in water at pH 7, they may be precipitated from the aqueous solution by neutralization.

3.2   Technique of Simple Extraction

In the undergraduate organic laboratory, the student will often need to use extraction in the purification of the products of his reactions. Some of the common applications of extraction that will be encountered are (1) removal of organic reaction products from a water solution in which they are formed by extraction with an organic solvent; (2) removal of acid or base catalysts or inorganic salts from a reaction mixture by extraction with water; (3) separation of organic acids or bases from other organic compounds in an organic solvent by extraction with dilute aqueous base or mineral acid. Accordingly, the technique of simple extraction with a separatory funnel is described here in some detail.

Separatory funnels are available in several different shapes, from almost spherical to elongated pear shape (Figure 3.4). The more elongated the funnel, the longer the time required for the two liquid phases to separate once shaken together. When the liquids have similar densities, the flatter (more spherical) separatory funnels are preferable, for otherwise a needless amount of time may be wasted in waiting for the phases to separate. The funnels have a stopcock at the bottom through which the contents may be drained.

To perform an extraction, the solution is placed in a separatory funnel (with the stopcock closed!) and to this is added a quantity of the extracting solvent. The funnel should not be too full (three-fourths of the height of the funnel is a reasonable maximum). The upper opening of the funnel is stoppered either with a ground-glass stopper,

FIGURE 3.4 Separatory funnels.

for which most separatory funnels are fitted, or with a rubber stopper. The funnel is held during the shaking process in a rather specific manner, which will, when mastered, be found to be quite efficient. If right handed, set the stopper at the top of the funnel against the base of the index finger of the left hand and grasp the funnel with the first two fingers and the thumb. The funnel should be turned so that the first two fingers of the right hand can be curled around the handle of the stopcock. In such a manner, the stopper and stopcock can be held tightly in place during the shaking process (see Figure 3.5). Left-handed people would, of course, find it easier to use the opposite hand for each position.

FIGURE 3.5 Efficient method of holding separatory funnel.

The funnel and its contents are shaken in a vigorous manner, so that the two immiscible liquids are mixed as intimately as possible. The purpose of the shaking process is to increase greatly the surface area of contact between the solvents so that the equilibrium distribution of solute between the solvents will be attained in a relatively short time. Every few seconds it is important to "vent" the funnel by inverting it (stopcock up) and carefully opening the stopcock to release any pressure that may have developed inside. This is particularly important when solvents of low boiling points are used, or if an acid solution is extracted with sodium bicarbonate solution (CO_2 is released). If this is not done, the stopper may well be blown free, and the contents of the funnel lost. If the funnel is held as described, the stopcock may be opened by twisting the fingers curled around it, without readjusting the grip on the funnel. At the end of the shaking period (if the shaking is vigorous, one or two minutes is usually sufficient), the funnel is vented a final time, supported on a ring [3] and the layers allowed

[3] When using an iron ring to support a funnel, it is best to cover the ring with a length of rubber tubing to prevent breakage. This may be accomplished by slicing the tubing on one side and then slipping the tubing over the ring. Copper wire may be used to fix the tubing permanently in place.

to separate. The two liquid layers then are separated by carefully drawing the lower layer into a flask through the stopcock.

As a general rule, the layers will separate so that the solvent of greater density will be on the bottom. Thus knowledge of the densities of solvents to be employed is useful in identification of the layers. This, however, is not foolproof, because the nature and concentration of solute may be such as to invert the relative densities of two solvents. A common mistake of students is to confuse the identity of the two layers in the funnel. Often one layer is discarded and it is later found that the wrong layer was saved. *It is suggested, as a point of caution, that both layers always be saved until there is no doubt about the identity of each.*

Occasionally, after shaking, the two immiscible liquids will not separate cleanly into two layers, but will instead form an emulsion. If one of the layers is water, often, but not always, the emulsion can be "broken" by adding a small amount of acid or base or saturated aqueous sodium chloride to the mixture and reshaking gently. This lowers the surface tension of the water and may allow the droplets to collect into separate layers. This method can be used only if the added material does not change the distribution coefficients in an unfavorable manner (which it usually does not unless pH is an important factor). In the case of difficult emulsions, it may be necessary to choose a different extracting solvent to avoid this problem. Small quantities of insoluble solid material often collect at the phase interface, making it difficult to observe the layer boundary. Filtration of the heterogeneous mixture before separating the layers is the usual remedy for this problem.

EXPERIMENTAL PROCEDURE

This experiment will demonstrate the ability to separate various types of organic compounds by control of the pH of the extracting aqueous medium.

Prepare a *mixture* of 2 g each of the following compounds: benzoic acid, *p*-nitroaniline, naphthalene, and 2-naphthol. Dissolve this mixture in about 100 ml of dichloromethane. Extract this solution two times with 30-ml portions of 6 *M* hydrochloric acid. Combine the aqueous extracts and set them aside in a labeled flask. Reextract the organic layer from above two times with 30-ml portions of 10% sodium *bicarbonate* solution. Combine the aqueous extracts in a labeled flask and set them aside. Again extract the organic layer two times with 30-ml portions of 3 *M* sodium hydroxide solution and combine these two aqueous extracts.

Dry the organic layer that is left at this point by adding 4–5 g of *anhydrous* sodium sulfate to the liquid contained in an Erlenmeyer flask;[4] let it stand for about 0.5 hour, swirling occasionally.

While the organic layer is drying, neutralize each of the three aqueous extracts with either 6 *M* hydrochloric acid or 6 *M* sodium hydroxide, whichever is necessary. The 6 *M* hydrochloric acid extract should be cooled with an ice bath before neutralization. Upon neutralization, precipitates will be observed for each of the three extracts. Cool each of the mixtures with an ice bath and then filter each by vacuum filtration. Each of the filtered solids should be washed *within the Büchner funnel* with *cold* distilled water. Pre-weigh (tare) four small labeled beakers, and place the three solid materials thus far obtained in three of them. Either allow these to air-dry until the next laboratory period[5] or place them in an oven held at 90° for 1.5 hours.

---

[4] For a discussion of drying agents, see Appendix I.

[5] If these materials are allowed to air-dry in the laboratory desk until the next period, the beakers should be covered loosely with a piece of paper to prevent contamination of the compounds with dust.

Filter the organic layer which has been drying over sodium sulfate by gravity filtration and remove the solvent on a steam bath *in the hood.* Collect any solid residue, transfer it to the fourth tared beaker, and allow it to *air-dry* (do not use oven).

Reweigh each of the four beakers and obtain the weights of the four solid materials. Determine the melting points of each of these four materials obtained and identify each as one of the original compounds; benzoic acid, mp 121–2°; 2-naphthol, mp 122–3°; *p*-nitroaniline, mp 146–7°; naphthalene, mp 80–1°. Due to the closeness of the melting points of benzoic acid and 2-naphthol, mixture-melting points will be needed to achieve certain identification. Intimately mix small samples of each of the two materials melting in the vicinity of 120° with small quantities of benzoic acid and determine the two melting points. You should now be able to identify each of the four materials.

EXERCISES

1. From the results of the above experiment, what can you conclude about the properties of the four compounds used? Write chemical equations showing any changes that occurred during the extraction procedure.
2. If, when extracting an aqueous solution with an organic solvent, you were uncertain of which layer in the separatory funnel was the organic layer, how could you quickly settle the issue?
3. Which layer (upper or lower) will each of the following solvents usually form if used to extract an aqueous solution: diethyl ether? chloroform? acetone? *n*-hexane? benzene?
4. Given 500 ml of an aqueous solution containing 8 g of compound *A*, from which it is desired to separate *A*, how many grams of *A* could be removed in a single extraction with 150 ml of diethyl ether? (Assume the distribution coefficient, diethyl ether: water, to equal 3.0.) How many *total* grams could be removed with three successive extractions of 50 ml each?

3.3 Chromatography

As mentioned in the introductory section, chromatography is based upon the general principles of phase distribution. Reduced to its fundamentals, the method involves the *selective removal* of the components of one phase from that phase as it is *flowing* past (or through) a second *stationary* phase. The removal of a component by the stationary phase is an equilibrium process, and the molecules of that component reenter the moving phase. Separation of two or more components in the moving phase will result when the equilibrium constants for the distribution of these components between the two phases differ. Expressed simply, the more tenaciously one component is held by the stationary phase, the higher the percentage of molecules of that component that are held *immobile.* A second component, less strongly held, will have a higher percentage of molecules in the *mobile* phase than will the first component. Therefore on the average, the molecules of the component that is held less strongly will move over the stationary phase (in the direction of flow) at a higher rate than the other, resulting in a migration of the components into separate regions (bands) of the stationary phase (see, for example, Figure 3.10).

The separation between the bands is linearly related to the distance traveled on the column. In general, the longer the distance, the greater the separation will be. For example, if one band moves 10 cm along the pathway and a second band moves 5 cm during the same time interval, the separation between the centers of the bands will be 5 cm. At a later time, if the first band has moved 100 cm, the second will have moved only 50 cm, and the separation will have increased to 50 cm. It should be remembered that the separation of a mixture by phase distribution requires that the components of the mixture have different distribution coefficients; if these coefficients are similar, only partial separation of the components into individual bands will occur, unless, of course, the path length is increased to give the components "time" to migrate apart.

There are four important types of chromatography based upon the principles discussed above. These are gas chromatography (more specifically, gas-liquid partition chromatography, glpc), column chromatography, thin-layer chromatography (tlc), and paper chromatography. These are discussed individually in the following sections.

3.4   Gas Chromatography

In gas chromatography, the mixture to be separated is *vaporized* and carried along a column by a flowing *inert* gas such as nitrogen or helium (the carrier gas). The gaseous mixture is the *mobile* phase. The column is packed with a solid, finely divided substance, on the surface of which is adsorbed a liquid of relatively low volatility. This liquid serves as the *stationary* phase. Due to selective phase distribution of the components of the mixture between the mobile and stationary phases, these components may move through the column at different rates, and thus be separated. The physical process involved in the separation of the components of a mixture in the glpc column is the *partitioning* of the components between the gas and liquid phases.

The solubility of a gas in a liquid depends to some extent on its vapor pressure. In general, lower boiling, more highly volatile components will move through the column faster and exit sooner than those of lower volatility because, to a rough approximation, the higher the vapor pressure of a gas the lower its solubility in a liquid. Factors other than volatility (and thus indirectly, molecular weight) are important, however, in determining the rate at which a compound will move through the column. The solubility of a substance in a liquid is also influenced by polar interactions between the molecules of solute and solvent, *e.g.*, hydrogen bonding and other electrostatic interactions.

1. Carrier Gas Supply
2. Pressure Reducing Valve
3. Dessicant
4. Fine-Control Valve
5. Flowmeter
6. Heated Injection Port
7. Column
8. Oven
9. Detector
10. Electronic Recorder
11. Exit Port

FIGURE 3.6   Schematic diagram of apparatus for gas chromatography.

A large variety of gas chromatographs of different types are commercially available. However, the basic features of these instruments are quite similar and are represented by the schematic shown in Figure 3.6. Parts 1–5 are needed to supply dry carrier gas at a controlled flow rate. The column (7) is connected to the gas supply and is contained within an oven (8); the temperature within the oven is controlled by a thermostat and heating elements. The sample to be separated is introduced into the flow system at the injection port (6), which is individually heated to facilitate vaporization of the sample. The vaporized sample is then swept into the column by the carrier gas. As the sample passes through the column, its components separate into individual "bands" in the carrier gas, which then pass through the detector (9). The detector produces an electronic signal whose voltage is proportional to the amount of material different from the carrier gas itself present in the gas stream. The recorder (10) plots this voltage as a function of time to give the gas chromatogram (see Figure 3.8, for example). The vapors then pass from the detector into either the atmosphere or a collecting system at the exit port (11).

The elapsed time necessary for a given component to pass from the injection port to the detector is known as its *retention time*. Four important experimental factors affect the retention time of a compound: (1) the *nature* of the stationary liquid phase; (2) the *length* of the column; (3) the *temperature* at which the column is maintained; (4) the *rate of flow* of the inert gas. *The retention time is independent of the presence (or absence) of other components in the mixture.* For any given column and set of conditions (temperature and flow rate), the retention time is a property of the compound in question, and may be used to identify it (see below).[6]

The choice of a stationary liquid phase best suited for a glpc experiment depends for the most part on the types of compounds to be separated. Although there are a very large number of liquid phases available, only a few (Table 3.1) are widely used. The best choice will be that liquid giving rise to the largest differences in the partition coefficients of the components to be separated. Obviously, then, the solubility principles of organic compounds need to be considered.

An age old rule of chemistry is "like dissolves like." This would suggest that if it were desired to separate nonpolar types of compounds, a nonpolar liquid substrate should be used. On the other hand, if the components to be separated were of the polar variety, a nonpolar substrate would not be a very good choice, for none of the components would be very soluble and all would tend to pass unhindered through the column with little, if any, separation. Thus, for polar samples it is best to use polar liquid phases.

The most common means of supporting the liquid phase within the column is as a thin film coating on an inert solid support. The support should be of small, evenly meshed granules, so that a large surface area is available. This provides a correspondingly large film area in contact with the vapor phase, which is necessary for efficient separation. Some common types of solid supports are given in Table 3.2. The liquid is coated on the solid support by dissolving the liquid in a suitable low-boiling solvent and mixing this solution with the solid. The low-boiling solvent is then evaporated,

---

[6] For the accurate determinations necessary in many research laboratories, *retention volumes* are used rather than retention times. These measurements are more difficult to perform, but are more reproducible than retention times. They are a measure not of the time necessary for a component to pass through the apparatus, but instead, a measure of the *volume* of inert gas needed to carry the component through.

TABLE 3.1   GLPC STATIONARY PHASES

Liquid Phase	Type	Property	Maximum Temp. Limit, °C	Used for Separating
Squalane	Hydrocarbon grease	Nonpolar	100	Hydrocarbons, general application
Apiezon-L	Hydrocarbon grease	Nonpolar	300	Hydrocarbons, general application
Carbowax 20M	Hydrocarbon wax	Polar	250	Alcohols, C_5-C_{18} aldehydes, sulfur compounds
DC-550	Silicone oil	Intermediate polarity	275	C_1-C_5 aldehydes, C_6-up hydrocarbons, halogen compounds
QF-1	Silicone (fluoro)	Intermediate polarity	250	Polyalcohols, alkaloids, halogen compounds, pesticides, steroids
SE-30	Silicone gum rubber	Nonpolar	375	C_5-C_{10} hydrocarbons, pesticides, steroids
Diethyleneglycol succinate (DEGS)	Polyester	Polar	190	Esters, fatty acids
Butanediol succinate	Polyester	Intermediate polarity	225	Esters, fatty acids

TABLE 3.2   SOLID SUPPORTS

Chromosorb "P"	Pink diatomaceous earth (surface area: 4–6 m²/g)
Chromosorb "W"	White diatomaceous earth (surface area: 1–3.5 m²/g)
Crushed Firebrick	
Chromosorb "T"	40/60 mesh Teflon 6
Porapak	Porous polymer beads

leaving the solid granules evenly coated, and the column is then filled with these coated granules.

An alternative method of supporting the liquid phase is used in capillary columns of the Golay type. These are very long (300 ft is not unusual) and very small diameter (0.1–0.2 mm) columns. In these columns, the liquid is coated directly on the inner walls of the tubing. These types of columns are highly efficient (and relatively expensive).

In general, column efficiency increases with increasing path length and decreases with increasing diameter. As suggested in the introductory discussion above, the separation between bands increases with increasing path length. Therefore, with a longer column, the likelihood of separating two components will be increased (the ratio of their retention times will be larger). The diameter of the column will affect the band *width*. A smaller diameter column will give rise to narrower bands and greater efficiency. It is evident that with a small band separation (measured from the band centers), wide bands are more likely to *overlap* [Figure 3.7(a)] than narrow bands [Figure 3.7(b)]. Therefore a smaller diameter column will effect a more efficient separation, i.e., the *resolution* will be greater.

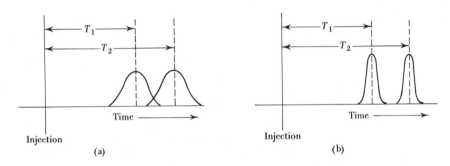

FIGURE 3.7 Effect of band width on resolution.

The temperature at which the column is maintained is another important factor in a glpc experiment. Retention times may be shortened by using higher temperatures. This is because the solubility of gases in liquids decreases with increasing temperature. The partition coefficients are thus affected, and the bands move through the column at a faster rate. This is sometimes desirable because of the convenience of a faster experiment, but it will also usually mean that the band separation will decrease, resulting in lower resolution.

For each liquid phase there is a maximum temperature limit. This point is determined by the stability and volatility of the liquid being used. At higher temperatures, the liquid phase may volatilize to some extent and be carried off the column by the carrier gas, which is of course undesirable. With some liquids there is a minimum temperature, governed by the melting point of the substance or its viscosity. Should the liquid be partially solidified or quite viscous, it will be very inefficient at dissolving the components of the gaseous mixture.

Just as with higher temperatures, higher flow rates of the carrier gas will also cause retention times to decrease. In spite of the decreased resolution obtained at higher temperatures and flow rates, these conditions are sometimes necessary for substances otherwise having very long retention times.

There are two basic types of experiments to which gas chromatography may be applied: (1) qualitative and/or quantitative analysis of the sample; (2) preparative experiments for the purpose of separating and purifying the components of the sample (preparative glpc).

Under a given set of conditions (column, temperature, flow rate, etc.), the retention time of a given compound is a property of that compound and may be used to identify it. Often, identification may be made by injecting and obtaining the retention time, under the same conditions, of material *known* to be the compound in question. If the retention time of the known compound matches that of a peak of the chromatogram, this is necessary evidence that the two compounds may be identical; it is not sufficient evidence, however, for more than one compound may exhibit the same retention time.

(I = Injection Point; t = time)

FIGURE 3.8   Gas chromatographic analysis of the distillation fractions from the distillation experiment of Chapter 2.

Since one usually has some indication of what compounds are present in the mixture, the identification is reasonably certain. Furthermore, if the voltage output of the detector is proportional to the mole fraction of the material being detected in the vapor, the relative *areas* under two peaks of the chromatogram will equal their percentage ratio in the mixture. Thus, from retention times and peak areas, the qualitative and quantitative composition of a mixture can be obtained.

An example of the analytical use of glpc is shown in Figure 3.8. These sets of peaks represent gas chromatographic analyses of the distillation fractions of a benzene-toluene mixture similar to those obtained in the distillation experiment of Chapter 2. The notations *A*, *B* and *C* refer to the three fractions taken in that experiment. Comparison of retention times with those of pure benzene and pure toluene indicates that the first peak (lower retention time) in each case is benzene and the other is toluene. Comparison of the relative areas for the peaks of fractions *A* and *C* with the packed column with the results for fractions *A* and *C* for the unpacked column demonstrates the greater efficiency of the packed fractional distillation column.

In preparative glpc, one obtains the purified components of the mixture by collecting (and condensing) the vapors as they come from the exit port. The recorder indicates when the material is passing through the detector; by making allowance for the time required for the material to pass from the detector to the exit port, one can estimate when to collect the condensed vapor. As different peaks are observed, different collection vessels may be used.

Many instruments are designed specifically for analytical determinations. Relatively small columns are used, and thus only very small samples may be injected (*ca.* 0.1–5 *micro*liters). With such sample sizes, it is evident that a very large number of repetitive injections would be needed before any sufficient quantity of the components could be collected in a preparative experiment. Instruments designed specifically for preparative experiments are also available; these utilize much larger columns (some of the larger ones are 12″ in diameter), in order to handle larger samples. Larger quantities of purified materials can thus be obtained with fewer repetitions of the injection-separation-collection cycle.

3.5 Column Chromatography

This technique is a type of liquid-solid adsorption chromatography. The stationary phase is a solid, which separates the components of a liquid passing through it by selective adsorption on its surface. The types of interactions causing adsorption are the same as those that cause attractions between any molecules, *i.e.*, electrostatic attraction, complexation, hydrogen bonding, van der Waals' forces, etc.

A column such as that shown in Figure 3.9 is used to separate a mixture by column chromatography. The column is packed with an "active" solid (the stationary phase) such as alumina or silica gel and a small liquid sample [7] is applied at the top. The sample will become adsorbed initially at the top of the column. An eluting solvent is then allowed to flow through the column; this mobile liquid phase will carry with it the components of the mixture. However, due to the selective adsorption power of the solid phase, the components may move down the column at different rates. A more weakly adsorbed compound will be eluted more rapidly than a more strongly adsorbed

---

[7] If the mixture is a solid, it must be dissolved in a minimum quantity of solvent and then applied.

FIGURE 3.9   Chromatography column.

compound because the former will have a higher percentage of molecules in the mobile phase. The process is seen to be quite analogous to that operating in glpc. The progressive separation of the components is depicted in Figure 3.10.

The separated components may be recovered in two ways: (1) the solid packing may be extruded and the portion of the solid containing the desired band cut out and extracted with an appropriate solvent; (2) solvent can be passed through the column until the bands are eluted from the bottom of the column and collected in different con-

FIGURE 3.10   Development of the chromatogram.

tainers, since they will be eluted at different times. The second method is the more commonly used; the first is rather difficult because the solid must be extruded from the column in one piece.

With colored materials, the bands may be directly observed as they pass down the column. The word "chromatography" was originally coined to describe this technique from such observations (Gr. *chromatos*, a color). With colorless materials, however, the changes cannot be observed directly. Many materials fluoresce when irradiated with ultraviolet light, however, and this provides a method of observing the bands in these cases. Usually, the progress of a column chromatographic experiment is followed by collecting a series of fractions of eluent of constant volume, for example, 25 ml. The solvent is then evaporated from each to see if any solute is present. If the volume of each fraction is kept relatively small—for example, less than 10% of the volume of the column—the different bands will usually be obtained in different flasks, although each component may be distributed among several flasks. Another convenient method of following the separation is to analyze the eluent at intervals by thin-layer chromatography (Section 3.6).

The composition of the material eluted from the column may be ascertained by determination of its physical constants (Chapter 1) or its spectra (Chapter 4) and comparison of these with those of known materials.

A few of the solid adsorbents commonly used include: alumina, silica gel, Florisil, charcoal, magnesia, calcium carbonate, starch, and sugar. The organic chemist usually finds alumina, silica gel and Florisil (fluorinated silicone polymer) of the greatest utility.

Alumina (Al_2O_3) is a highly active, strongly adsorbing, polar compound coming in three forms: neutral, and base or acid washed. Basic and acidic alumina offer good separating power for acids and bases, respectively. For compounds that are sensitive to chemical reaction under acidic or basic conditions, neutral alumina should be used. Being highly polar itself, alumina adsorbs polar compounds quite tenaciously, so that they may be difficult to elute from the column. The activity (adsorptivity) of alumina may be reduced by additions of small amounts of water; the weight percentage of water present determines the activity grade of the alumina. Silica gel and Florisil are also polar, but less so than alumina.

For the greatest effectiveness, the solid adsorbent should be of uniform particle size and of high *specific area*, a property which promotes more rapid equilibrium of the solute between the two phases. This is important for producing narrow bands. Good grades of alumina and silica gel have very high specific areas, of the order of several hundred m²/g.

The strength of adsorption depends on the adsorbate as well as on the adsorbent. It has been found that the strength of adsorption for compounds having the following types of polar functional groups increases in the order noted on any given adsorbent:

$$Cl-, \; Br-, \; I- \; < \; \diagup\!\!\!C\!\!=\!\!C\diagdown \; < \; -OCH_3 \; < \; -CO_2R \; < \; \diagup\!\!\!C\!\!=\!\!O \; < \; -CHO$$

$$< \; -SH \; < \; -NH_2 \; < \; -OH \; < \; -CO_2H$$

The nature of the liquid phases (solvents) to be used is an important consideration when designing a chromatographic experiment. The solvent may also be adsorbed on the solid, thereby competing with the solute for the adsorptive sites on the surface. If the solvent is more polar and more strongly adsorbed than the components of the mixture, these components will remain almost entirely in the mobile liquid phase and little separation will occur during the experiment. For effective separation, then, the eluting solvent must be significantly less polar than the components of the mixture. Furthermore, the components must be soluble in the solvent, for if they are not they will remain permanently adsorbed on the stationary phase of the column. The eluting powers of various solvents, *i.e.*, their ability to move a given substance down a column, are generally found to occur in the order shown.

hexane	
carbon tetrachloride	
toluene	increasing
benzene	eluting
dichloromethane	power
chloroform	
diethyl ether	
ethyl acetate	
acetone	
propanol	
ethanol	
methanol	
water	

In a *simple elution* experiment, the sample [8] is placed on the column and a *single* solvent is used throughout the separation. The optimum solvent choice will be that which produces the greatest band separation. Since the best solvent will most likely be found only by trial and error, it is sometimes convenient to use the techniques of thin-layer chromatography (Section 3.6) in the selection of a solvent for column chromatography. A series of thin-layer chromatographic experiments using various solvents may be performed in a relatively short time. The best solvent, or mixture of solvents, found in this way will usually be appropriate for the column chromatography.

A procedure known as *stepwise* (or *fractional*) *elution* is most commonly used. In this method, a *series* of increasingly more polar solvents is used to develop the chromatogram. Starting with a nonpolar solvent (usually hexane), one band may move down and off the column while the others remain very near the top. Ideally, then, the solvent would be changed to one of slightly greater polarity in the hope that one more band will be eluted while the others remain behind. If too large a jump in polarity is attempted, all of the remaining bands may come off at once. Therefore, small systematic increases in solvent polarity should be effected at each step. This is best accomplished not by changing solvents entirely, but by using *mixed* solvents. For example, an appropriate quantity [9] of hexane could be passed through the column, followed by a quantity of solvent having the composition 95% hexane—5% benzene. Solvent mixtures containing 10, 15, 20, 40, and 80% benzene could then be used in succession, followed by pure benzene. A similar series utilizing benzene with diethyl ether or chloroform could then be employed. This technique has proven to be quite efficient with regard both to column resolution and to time invested, and is very commonly employed in research laboratories.

The method of packing the column is exceedingly important, because a poorly packed column will have minimal resolution. The packing should be homogeneous and should not have entrapped air or vapor bubbles. The best way to achieve an evenly packed column is described in the experiment below. Column sizes are variable, but the same considerations hold as for glpc: the larger the diameter, the longer the column should be.

---

[8] A sample size of approximately 1 g of sample per 25 g of adsorbent is satisfactory.

[9] A "rule-of-thumb" to use in determining the amount of solvent to pass through before changing to one of higher polarity is to use a volume equal to approximately three times the packed volume of the column. There are exceptions, but this is a reasonable suggestion in the absence of more specific knowledge

EXPERIMENTAL PROCEDURE

A. PREPARATION OF COLUMN

Clamp in a vertical position a 50-ml buret with its stopcock closed but ungreased. Fill the buret to approximately the 40-ml mark with 30–60° petroleum ether, and insert a small plug of glass wool into the bottom of the buret by means of a long piece of glass tubing. Introduce into the buret enough clean sand to form a 1-cm layer on top of the glass wool plug. As the sand settles through the liquid, any entrapped air bubbles will be allowed to escape. *Slowly* add 15 g of alumina to the buret while continually tapping the column with a "tapper" made from a pencil and a one-holed rubber stopper. The agitation of the column while the alumina is being sifted in assures an evenly packed column. With a little additional petroleum ether, wash down the inner walls of the buret to loosen any adhered alumina. In order to protect the packed alumina, introduce a 1-cm layer of sand at the top of the column. Open the stopcock and allow the solvent to drain just to the top of the top layer of sand.[10] The column, as shown in Figure 3.9, is now ready to receive the sample mixture to be separated.

B. SEPARATION OF SYN- AND ANTI-AZOBENZENES

Obtain from your instructor 1–2 ml of a half-saturated petroleum ether solution of *syn*- and *anti*-azobenzene. Prepare a pipet by drawing out a piece of soft glass tubing.

$$
\begin{array}{cc}
\text{N=N} & \text{C}_6\text{H}_5 \\
& \text{N=N} \\
\text{C}_6\text{H}_5 \qquad \text{C}_6\text{H}_5 & \text{C}_6\text{H}_5 \\
\textit{syn}\text{-azobenzene} & \textit{anti}\text{-azobenzene}
\end{array}
$$

Apply 1.0 ml of the sample to the column with the pipet, allowing the solution to run evenly down the inside surface of the buret. Open the stopcock and allow the liquid to drain to the top of the sand. Introduce, in a like manner, 1–2 ml of fresh petroleum ether to wash down the inner surface of the buret. Again allow the liquid to drain to the top of the sand. Fill the buret with fresh petroleum ether. Open the stopcock; as the solvent runs through the column, a broad orange band of the *anti*-isomer should be observed to move slowly down the column. A narrow yellow band of the *syn*-isomer should remain very near the top of the column. Allow petroleum ether to run through the column until the orange band is eluted and collected at the bottom of the column in a small Erlenmeyer flask.[11] Remove the solvent from the solution of the *anti*-isomer on a steam bath in the hood. Air-dry the material obtained and determine its melting point.

After eluting the *anti*-isomer, allow the liquid in the buret to drain to the top of the sand and fill the buret with 5% methanol (by volume) in petroleum ether. Elute and collect the *syn*-isomer, remove the solvent, and, after air-drying, determine its melting point.

If desired by your instructor, obtain infrared spectra (Chapter 4) of the purified isomers and carefully note the differences.

---

[10] At no time during the experiment should the solvent be allowed to drain below this level. Air bubbles and channels would be introduced into the alumina which would give rise to ragged bands during the development of the column.

[11] If it is desired to elute the *anti*-isomer more rapidly, 20% benzene (by volume) in petroleum ether may be used for the second filling of the buret.

3.6 Thin-Layer Chromatography

Thin-layer chromatography involves the same principles as column chromatography; it is a form of solid-liquid adsorption chromatography. However, in this case the solid adsorbent is spread as a thin layer (*ca.* 250μ) on a piece of glass or rigid plastic. A drop of the solution to be separated is placed near one edge of the plate and the plate is placed in a container (developing chamber) with enough of the eluting solvent to come to a level just below the "spot." The solvent migrates *up* the plate carrying with it the components of the mixture at different rates. The result may then be a series of spots on the plate, falling on a line perpendicular to the solvent level in the container [see, for example, Figure 3.11(b)].

This chromatographic technique is very easy and rapid to perform. It lends itself well to the routine analysis of mixture composition, and may also be used to advantage in determining the best eluting solvent for subsequent column chromatography.

The same solid adsorbents used for column chromatography may be employed for tlc, with silica and alumina being the most widely used. The adsorbent is usually mixed with a small amount of "binder," for example, plaster of paris, calcium sulfate or starch, to insure proper adherence of the adsorbent to the plate. The plates may be prepared before use, or commercially available prelayered plastic sheets may be used.

The relative eluting abilities of the various solvents that may be used are the same as those given earlier in Table 3.2. It should be remembered that the eluting power required of the solvent is directly related to the strength of adsorption of the components of the mixture on the adsorbent.

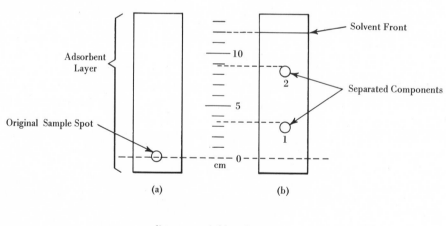

$$R_f = \frac{\text{distance traveled by substance}}{\text{distance traveled by solvent}}$$

$$R_f (\text{compound 1}) = \frac{3.4\ \text{cm}}{12\ \text{cm}} = 0.28$$

$$R_f (\text{compound 2}) = \frac{8.8\ \text{cm}}{12\ \text{cm}} = 0.73$$

FIGURE 3.11 Thin-layer chromatogram: (a) original plate; (b) developed chromatogram.

A distinct advantage of tlc is the very small quantity of sample required. A lower limit of detection of 10^{-9} grams is possible in some cases. However, sample sizes as large as 500 micrograms may be used. With the larger samples, preparative experiments may be conducted by scraping the individual spots from the plate and eluting (extracting) them with an appropriate solvent. Of course, it would be necessary to repeat this process a number of times in order to obtain even several *milli*grams of material. Such a procedure would offer a method of identification of the various spots, however, for enough could be collected to obtain spectra of the individual components (see Chapter 4).

Detection of spots on the chromatogram is easy for colored materials, and a number of procedures are available for locating spots of colorless materials. For example, irradiation of the plate with ultraviolet light will permit location of the spots of compounds that fluoresce. Alternatively, the solid adsorbent may be impregnated with an otherwise inert, fluorescent substance. Spots of materials that absorb ultraviolet light, but do not fluoresce, will show up as black spots against the fluorescing background when the plate is irradiated with ultraviolet light. Other detecting agents are more often used. These agents may be sprayed onto the chromatograms, causing the spots to become readily apparent. Examples of detecting agents used in this way are sulfuric acid, which causes many organic compounds to char, and potassium permanganate solution. Iodine is another popular detecting agent. In this case, the plate is placed in a vessel whose atmosphere is saturated with iodine vapor. Iodine is adsorbed by many organic compounds, and their spots on the chromatogram become colored (usually brown).

Under a given set of conditions (adsorbent, solvent, layer thickness and homogeneity) the rate of movement of a compound with respect to the rate of movement of the solvent front, R_f, is a property of that compound. The value is determined by measuring the distance traveled by a substance from a starting line to the front (top) of the spot, and dividing this distance by the distance the solvent has traveled, as measured from the same starting line (Figure 3.11). The property has the same significance as retention volume [6] in a glpc experiment.

EXPERIMENTAL PROCEDURE

A. Separation of Syn- and Anti-Azobenzenes

Obtain from your instructor a 10-cm strip of silica gel chromatogram sheet (without fluorescent indicator) and about 0.5 ml of a 10% benzene solution of commercial azobenzene. Place a spot of this solution on the tlc plate about 10 mm from one edge and about 20 mm from the bottom, using a capillary tube to apply the spot. The spot should be 1 or 2 mm in diameter. Allow the spot to dry and then expose the plate to sunlight for one or two hours. Alternatively, the plate may be placed beneath a sunlamp for about 20 min. After this time, apply another spot of the *original* solution on the plate at the same distance from the bottom as the first and leave about 10 mm between the spots. Again, allow the plate to dry.

A wide-mouth bottle with a tightly fitting screw top cap may be used as a developing chamber. Alternatively, a beaker covered with a watch glass may be used. Prepare an 8:1 (by volume) mixture of cyclohexane: benzene and place a 1-cm layer of this mixture in the bottom of the developing chamber. Fold a piece of filter paper into the shape of a round-ended rectangle [Figure 3.12(a)], and place it into the developing chamber as shown in Figure 3.12(b). Saturate the chamber with the vapors of the solvent by shaking. This inhibits the evaporation of solvent from the plate during the development of the chromatogram. The piece of filter paper aids in the maintenance of this saturated state.

(a) (b)

FIGURE 3.12 TLC chamber.

Place the chromatogram plate in the chamber, being careful not to splash the solvent onto the plate. Allow solvent to climb to within 15 mm of the top of the plate and then remove the plate and allow it to air-dry. Note the number of spots arising from each of the two original spots and compare the intensities of the two spots corresponding to *syn*-azobenzene (the spots nearest the starting point).

B. SEPARATION OF GREEN LEAF PIGMENTS

Place a few milliliters of a 2:1 mixture of petroleum ether and ethanol, along with a few green leaves [12] in a mortar and crush the leaves with a pestle. By means of a pipet, transfer the liquid extract to a small separatory funnel and swirl with an equal volume of water. Separate and discard the lower aqueous phase. Repeat the water washing twice, discarding the aqueous phase each time. Transfer the petroleum ether layer to a small Erlenmeyer flask and add 2 g of *anhydrous* sodium sulfate. The solution may be concentrated, if necessary, by evaporation with a stream of dry nitrogen.

Obtain from your instructor a 10-cm strip of silica gel chromatogram sheet (without fluorescent indicator). Place a 1–2 mm spot of the pigment solution on the sheet about 1.5 cm from one end, using a capillary tube to apply the spot. Allow the spot to dry. Using 7:3 (v/v) benzene: acetone as a developing solvent, develop the chromatogram according to the general directions given in part A.

It is possible to observe as many as eight colored spots. In order of decreasing R_f values, these spots may be identified as the carotenes (two spots, orange), chlorophyll *a* (blue-breen), chlorophyll *b* (green), and the xanthophylls (four spots, yellow).

EXERCISES

1. Which of the two isomers of azobenzene would you expect to be the more thermodynamically stable and why?

---

[12] Green leaves from a variety of sources might be used, such as grass and various types of trees and shrubberies. The interested student may wish to complete this experiment with more than one type of leaf and compare the results.

2. From the results of the tlc experiment with the azobenzenes, describe the role of sunlight.
3. In a tlc experiment, why must the *spot* not be immersed in the solvent in the developing chamber?
4. Explain why the solvent must not be allowed to evaporate from the plate during the development.

3.7 Paper Chromatography

Paper chromatography bears a resemblance to thin-layer chromatography but is slightly different in principle. Small spots of the mixture to be separated are placed near the bottom of a strip of paper; the end of the paper strip (but not the spot) is placed in solvent, and, as the solvent ascends up the paper, the components of the mixture are separated into individual spots. Paper chromatography involves distribution of the sample between a polar liquid phase (normally water), which is strongly adsorbed on the cellulose fibers of the paper, and an eluting solvent. It is thus an example of liquid-liquid partition chromatography.

Compounds may be identified by comparison of their R_f values (defined in Section 3.6). R_f values are far more reproducible with paper chromatography than with tlc and thus are more valuable in this application.

The great utility of paper chromatography (and tlc as well) is in the rapid analysis of reaction mixtures rather than in separation of compounds on a useful preparative scale. The minute quantities of sample required to accomplish an analysis and the ease and rapidity with which qualitative analysis of reaction mixtures can be performed make this chromatographic technique a valuable tool to the organic chemist, and especially to the biochemist.

EXPERIMENTAL PROCEDURE

A. ANALYSIS OF INKS

Cut a 3 × 12-cm strip of filter paper. Apply spots of two different ink solutions (washable black, washable blue and permanent emerald green are provided) to the paper at a distance approximately 1 cm from the end and 5 mm from each edge. The amount of solution used should be such that spots of only 1–2 mm in diameter are formed. Allow the paper to dry. Place a pencil mark at the edge of the strip at the level of these spots. Thread the paper through a slot in a stiff piece of cardboard, and hang the strip in a vertical configuration within a wide-mouth bottle which contains a 1-cm layer of water. The stiff piece of paper should cover the mouth of the bottle making the bottle as air-tight as possible (Figure 3.13). Do not let the strip of paper touch the sides of the bottle; only the bottom of the strip (the end having the spots of ink) is to be in contact with solvent. The spots should be above the water surface and not in contact with it. Leave the strip in the solvent until the solvent has climbed about 10 cm up the paper. Then remove the strip, place a pencil mark on it at the solvent front and allow the paper to dry. The developed chromatogram should exhibit several colored spots, each of which corresponds to an organic dye which was a component of one or the other of the two inks. Calculate the R_f value for each dye.

Repeat this experiment with a *mixture* of the two inks. Calculate R_f values for each of the spots obtained. Using the R_f values obtained from the first part of the experi-

Solvent

Paper
Chromatogram

FIGURE 3.13   Paper chromatography chamber.

ment, identify each of the spots of the chromatogram of the mixture of inks with one or the other of the original ink solutions.

B. ALTERNATIVE EXPERIMENT

As an alternative to the use of inks in part A, aqueous solutions of various food coloring agents may be used.

REFERENCES

A. EXTRACTION

L. C. Craig and D. Craig, in *Techniques of Organic Chemistry*, 2nd Ed., A. Weissberger, editor, Interscience Publishers, New York, 1956, Vol. III, Part I, Chapter 2.

B. CHROMATOGRAPHY

1. R. Stock and C. B. F. Rice, *Chromatographic Methods*, Reinhold Publishing Corp., New York, 1963.
2. *Chromatography*, E. Heftmann, editor, Reinhold Publishing Corp., New York, 1961.
3. A. I. M. Keulemans, *Gas Chromatography*, C. G. Verver, editor, Reinhold Publishing Corp., New York, 1957.
4. H. G. Cassidy, in *Techniques of Organic Chemistry*, A. Weissberger, editor, Interscience Publishers, New York, 1957, Vol. X, Chapter 8.
5. J. G. Kirchner, in *Techniques of Organic Chemistry*, E. S. Perry and A. Weissberger, editors, Interscience Publishers, New York, 1967, Vol. XII.
6. Louis F. Fieser, "Thin Layer Chromatography," *Chemistry*, **37**, 23 (1964).

4

Spectroscopic methods of identification and structure proof

An important part of the science of organic chemistry is determination of the exact structures of products formed in chemical reactions. This formerly was an exceedingly time consuming and often impossible task for the classical organic chemist because his primary method of determining the structures of compounds usually involved conversion, by chemical reaction, of his "unknown" substance to a compound having a "known" structure. The procedure is now known to have often led to erroneous conclusions concerning the identity of the "unknown" because the reactions employed to convert the "unknown" to a "known" compound often produced structural rearrangements which went unrecognized by the investigator. The modern organic chemist fortunately has at his disposal various spectroscopic techniques which supplement and even replace the more classical chemical methods of proof of structure and which permit much more rapid analysis of "unknowns." It is noteworthy that the development of spectroscopy as a tool for the elucidation of structures of organic compounds has, more than anything else, contributed to the extremely rapid growth of organic chemistry in the past fifteen years, and, consequently, it is vitally important that the student learn how to analyze and to interpret the information available from infrared (ir), nuclear magnetic resonance (nmr) and ultraviolet (uv) spectroscopy and mass spectrometry (ms). Only the first two spectroscopic techniques will be discussed in this chapter because they are more readily available and of greater utility to the undergraduate. However, the interested student can profit by reading one or more of the references on ultraviolet spectroscopy and mass spectrometry cited at the end of the chapter.

4.1 Infrared Spectroscopy

The spectra most generally available for aid in elucidation of the structures of an organic compound are those obtained by irradiation of the compound with light from the infrared region, 5000 to 500 cm^{-1} (reciprocal centimeters or wave numbers),[1] of the

---

[1] The energy (in ergs) available from electromagnetic radiation is given by the familiar formula, $E = h\nu$, where h is Planck's constant (6.62×10^{-27} erg/second) and ν is the frequency of the radiation in cycles per second. The frequency of radiation expressed in cm^{-1} is obtained by dividing ν by the speed of light, c, expressed in cm/sec. By dividing 10,000 by the frequency in cm^{-1}, the wavelength, in microns (μ), of the radiation is obtained. The position of an absorption band can thus be expressed either in microns or in wave numbers; the latter designation will be used in this text.

electromagnetic spectrum. The interaction of infrared light with an organic molecule can be explained qualitatively by imagining that molecular bonds between atoms are analogous to springs. These molecular springs are constantly undergoing stretching and bending motions (Figure 4.1) at frequencies which depend upon the masses of the

Symmetric stretch   Asymmetric stretch   Symmetric in-plane bend   Asymmetric in-plane bend

FIGURE 4.1   Some vibrational modes of a group of atoms.

atoms involved and upon the type of chemical bond joining the atoms. Since the frequencies of the various vibrations of the molecule correspond to those of infrared radiation, absorption of the radiation occurs, producing an increase in the *amplitude* of the molecular vibrational modes. No irreversible change in the molecule results, however, because the energy gained by the molecule in the form of light is soon lost in the form of heat. By plotting the percent transmittance, the ratio of the intensity of the light passing through the sample, I, to the intensity of the light striking the sample, $I°$, multiplied by 100 [equation (1)], *vs.* the frequency of the radiation, an *infrared spec-*

$$\%T = \frac{I}{I_\circ} \times 100 \tag{1}$$

trum such as that shown in Figure 4.2 is obtained.

FIGURE 4.2   IR spectrum of a 3-methyl–2-cyclohexenone.

As a general rule, absorptions between 5000 and 1250 cm^{-1} involve vibrational excitations of particular functional groups, such as carbonyl, cyano, and nitro, present in the molecule. Thus, the stretching mode of the carbon-carbon double bond of an alkene usually appears in the region 1600–1680 cm^{-1} (see Figure 4.2). Analysis of the portion of the infrared spectrum from 5000–1250 cm^{-1} can therefore provide much information concerning the presence or absence of a number of functional groups. A tabulation of the normal frequency range of absorption for a number of groups is presented in Table 4.1.

TABLE 4.1 INFRARED ABSORPTION RANGES OF FUNCTIONAL GROUPS

Bond	Type of Compound	Frequency Range, cm^{-1}	Intensity
C—H	Alkane	2850–2970	Strong
		1340–1470	Strong
C—H	Alkenes $\left(\begin{array}{c} \diagdown \\ / \end{array} C{=}C \begin{array}{c} {}^{H} \\ \diagdown \end{array} \right)$	3010–3095	Medium
		675–995	Strong
C—H	Alkynes (—C≡C—H)	3300	Strong
C—H	Aromatic rings	3010–3100	Medium
		690–900	Strong
O—H	Monomeric alcohols, phenols	3590–3650	Variable
	Hydrogen-bonded alcohols, phenols	3200–3600	Variable, sometimes broad
	Monomeric carboxylic acids	3500–3650	Medium
	Hydrogen-bonded carboxylic acids	2500–2700	Broad
N—H	Amines, amides	3300–3500	Medium
C=C	Alkenes	1610–1680	Variable
C=C	Aromatic rings	1500–1600	Variable
C≡C	Alkynes	2100–2260	Variable
C—N	Amines, amides	1180–1360	Strong
C≡N	Nitriles	2210–2280	Strong
C—O	Alcohols, ethers, carboxylic acids, esters	1050–1300	Strong
C=O	Aldehydes, ketones, carboxylic acids, esters	1690–1760	Strong
NO$_2$	Nitro compounds	1500–1570	Strong
		1300–1370	Strong

The absorption maxima observed between 1250 and 500 cm^{-1} in the infrared spectrum cannot usually be associated with vibrational excitation of a particular functional group, but rather are the result of a complex vibrational-rotational excitation of the *entire* molecule. Accordingly the spectrum from 1250–500 cm^{-1} is characteristic and unique for every compound, and the apt description, "fingerprint region," is often applied to this portion of the spectrum. It is considered highly unlikely that two different organic compounds would exhibit identical spectra in the fingerprint region even though their spectra might be quite similar from 5000–1250 cm^{-1} (see Figure 4.3).

The apparently unique character of the infrared spectrum of each different organic compound has led to the acceptance of the hypothesis that if two spectra are completely identical as to positions and intensities of absorption maxima, that is, the spectra are completely superimposable, the spectra must be of the same compound. Comparison of spectra of "unknown" compounds with those of known compounds can therefore be one very useful technique for structure elucidation.

(a) 2,3-dimethylbutane

(b) 2-methylbutane

FIGURE 4.3   IR spectra of methylbutanes.

When the student is confronted with the infrared spectrum of a compound which he is attempting to identify, he should first attempt to determine what functional groups are present in the compound by careful analysis of the region, 5000–1250 cm^{-1}. Subsequently, he may be able to assign a few of the bands in the fingerprint region of the spectrum in order to define further the structure of the unknown substance. At this point, he should be in a position to formulate several possible structures for the unknown and may be able to differentiate between the several possibilities by comparison of their published spectra with that of the unknown. Several catalogs of infrared spectra are currently available, and some of the more useful and comprehensive are listed in the references at the end of this chapter.

A majority of the possibilities may also be eliminated by a comparison of the physical properties of the known, for example, mp, bp, index of refraction, with those of the unknown, or by their nuclear magnetic resonance spectra (see below). If, however, reference spectra of some of the substances considered as possibilities are not available, the information derived from the analysis of the infrared spectrum of the unknown may be utilized to suggest a specific chemical reaction which would provide a solid derivative, the melting point of which is published in the chemical literature. The identity of the derivative of the unknown with that of the known would then depend on a comparison of their melting points, or, ideally, a melting point of a mixture of the derivative of the unknown and an authentic sample of the same derivative obtained from the known compound.

Experimentally, infrared spectra can be obtained of samples contained in the gas, liquid (either of a pure, "neat," substance or of a solution) or solid phases. Only the general techniques necessary for the preparation for infrared spectra of samples in the liquid phase will be considered here since the apparatus required for obtaining spectra of gaseous or solid samples is somewhat too specialized for adaptation to the basic organic chemistry course. The student should see reference 2 for a thorough discussion of various ways of preparing samples for infrared analysis.

A brief discussion of the factors which determine the intensity of an infrared spectrum is appropriate here. The amount of absorption, termed absorbance, A, at a given frequency is defined by equation (2) in which I and I_o are, respectively, the

$$A = \log_{10} \frac{I_o}{I} = kcl \tag{2}$$

amounts of light of the given frequency transmitted by and impinging on the sample, k is the absorptivity of the sample at that frequency, l is the path length, in cm, of the cell in which the sample is contained, and c is the concentration, in g/l, of the solute in the solution. Although the absorptivity is a constant, its magnitude is frequency dependent which results in the observation of maxima and minima in the spectrum and which is somewhat characteristic of the functional group absorbing the radiation. Thus, the absorptivity of the stretching mode of a carbonyl group is generally greater than that of a carbon-carbon double bond, as can be seen by comparison of the intensities of the corresponding bands in the spectrum reproduced in Figure 4.2.

As the absorptivity of a given molecule is not subject to experimental modification, there are only two variables which can be changed in order to control the intensity of a spectrum: the path length, l, of the cell containing the sample and the concentration, c, of the solute in a solution. For spectra of neat samples, of course, only the former alternative is available, and it is found that path lengths of 0.025 to 0.030 mm are appropriate. As a general rule, solutions having 5 to 10 weight percent of solute and contained in a 0.1 mm cell yield suitable spectra.

The windows of most types of infrared cells are constructed of a clear fused salt such as sodium chloride or potassium bromide. As a result, contact of the cells with moisture must be scrupulously avoided or the windows will soon become cloudy and will transmit very little light. Thus, one should avoid breathing directly onto the windows, touching them with fingers, or filling them with samples which contain moisture. Furthermore, the cells should be cleaned with *dry* solvent (it is unwise to use any hydroxylic solvents, even alcohols, to wash the cells) and flushed with a slow stream of dry nitrogen or some other dry gas. The dried infrared cells should then be stored in a desiccator.

Infrared spectroscopy provides valuable information concerning the functional groups present in molecules and serves as a means of identifying an unknown compound if a reference spectrum of the compound is available.[2] In those cases, and there

---

[2] In addition to its use as a tool for *qualitative analysis* of organic compounds, infrared spectroscopy is also utilized for *quantitative analysis* of mixtures of known substances. A discussion of this latter technique can be found in reference 2 at the end of this chapter.

are many, in which the infrared spectrum alone does not provide sufficient information to permit determination of structure, other chemical or spectroscopic techniques must be employed. One such spectroscopic technique is discussed below.

EXERCISE

Assign as many bands as possible in the spectra presented in Figures 4.2 and 4.3 to the functional groups responsible for the absorptions.

4.2 Nuclear Magnetic Resonance Spectroscopy

Although infrared spectroscopy has been used profitably by organic chemists since the early 1950s, nuclear magnetic resonance (nmr) spectroscopy became generally available only in the early 1960s. This extremely important experimental technique is based upon the interesting property of *nuclear spin* exhibited by hydrogen nuclei.[3,4]

The hydrogen nucleus behaves as a spherical body having a uniform distribution of charge and a mechanical spin about an axis. The spin of the nucleus produces circulation of charge which, in turn, generates a nuclear magnetic dipole, the direction or moment of which is *colinear* with the axis of nuclear spin [Figure 4.4(a)].

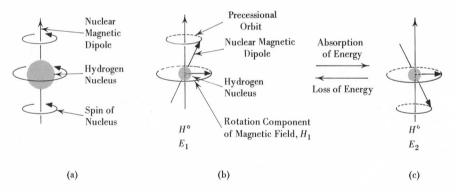

(a)   (b)   (c)

FIGURE 4.4 Spin properties of the hydrogen nucleus.

When the spinning nucleus is placed in an external magnetic field, $H°$, the magnetic dipole, and hence the axis of spin, becomes oriented either *with* the applied field, state E_1, or *against* it, state E_2 [Figure 4.4(b)]. The energy state E_1 represents the more stable alignment. It is very important to realize that the magnetic dipole does not become aligned either precisely parallel or antiparallel to $H°$ but rather traces out a circular path about the axis defined by $H°$, in a process called *precession*. The phenomenon of nuclear precession can be likened to the motion observed for a spinning gyroscope or toy top that has not been oriented exactly parallel to the earth's gravitational field.

---

[3] It will be convenient, during the presentation of this discussion, occasionally to refer to the hydrogen nucleus simply as a hydrogen or as a proton. These latter designations are not strictly correct since we are referring to the nucleus of a covalently bound hydrogen and not to the free atom or to the charged ion, H^+.

[4] The phenomenon is not limited to hydrogen nuclei but is observed for other nuclei, such as ^{19}F, ^{13}C, ^{31}P, as well. Our discussion, however, will treat only the case of hydrogen.

The frequency, $\omega°$, of the precession is linearly related to the strength of the applied field, $H°$, as is expressed in equation (3), in which the constant, γ_H, is the magnetogyric

$$\omega° = \gamma_H H° \tag{3}$$

ratio for the hydrogen nucleus. Thus, $\omega°$ increases as $H°$ increases. Now, a transition between the energy states E_1 and E_2 can be accomplished if the nucleus, precessing at a frequency, $\omega°$, in a given field, $H°$, is subjected to a second, oscillating magnetic field, H_1, which has a rotation component perpendicular to $H°$ [Figure 4.4(b)]. The frequency of rotation of the magnetic component can be varied by changing the frequency, ν, of the oscillator used to produce the oscillating magnetic field, H_1.

When the frequency of rotation becomes equal to the precessional frequency, $\omega°$, *resonance* between the precessing nucleus and H_1 will be achieved. At this point, the nucleus in state E_1 can absorb the electromagnetic energy provided by the oscillator and undergo a "flip" of its spin into the higher energy state, E_2. The absorption of energy can be detected electronically and recorded as a peak on a chart.

The oscillator frequency required to produce the desired nuclear magnetic transition depends upon the strength of the applied field, $H°$, as is expressed in equation (4). It

$$\nu = \frac{\gamma_H H°}{2\pi} \tag{4}$$

turns out that if $H°$ is of an experimentally convenient magnitude, about 14,000 gauss, ν must be of the order of 60 megacycles per second (Mcps),[5] in order to produce resonance. This value is in the radiofrequency (rf) region of the electromagnetic spectrum so that an rf oscillator is required.

Inspection of equation (4) should make it apparent that the condition of resonance for a proton could be achieved either by holding $H°$ constant and varying ν, as is described above, or by maintaining ν at a constant value and changing $H°$. The latter approach is more practical and is employed in all commercially available spectrometers.

A typical nmr spectrum, that of 1-nitropropane, is presented in Figure 4.5. In the spectrum, the strength of the applied magnetic field, $H°$, *increases* as one goes from left to right, the *upfield* direction. Since the frequency of the radio oscillator is held constant, the *energy* required to produce the transition between the spin states, E_1 and E_2, also *increases* from left to right.[6] The three groups of peaks, *A*, *B*, and *C*, all result from absorption of energy by hydrogens, but the amount of energy required to realign the nuclei in the magnetic field, *i.e.*, the energy necessary to produce *resonance*, is clearly not identical for all the protons in the molecule. This is surprising as equation (4) indicates that all hydrogen nuclei should undergo spin flip at the same $H°$ (at a given value of ν) so that only a *single* peak would be expected.

The qualitative explanation for the observation of more than one peak is as follows. The three groups, *A*, *B*, and *C*, represent three types of hydrogens which are rendered nonequivalent by virtue of the slightly different magnetic environments in which they

---

[5] One megacycle is 10^6 cycles. Another designation, Hertz (Hz), is synonymous with cycles per second (cps) and is often used. The term, megaHertz (MHz), which is equivalent to Mcps, is also encountered.

[6] Both ν and $H°$ are linearly related to energy through the expression, $E = h\nu = \gamma_H H°/2\pi h$.

FIGURE 4.5   NMR spectrum of 1-nitropropane.

exist. The magnetic environment of each proton in the molecule is dependent upon two major factors: (1) The externally applied field, $H°$, which has a uniform strength over the entire molecule and therefore cannot produce nonequivalence of the different hydrogens; (2) a magnetic field which is induced by circulation of the electrons of the molecule and which, because the electron density varies over the molecule, is *not* uniform. This latter field results in the magnetic nonequivalence of the protons. Those hydrogen nuclei which experience higher fluxes of the induced magnetic field of the molecule will, because this field will act to oppose the external, applied field, be more highly shielded from the external field. In other words, more highly shielded nuclei will experience an *effective* field, H_e, which is *less* than the applied field so that *higher* applied fields will be necessary to produce the desired nuclear spin transitions. The appearance of more than one group of peaks in the spectrum therefore reflects the difference in the flux of the magnetic field induced within the molecule by circulation of electrons.

The positions of peaks in the nmr spectrum of a compound are routinely measured relative to the position of the peak produced by the protons of a standard compound, tetramethylsilane (TMS), $(CH_3)_4Si$. In reference to the spectrum of Figure 4.5, the groups of peaks are found to be centered at 60, 123 and 263 cycles per second (cps) *downfield* from the TMS peak when an rf oscillator set at 60 Mcps is employed. These values represent the *chemical shifts* relative to TMS of the three types of hydrogen nuclei in 1-nitropropane.

The chemical shift of a hydrogen determined in units of cps is dependent on the frequency at which the oscillator is operating. It has been found useful to convert such shifts into units independent of the oscillator frequency. This is accomplished by dividing the chemical shift, expressed in cps, by the frequency, in cps, of the oscillator and then, to arrive at more convenient numbers, multiplying the result by 10^6 [equation (5)]:

$$\delta = \frac{\text{chemical shift in cps}}{\text{oscillator frequency in cps}} \times 10^6 \tag{5}$$

Chemical shifts obtained in the above manner are values on a standard scale, termed the delta (δ) scale, which has the peak for TMS at 0δ, peaks downfield from TMS at values of $+\delta$, and peaks upfield from TMS at values of $-\delta$.[7] The oscillator frequency-independent chemical shifts, δ, of the groups of peaks observed for 1-nitropropane are as follows:

Group A: $(60 \text{ cps})/(60 \times 10^6 \text{ cps}) = 1.00 \times 10^{-6}$ or 1 part per million (ppm)
$(1.00 \times 10^{-6})(10^6) = 1.00 \, \delta_A$
Group B: $(123)/(60 \times 10^6)(10^6) = 2.05 \, \delta_B$
Group C: $(263)/(60 \times 10^6)(10^6) = 4.38 \, \delta_C$

A second standard scale, values of which are also independent of oscillator frequency, has been derived in which the TMS peak is defined as appearing at 10 tau (τ).[8] To convert a chemical shift given on the delta scale to a value on the tau scale, one simply subtracts the shift as measured on the delta scale from 10. The groups of peaks in the spectrum of 1-nitropropane, for example, appear at $10 - 1.0 = 9.0 \, \tau$, $10 - 2.05 = 7.95 \, \tau$ and $10 - 4.38 = 5.62 \, \tau$. As both these standard scales are currently in use, it is important to know how they are derived and how they can be interconverted.

The magnitude of the chemical shift of a proton attached to a given type of functional group is relatively constant so that determination of the shifts of the various

TABLE 4.2   TYPICAL CHEMICAL SHIFTS OF HYDROGENS ATTACHED TO VARIOUS TYPES OF FUNCTIONAL GROUPS

Type of Hydrogen	*Chemical Shift, ppm* τ	*δ*	*Type of Hydrogen*	*Chemical Shift, ppm* τ	*δ*
Cyclopropane	9.6–10.0	0.0–0.4	ICH	6–8	2–4
RCH_3	9.1	0.9	OCH (alcohols,		
R_2CH_2	8.7	1.3	ethers)	6–6.7	3.3–4
			O—CH$\,|\,$O	4.7	5.3
R_3CH	8.5	1.5			
C=CH	4.1–5.4	4.6–5.9	OCH (esters)	5.9–6.3	3.7–4.1
CCH	7–8	2–3	RO_2CCH	7.4–8	2–2.6
ArH	1.5–4	6–8.5	R$\,|\,$O=CCH	7.3–8	2–2.7
ArCH	7–7.8	2.2–3			
C=CCH$_3$	8.3	1.7			
C≡CCH$_3$	8.2	1.8	O$\,\|\,$RCH	0–1	9–10
FCH	5–5.6	4–4.5			
ClCH	6–7	3–4	ROH	4.5–9	1–5.5
Cl_2CH	4.2	5.8	ArOH	−2 to 6	4–12
BrCH	6–7.5	2.5–4	RCOOH	−2 to −0.5	10.5–12
O_2NCH	5.4–5.8	4.2–4.6	RNH_2	5–9	1–5

[7] Actually, the sign convention should be exactly opposite to that given, so that peaks downfield from TMS would be at values of $- \delta$. However, common practice corresponds to the convention described above.

[8] The standard value of 10τ is chosen because the protons of most organic compounds absorb within 600 cps downfield of TMS when a 60-Mcps oscillator is used.

protons of an unknown compound will often provide valuable information concerning the functional groups present in the compound. Tables are available which correlate chemical shifts with structural features, and a partial compilation is presented in Table 4.2.

The peaks within each group of absorptions in the spectrum of 1-nitropropane are the result of *spin-spin splitting*, a phenomenon which results because there can be magnetic interaction, termed *coupling*, between magnetically nonequivalent hydrogen nuclei in the molecule. As a rule, only nonequivalent nuclei on the same carbon or on adjacent carbons will interact with one another to produce spin-spin splitting. Thus, if we label the three types of hydrogens as shown in Figure 4.6, we would expect the nuclei designated H_a to couple with those labeled H_b and these latter protons to interact with both the nuclei, H_a *and* H_c. However, the H_c protons should couple only with the H_b protons.

$$
\begin{array}{ccc}
H_c & H_b & H_a \\
| & | & | \\
H_c-C-&C-&C-NO_2 \\
| & | & | \\
H_c & H_b & H_a
\end{array}
$$

FIGURE 4.6   Types of hydrogens in 1-nitropropane.

The magnitude of the spin-spin splitting between hydrogen nuclei can be measured in cycles per second, and the values obtained are called *coupling constants, J*. In contrast to chemical shifts, the magnitude of coupling constants is *independent* of oscillator frequency, so units of cps or Hz[5] are used with such values. Some typical hydrogen-hydrogen coupling constants, J_{HH}, are given in Figure 4.7.

$J_{HH} = 10\text{--}15$ cps $J_{HH} = 5\text{--}8$ cps

$J_{HH} = 11\text{--}18$ cps $J_{HH} = 6\text{--}14$ cps

FIGURE 4.7   Some typical coupling constants.

The number of peaks into which the nmr absorption of a given hydrogen nucleus, H_i, will be divided due to spin-spin splitting depends in a linear manner on the number of protons coupling with H_i. Thus, if N protons, all of which have the same coupling constant, J, with a hydrogen, H_i, are coupled with this nucleus, the absorption of H_i will appear as a group of $N + 1$ peaks.[9] If the coupling constants are not identical, the spectrum may be considerably more complicated, but a detailed discussion of such cases is not appropriate here.

[9] This statement is an oversimplification and is strictly true only in those cases in which the ratio of the chemical shift, expressed in cps, between coupled protons and their coupling constant is greater than 5 to 10. For further discussion of this point, see, for example, references 3, 5, and 7.

Applying the $N + 1$ rule to the different hydrogens of our reference compound, 1-nitropropane, the H_a protons should appear as a triplet, the H_b protons as a sextet, *if* $J_{ab} = J_{bc}$, and the H_c protons as a triplet. These expectations are confirmed in the observed spectrum. It should be apparent that the multiplicity of the nmr absorption of a given hydrogen, H_i, can be utilized to determine the number of protons adjacent to it, a fact which is quite useful for purposes of analyzing spectra.

The relative numbers of different types of protons can be determined by integration, electronic or manual, of the absorption bands of the spectrum since the areas under the bands are a linear function of the number of hydrogen nuclei producing the band. For example, in the spectrum of Figure 4.5, the relative areas of the three bands are obtained by measuring the distances, *a–b*, *a–c* and *a–d*. The ratio of these heights is (if measured in millimeters) 28:28:44, so that the relative abundance of the corresponding protons is determined to be 2.0:2.0:3.1. This is within experimental error of the absolute ratio of 2.0:2.0:3.0 anticipated for 1-nitropropane.

It should now be clear that analysis of an nmr spectrum can yield a wealth of information. Determination of the *chemical shift* of each peak or group of peaks allows one to speculate about the *type of functional group* to which the hydrogen nucleus producing the absorption is attached; the *spin-spin splitting* pattern provides knowledge concerning the *number of nearest neighbors* of the hydrogen; and *integration* permits an evaluation of the *relative numbers of each type* of hydrogen present in the molecule.

When attempting to analyze the nmr spectrum of an unknown compound, it is profitable to begin by determining the relative numbers of different types of hydrogen present by measurement of the integrated peak areas. If the molecular formula of the unknown is available, the *relative* ratios can be converted to *absolute* numbers of the hydrogens present.

The second step in the analysis is an attempt to define the functional groups present in the molecule by determination of the chemical shifts of the different hydrogens. Further knowledge concerning the molecular environment of a given type of hydrogen can be derived by analysis of the spin-spin splitting patterns. At this point, it should be possible to propose for the unknown one or more complete, or at least partial, structures which are consistent with the spectrum.

If an infrared spectrum of the unknown is also available, it can be analyzed as described in Section 4.1. This analysis should provide additional information concerning the nature of functional groups present, and this information may well be used to supplement or to confirm that derived from analysis of the nmr spectrum.

The structures of many "unknowns" whose molecular formulae are available can often be derived immediately by combination of the information available from the ir and nmr spectra.[10] However, proper interpretation of the spectra requires considerable practice on the part of the beginning student of spectroscopy. In order to provide an opportunity for such practice in interpretation, the spectra of many of the organic compounds encountered as reagents or products in reactions described in this text have been reproduced at the end of each experiment. As these spectra are labeled, you will not be attempting to assign a structure to the substance producing them, but will be

---

[10] Structural assignments based solely on interpretation of spectra must be confirmed either by comparison of the spectra of the unknown with those of the known compound, which has been synthesized in an unequivocal fashion, or by more classical means such as determination of mixture-melting points of solid derivatives (see Section 1.2).

trying to interpret the spectra in terms of the known structure. As complete analysis as is possible should be included as part of the write-up of each experiment.

As an example of the kind of analysis which can be done, consider the spectra of isobutyl alcohol, $(CH_3)_2CHCH_2OH$, given in Figure 4.8. Looking first at the ir spectrum [Figure 4.8(a)], three prominent bands are present in the region of 5000–1250 cm^{-1}. By referring to Table 4.1, it is possible to assign the absorptions at about 2900 and 1465 cm^{-1} to the C—H bonds of the molecule. The broad and intense band at 3250 cm^{-1} undoubtedly is due to the O—H function in the alcohol. The position and breadth of the band indicates that the alcohol is hydrogen bonded. Finally, the very strong absorption at 1050 cm^{-1} is characteristic of a C—O single bond. As would be expected on the basis of the structure of isobutyl alcohol, no absorptions are seen for carbonyl functions, aromatic rings or carbon-carbon multiple bonds.

Turning now to the nmr spectrum [Figure 4.8(b)], the first point of interest is the integration. The areas of the peaks, going *upfield*, are in the ratio (measured in milli-

(a) IR of isobutyl alcohol

(b) NMR of isobutyl alcohol

FIGURE 4.8   Spectra of isobutyl alcohol.

meters), 2.8:5.6:3.0:15, which is equal to the ratio, 1.0:2.0:1.1:5.4. Since the total number of hydrogens in isobutyl alcohol is 10, the latter ratio is within experimental error of the absolute ratio.

It is now possible to complete the analysis of the spectrum by assigning absorptions to specific types of hydrogens, basing the assignment on chemical shift and integrated area. The doublet at about 0.85 δ has the appropriate relative area, 6, and chemical shift to be due to the six hydrogens of the two methyl groups of the alcohol. The multiplicity of the methyl absorption is due to coupling with the lone methine (tertiary) hydrogen, $(CH_3)_2CH$. The multiplet from 1.4–2.0 δ can be assigned to the methine hydrogen by consideration of the area, 1, the chemical shift, which is consistent with the nucleus being tertiary, and the multiplicity. The methine hydrogen is coupled with no less than eight other hydrogens. The doublet at about 3.3 δ has the area, 2, the chemical shift and multiplicity anticipated for the methylene group of isobutyl alcohol. This leaves the somewhat broad singlet at 4.2 δ to be assigned to the hydroxylic hydrogen.[11]

For purposes of analysis by nmr, approximately 0.3 ml of a 10–20 percent solution (by weight) is normally required. Carbon tetrachloride is an excellent choice for the solvent since it has no protons which would produce absorption peaks in addition to those of the solute. For samples which are insoluble in carbon tetrachloride, benzene, 7.67 δ (chemical shift), chloroform, 7.27 δ, nitromethane, 4.33 δ, acetonitrile, 2.00 δ, or acetone, 2.17 δ, may be suitable solvents if their proton resonances do not occur in a region of absorption of the solute.[12]

EXERCISES

1. The nmr spectra of the three fractions obtained in the distillation of a mixture of benzene and toluene consisted of a sharp singlet and a multiplet in the region of 7.2 δ, and a singlet at 2.3 δ. The integrated intensities of the downfield and upfield absorptions are given below for each fraction.

	Downfield	*Upfield*
Fraction 1	165	9
Fraction 2	192	30
Fraction 3	167	81

   Calculate the percentage of toluene present in each fraction.
2. The ir and nmr spectra of a compound, $C_{12}H_{18}$, are given in Figure 4.9(a), (b). Interpret both spectra as completely as possible and attempt to assign a structure or structures to the unknown on the basis of your interpretation.

[11] It may seem strange that the hydroxylic proton appears as a singlet, albeit a broad singlet, since one would expect it to be coupled to the methylene hydrogens. Such coupling can, in fact, be observed in highly purified alcohols, but for our purposes, one can assume that no spin-spin splitting will be detected between such a proton and protons on the carbon atom bonded to the oxygen.

[12] Many deuteriated solvents, such as chloroform-*d* ($CDCl_3$), acetone-d_6 (CD_3COCD_3), and benzene-d_6 (C_6D_6), are available. Use of these solvents circumvents the problem of resonance absorptions of solvent protons in the nmr spectrum because the chemical shift of deuterium (^2H) is vastly different from that of hydrogen.

FIGURE 4.9 Spectra of $C_{12}H_{18}$.

REFERENCES

1. L. J. Bellamy, *The Infra-red Spectra of Complex Molecules*, 2nd Ed., John Wiley & Sons, Inc., New York, 1958.
2. R. T. Conley, *Infrared Spectroscopy*, Allyn and Bacon, Inc., Boston, 1966.
3. J. R. Dyer, *Applications of Absorption Spectroscopy of Organic Compounds*, Prentice-Hall, Inc., Englewood Cliffs, N. J. (paperback).
4. L. M. Jackman, *Applications of Nuclear Magnetic Resonance Spectroscopy in Organic Chemistry*, Pergamon Press, New York, 1959.
5. J. D. Roberts, *An Introduction to the Analysis of Spin-Spin Splitting in High-Resolution Nuclear Magnetic Resonance Spectra*, W. A. Benjamin, New York, 1962.
6. J. D. Roberts, *Nuclear Magnetic Resonance. Applications to Organic Chemistry*, McGraw-Hill Book Co., New York, 1959.
7. R. M. Silverstein and G. C. Bassler, *Spectrometric Identification of Organic Compounds*, 2nd Ed., John Wiley & Sons, Inc., New York, 1967, Infrared, nuclear magnetic resonance and ultraviolet spectroscopy and mass spectrometry.
8. I. Fleming and D. H. Williams, *Spectroscopic Methods in Organic Chemistry*, McGraw-Hill Book Co., Inc., New York, 1966.

CATALOGS OF SPECTRA

1. Sadtler Research Laboratories, Inc., "Nuclear Magnetic Resonance Spectra," Vol. 1–3.
2. Sadtler Research Laboratories, Inc., "Sadtler Standard Spectra, Midget Edition," Vol. 1–26 (IR Spectra).
3. Varian Associates, "High Resolution NMR Spectra Catalogue," Vol. 1 (1962) and Vol. 2 (1963).

5

Alkanes
Free radical halogenation

5.1 *Chlorination by Means of Sulfuryl Chloride*

In general, saturated hydrocarbons are chemically inert even in the presence of very strong acids or bases, and the organic chemistry of these substances is rather limited as a consequence. Some chemical reactions have been developed, however, by which hydrocarbons can be converted to organic substances suitable for use in a variety of reactions. For example, a hydrocarbon, RH, can be transformed to a nitroalkane, RNO_2, or to a hydroperoxide, ROOH, upon reaction with nitrogen tetroxide or molecular oxygen, respectively.

This experiment involves the production of alkyl chlorides, RCl, from hydrocarbons. No reaction occurs when chlorine gas and a hydrocarbon are simply mixed at room temperature, but ultraviolet irradiation or heating (200–400°) of the mixture affords an alkyl chloride and hydrogen chloride [equation (1)]. The photochemical or thermal

$$RH + Cl_2 \xrightarrow[h\nu]{\text{heat or}} RCl + HCl \tag{1}$$

activation of the mixture is required in order to convert some molecular chlorine into chlorine atoms. The generation of chlorine atoms, which are free radicals, is essential to the initiation of the reaction between the hydrocarbon and chlorine to give the alkyl chloride.

It is somewhat easier experimentally to generate chlorine atoms by use of an initiator, a substance which will decompose to free radicals under relatively mild conditions [equation (2)]. The free radicals, In·, arising from decomposition of the initiator will

$$In\!-\!In \xrightarrow[h\nu]{\text{heat or}} In\cdot + In\cdot \tag{2}$$

$$In\cdot + Cl\!-\!Cl \longrightarrow InCl + Cl\cdot \tag{3}$$

react with molecular chlorine to produce InCl and chlorine atoms [equation (3)], and these chlorine atoms will then react with the hydrocarbon to produce an alkyl chloride.

A second modification of the usual manner of chlorinating hydrocarbons will be made to make the reaction more suitable for study in the undergraduate laboratory. This is the use of sulfuryl chloride, SO_2Cl_2, in place of chlorine gas.

The mechanism of the free radical chlorination of hydrocarbons is fundamentally the same whether sulfuryl chloride or molecular chlorine is employed and can be divided into three different phases: *initiation, propagation,* and *termination* [equations (4)–(13)]. In our experiment, the first step of the initiation phase is the homolytic

INITIATION

$$C_6H_5\overset{\|}{\underset{O}{C}}O\!-\!O\overset{\|}{\underset{O}{C}}C_6H_5 \xrightarrow{\;80°\;} 2\, C_6H_5\overset{\|}{\underset{O}{C}}O\cdot \tag{4}$$

$$C_6H_5\overset{\|}{\underset{O}{C}}O\cdot \longrightarrow C_6H_5\cdot + CO_2 \tag{5}$$

$$(C_6H_5\overset{\|}{\underset{O}{C}}O\cdot \quad \text{or} \quad C_6H_5\cdot = In\cdot)$$

$$In\cdot + Cl\!-\!\overset{O}{\underset{O}{\overset{\|}{\underset{\|}{S}}}}\!-\!Cl \longrightarrow InCl + \cdot\overset{O}{\underset{O}{\overset{\|}{\underset{\|}{S}}}}\!-\!Cl \tag{6}$$

$$\cdot\overset{O}{\underset{O}{\overset{\|}{\underset{\|}{S}}}}\!-\!Cl \longrightarrow \overset{O}{\underset{O}{\overset{\|}{\underset{\|}{S}}}} + Cl\cdot \tag{7}$$

PROPAGATION

$$Cl\cdot + RH \longrightarrow R\cdot + HCl \tag{8}$$

$$R\cdot + ClSO_2Cl \longrightarrow RCl + \cdot SO_2Cl \tag{9}$$

$$\cdot SO_2Cl \longrightarrow SO_2 + Cl\cdot \tag{10}$$

TERMINATION

$$Cl\cdot + Cl\cdot \longrightarrow Cl_2 \tag{11}$$

$$R\cdot + Cl\cdot \longrightarrow RCl \tag{12}$$

$$R\cdot + R\cdot \longrightarrow R\!-\!R \tag{13}$$

cleavage [1] of benzoyl peroxide at the oxygen-oxygen bond [equation (4)]. The necessary energy for the rupture of this bond (about 30 kcal/mole) is obtained by heating the peroxide to 60–80°. Either the resulting benzoyl radicals ($C_6H_5CO_2\cdot$) or the phenyl radicals ($C_6H_5\cdot$) produced from them by loss of CO_2 [equation (5)] may attack sulfuryl chloride to produce SO_2 and chlorine atoms [equations (6) and (7)].

The propagation steps involve abstraction of a hydrogen atom from the hydrocarbon by a chlorine atom to produce a new free radical, $R\cdot$, which can attack the sulfuryl chloride to produce the alkyl chloride and to regenerate the radical, $\cdot SO_2Cl$, the precursor of chlorine atoms [equations (8)–(10)]. A chain reaction is thus produced.

Once initiated, the chain reaction could in principle continue until either of the

---

[1] Homolytic cleavage may be defined as rupture of a bond in a fashion such that the bonding elections become distributed equally between the atoms originally linked by the bond. The alternate manner of distribution in which the bonding electrons are not equally divided between the atoms is termed heterolytic cleavage.

principal reagents, sulfuryl chloride or the hydrocarbon, depending on which is the *limiting reagent*,[2] is exhausted. In practice, however, the termination reactions [equations (11)–(13), for example] intervene to interrupt the free radical-chain process so that initiation must be continued throughout the reaction.

It is worthwhile to note some characteristics of the various steps in a free radical-chain mechanism. In general, an *initiation step* involves the formation of a low concentration of free radicals from molecules and results in an increase in the concentration of free radicals present in the system. A *propagation step* produces no *net* change in the concentration of radicals present, whereas a *termination step* gives a decrease in their concentration.

As free radical halogenation of hydrocarbons generally produces mixtures of several isomers and polyhalogenated products which are difficult to separate into pure components, other methods of preparation of pure alkyl halides are commonly used. The reaction of a hydrogen halide with an alkene [equation (14)] or an alcohol [equation (15)] provides the corresponding alkyl halide. Alcohols can also be transformed to

$$\underset{\substack{R \\ R}}{\overset{R}{\diagdown}}C=C\underset{R}{\overset{R}{\diagup}} + HX \longrightarrow R-\underset{\substack{| \\ X}}{\overset{\overset{R}{|}}{C}}-\underset{\substack{| \\ H}}{\overset{\overset{R}{|}}{C}}-R \tag{14}$$

$$R-OH + HX \longrightarrow R-X + H_2O \tag{15}$$

alkyl halides by reaction with a thionyl halide, SOX_2 ($X = Cl$ or Br), followed by thermal decomposition of the intermediate halosulfite ester [equation (16)]. These reactions of alcohols are discussed further in Chapter 11.2.

$$RO-H + X-\underset{\substack{\| \\ O}}{S}-X \xrightarrow{-HX} RO-\underset{\substack{\| \\ O}}{S}-X \xrightarrow{heat} R-X + SO_2 \tag{16}$$

EXPERIMENTAL PROCEDURE

This experiment provides for the chlorination of three compounds: (A) cyclohexane, (B) *n*-heptane and (C) 1-chlorobutane. Optimally, each student may do all three parts, since the number, type and proportion of products from each of the three compounds illustrate important principles. Alternatively, the three compounds may be distributed among the class, and the students may compare their results.

Caution: Sulfuryl chloride is a lachrymator and should be weighed in a hood. Benzoyl peroxide has been known to explode from shock or static electricity. This chemical should be dispensed from cardboard containers and transferred with a folded piece of stiff paper (filter paper is suitable).

A. CYCLOHEXANE

Fit a 100-ml round-bottomed flask with a water-cooled reflux condenser which is connected at the top to a gas trap by means of a one-hole rubber stopper or a thermometer adapter and rubber sleeve, a bent glass tube, and a rubber (better, Tygon)

[2] See Appendix II for explanation of this term.

tube (see Figure 5.1).[3] Place 33.6 g (43 ml, 0.4 mole) of cyclohexane,[4] 27.0 g (10 ml, 0.2 mole) of sulfuryl chloride and 0.1 g of benzoyl peroxide in the flask, and weigh the flask and its contents. Connect the condenser and trap and heat the mixture to gentle reflux with a micro burner until the evolution of gases appears to be complete (40–60 min).★ Cool the reaction mixture, disconnect the flask from the condenser and weigh it and its contents. If the loss of weight is less than theoretical,[5] add another small portion of the benzoyl peroxide and heat the mixture again to reflux.

After the theoretical amount of weight has been lost, cool the reaction mixture and pour it into 50 ml of ice water. Transfer the resulting two-phase solution to a separatory funnel and separate the layers.★ If separation into two layers does not occur readily, add some sodium chloride to the separatory funnel and shake. Wash the organic layer (which is it?) with 5% sodium carbonate solution until the washes are basic to litmus paper. Wash it once again with water and then dry it over about 3 g of *anhydrous* calcium chloride.[6]★ Filter the dried solution into a 100-ml flask fitted with an unpacked column for fractional distillation and carefully distil in order to separate the chlorinated products from unchanged starting material. Suggested boiling ranges are as follows:

Fraction 1: Ambient-85°;
Fraction 2: 85–145°.

Weigh each of the distillation cuts and submit them for analysis by gas chromatography. Estimate the approximate yields of the chlorinated products, taking into account the amount of unchanged starting material recovered.

FIGURE 5.1   Details of gas trap arrangement.

[3] The water flow in the trap must be regulated carefully to prevent sucking water back into the reaction mixture, which would cause a violent reaction.

[4] If the starting material has been stored in a metal container, it may be necessary to perform a simple distillation to remove traces of metal ions, which can inhibit the reaction.

[5] The theoretical loss of weight is calculated on the basis that one mole each of HCl and of SO_2 is evolved for each mole of sulfuryl chloride consumed. This calculation should be made before coming to laboratory.

[6] See Appendix I for a discussion of drying agents.

B. *n*-Heptane

Follow the directions given in part A, using 40.0 g (59 ml, 0.4 mole) of *n*-heptane [4] in place of cyclohexane. Suggested boiling ranges are as follows:

Fraction 1: Ambient–105°;
Fraction 2: 105–160°.

C. 1-Chlorobutane

Follow the directions given in part A, using 43.2 g (49 ml, 0.4 mole) of 1-chlorobutane [4] in place of cyclohexane. Suggested boiling ranges are as follows:

Fraction 1: Ambient–105°;
Fraction 2: 105–160°.

EXERCISES

1. Suggest at least one termination process for halogenation with sulfuryl chloride reagent which has no counterpart in halogenation with molecular chlorine.
2. What is the reason for using less than the amount of sulfuryl chloride theoretically required to convert all of the starting materials to monochlorinated products?
3. From which of the three starting materials should it be easiest to prepare a single pure monochlorinated product? Why?
4. What factors determine the proportion of monochlorinated isomers of *n*-heptane? Of 1-chlorobutane?
5. Why is only a catalytic amount of initiator used?
6. Calculate the heat of reaction for the reaction between cyclohexane and chlorine

   to yield chlorocyclohexane and hydrogen chloride, using the following bond energies (in kcal/mole):

$$C-H, 98.7; \quad C-Cl, 81; \quad Cl-Cl, 58.0; \quad H-Cl, 103.2.$$

7. Calculate the percent of each monochlorination product expected from *n*-heptane based on a relative reactivity of primary (1°): secondary (2°): tertiary (3°) hydrogens of 1.0:3.3:4.4. By referring to Figure 5.2, calculate the observed ratio of 1°:2° chloroheptanes and compare the result with the anticipated theoretical ratio.
8. Chlorination of propene leads to high yields of 3-chloro-1-propene to the exclusion of products of substitution of the vinyl hydrogens. Explain why this is so.
9. Write out the expected products of free radical reaction of SO_2Cl_2 with 1-methylcyclohexane. Predict which chlorinated isomer would be formed in highest yield.
10. By reference to Figure 5.3, calculate the percentage of each dichlorobutane present in fraction 2 (do not include the area of the peak due to 1-chlorobutane when making the calculation). Explain why the observed ratio of products does

FIGURE 5.2 Gas chromatogram of fraction 2 resulting from chlorination of *n*-heptane. Peak 1: *n*-Heptane; Peak 2: 2-, 3-, 4-Chloroheptanes; bp, 46° (19.5 mm), 48.3 21 mm), and 48.9° (21 mm), respectively; Peak 3: 1-Chloroheptane bp 61.4 (27 mm); 158.5–159.5 (760 mm); Column and conditions: 5 ft., 5% silicone elastomer or Chromosorb *W;* 80°, 40 ml/minute.

← Time

FIGURE 5.3 Gas chromatograms of fractions 1 and 2 resulting from chlorination of 1-chlorobutane. Peak 1: 1-Chlorobutane; Peak 2: 1,1-Dichlorobutane (bp 114–115°); Peak 3: 1,2-Dichlorobutane (bp 121–123°); Peak 4: 1,3-Dichlorobutane (bp 131–133°); Peak 5: 1,4-Dichlorobutane (bp 155–156°); Column and conditions: 5 ft, 5% silicone elastomer on Chromosorb *W;* 60°, 40 ml/minute.

not agree with that predicted by use of the relative reactivities given in question 7.

11. In Figure 5.4 is given the nmr spectrum of a mixture of dichlorobutanes obtained by free radical chlorination of 1-chlorobutane.

(a) Assign the multiplet at 5.80 δ, the doublet at 1.55 δ, and the triplet at 1.07 δ to the hydrogen nuclei of the three different isomers from which they arise. (Hint: Write the structures of the isomeric 1, x-dichlorobutanes and predict the multiplicity and approximate chemical shift of each group of hydrogens.)

(b) Calculate the approximate percentage of each isomer present in the mixture.

FIGURE 5.4 NMR spectrum of mixture of 1-*x*-dichlorobutanes.

REFERENCE

M. S. Kharasch and H. C. Brown, *Journal of the American Chemical Society*, **61**, 2142 (1939).

SPECTRA OF STARTING MATERIALS AND PRODUCTS

FIGURE 5.5 IR spectrum of 1-chlorobutane.

FIGURE 5.6 NMR spectrum of 1-chlorobutane.

FIGURE 5.7 IR spectrum of cyclohexane.

FIGURE 5.8 IR spectrum of chlorocyclohexane.

5.2   Bromination. Relative Ease of Substitution of Hydrogens in Different Environments

The preceding experiment allows a comparison of the reactivity of hydrogen atoms bonded to primary (1°) and secondary (2°) carbon atoms [7] toward substitution by chlorine atoms on the basis of the relative amounts of primary and secondary chloroheptanes produced from *n*-heptane. In the present experiment, the relative rates of reaction of bromine with several different hydrocarbons containing primary, secondary, and tertiary hydrogens will be determined. The hydrogens may also be described as being aliphatic, aromatic, or benzylic.[8] We may combine these terms and speak of primary, secondary and tertiary aliphatic or benzylic hydrogens. Only one type of aromatic hydrogen is possible. By careful consideration of your results from this experiment, you should be able to deduce an order of reactivity for seven different types of hydrogens.

Experimentally, the relative rates of bromination of the various hydrocarbons are measured by the lengths of time required for the bromine color to be discharged in the reactions, which are conducted under similar experimental conditions. Bromine is a less reactive and hence a more selective reagent than chlorine, so that the rates of bromination of the various hydrocarbons are sufficiently different to allow a qualitative order of reactivity to be determined quite easily.

The reactions are to be carried out in carbon tetrachloride solution under three different conditions: (1) at room temperature without special illumination, (2) at room temperature with strong illumination, and (3) at about 50° with strong illumination. Under all of these conditions, substitution of hydrogen by bromine occurs by a free radical mechanism analogous to that of the chlorination reactions of Section 5.1. In this case the initiation step is the thermal and/or photochemical dissociation of bromine molecules into bromine atoms (radicals) as shown in equation (17). The propagation steps are those of equations (18) and (19). The termination steps are exactly analogous to those of equations (11)–(13) of Section 5.1.

$$Br_2 \xrightarrow[\text{or h}\nu]{\text{heat}} 2\ Br\cdot \tag{17}$$

$$Br\cdot + RH \rightarrow HBr + R\cdot \tag{18}$$

$$R\cdot + Br_2 \rightarrow RBr + Br\cdot \tag{19}$$

EXPERIMENTAL PROCEDURE

1. Construct a table in your notebook with the following headings: *Hydrocarbon, Types of Hydrogen* (1°, 2°, etc.), *Conditions* (with the subheadings, 25°; 25°, hν; 50°, hν).

2. In each of seven 18 × 100-mm test tubes place a solution of 1 ml of hydrocarbon in 5 ml of carbon tetrachloride, using each of the following hydrocarbons: benzene, toluene, ethylbenzene, isopropylbenzene (cumene), *t*-butylbenzene, cyclohexane, and

[7] In the following discussion we shall follow the slightly inaccurate but convenient practice of referring to a hydrogen atom bonded to a primary carbon atom as a "primary hydrogen," etc.

[8] A benzylic hydrogen is one attached to a carbon atom bonded to an aromatic ring.

methylcyclohexane. In each of seven other test tubes place 1 ml of $1M$ bromine in carbon tetrachloride.[9] Add the bromine solutions to the hydrocarbon solutions in rapid succession, and shake each mixture after the addition.[10] Note and record the time required for the red bromine color to be discharged in each reaction mixture.

3. Carry out the experiment as in paragraph 2 above, but place the reaction tubes, immediately after mixing the reactants, about 4–5 in from an unfrosted 100- or 150-watt light bulb. Record the times of disappearance of color.

4. Carry out the experiment as in paragraph 2 above, but before mixing the hydro-carbon and bromine solutions, heat them all to about 50° in a large beaker of water placed on a steam cone or hot plate. Also, position the light bulb directly above the beaker containing the test tubes so that the reaction mixtures are all about 4–5 inches from the bulb. After mixing, keep the tubes containing the reaction mixtures in the beaker of warm water, and maintain the temperature of the water as constant as pos-sible. Record the times of disappearance of color as before.

EXERCISES

1. Arrange the seven hydrocarbons in increasing order of reactivity toward bromina-tion.
2. On the basis of the order of reactivity of the hydrocarbons, deduce the order of reactivity of the seven different types of hydrogens found in these compounds; *i.e.*, (1) primary aliphatic, (2) secondary aliphatic, (3) tertiary aliphatic, (4) primary benzylic, (5) secondary benzylic, (6) tertiary benzylic, and (7) aromatic.
3. Clearly explain how you arrived at your sequence in question 2.

SPECTRA OF STARTING MATERIALS

FIGURE 5.9 IR spectrum of ethylbenzene.

[9] Bromine, even in solution, is a potentially dangerous chemical. Do not breathe vapors of the car-bon tetrachloride solution or spill the solution on your skin. In case of accident, see page 5.

[10] Because the rates of some of the brominations are quite high, it may be advisable to start only three or four of the reactions at one time and the remaining three or four at another time.

FIGURE 5.10 NMR spectrum of ethylbenzene.

FIGURE 5.11 IR spectrum of isopropylbenzene.

FIGURE 5.12 NMR spectrum of isopropylbenzene.

FIGURE 5.13   IR spectrum of *t*-butylbenzene.

FIGURE 5.14   NMR spectrum of *t*-butylbenzene.

5.3   *Triphenylmethyl Bromide, Triphenylmethyl Radical, Hexaphenylethane, and Triphenylmethyl Peroxide*

This experiment demonstrates the use of direct halogenation for the production of an organic halide in practical yield. The starting material, triphenylmethane, is a compound containing one tertiary benzylic hydrogen [8] and fifteen aromatic hydrogens, thus assuring good selectivity (see Section 5.2) and freedom from significant amounts of isomers of the bromination product, triphenylmethyl bromide (**1**).

$$\text{Ph}_3\text{C}-\text{H} + \text{Br}_2 \longrightarrow \text{Ph}_3\text{C}-\text{Br} + \text{HBr} \qquad (20)$$

1

The first authentic free radical was identified by Moses Gomberg in 1900 as a product from triphenylmethyl chloride. Similar results can be obtained from the corresponding bromide. When triphenylmethyl bromide is treated with powdered zinc, the carbon-bromine bond is broken homolytically, and triphenylmethyl radicals (2) are produced [equation (21)]. Some of these radicals combine (dimerize) to form hexaphenylethane (3),[11] but this reaction does not go to completion for two reasons: first, the bond between the two aliphatic carbon atoms in hexaphenylethane is weak because of the steric interaction between the bulky phenyl groups and, second, the triphenylmethyl radicals are stabilized by delocalization of the unpaired electron by the three benzene rings [equation (22)].

In the presence of oxygen, triphenylmethyl radicals react to form a stable peroxide [4, equation (23)]. It should be noted that in the peroxide the bulky triphenylmethyl groups are spaced farther apart by the intervening oxygen atoms than in hexaphenylethane.

$$(21)$$

2

$$(22)$$

yellow   white

3

$$(23)$$

yellow   white

4

[11] Recently, a different structure for the dimer in equilibrium with triphenylmethyl radicals has been proposed on the basis of spectroscopic evidence, but there is no doubt about the nature of the triphenylmethyl radicals. See H. Lankamp, W. Th. Nauta and C. MacLean, *Tetrahedron Letters*, 249 (1968).

EXPERIMENTAL PROCEDURE

A. PREPARATION OF TRIPHENYLMETHYL BROMIDE

Caution: Bromine is a hazardous chemical. Do not breathe its vapors or allow it to come in contact with the skin. All operations involving transfer of the pure liquid into solvents or reaction flasks should be carried out in an efficient hood. In case of accident, see page 00.

The preparation of triphenylmethyl bromide is carried out in apparatus such as that pictured in Figure 5.15. Insert a small pad of glass wool into a vacuum adapter and then fill the adapter with sodium hydroxide pellets. In a 50-ml round-bottomed flask, dissolve 2 g (0.009 mole) of triphenylmethane in 10 ml of dry carbon tetrachloride.

Twisted Wire

Standard Taper Joint

FIGURE 5.15 Apparatus of bromination of triphenylmethane.

Attach the vacuum adapter and close its two openings with rubber serum caps as shown in the figure. Insert an open syringe needle (21 gauge or larger) in the upper serum cap. Measure 0.5 ml (1.38 g, 0.009 mole) of bromine into a syringe and dispense the bromine into 10 ml of dry carbon tetrachloride. Take up the whole solution of bromine in carbon tetrachloride in a syringe, insert the needle through the lower serum cap and add the solution dropwise to the hydrocarbon, keeping the flask near (4–6 in) a 150-watt unfrosted light bulb. If the reaction mixture is orange or yellow when the reaction is complete (10 to 45 min, depending on the intensity of the light and its distance from the flask), add a drop or two of cyclohexene with a syringe.

Remove the upper needle and the lower serum cap and attach the side arm to an aspirator. Evaporate the solvent under aspirator vacuum, warming the flask by placing it in a pan or beaker of warm water. Swirl the flask when applying the vacuum to keep the liquid from bumping and be careful not to warm the flask too much at first. Recrystallize the crystalline residue from petroleum ether to give pale yellow crystals of triphenylmethyl bromide (reported mp 153–155°). The yield should be 40–60%.

B. PREPARATION AND REACTIONS OF TRIPHENYLMETHYL RADICALS

In a 50-ml Erlenmeyer flask, dissolve 1 g of triphenylmethyl bromide in 15 ml of benzene. Add 3 g of powdered zinc, stopper the flask, and shake it vigorously for about 5 min. Filter the mixture through a fluted filter, collecting the filtrate in a 125-ml Erlenmeyer flask. Stopper the flask and shake the solution vigorously. Allow the flask to stand quietly for a few minutes. Note any color changes observed.

Remove the stopper and allow the flask to stand open for several minutes. Replace the stopper and shake the solution. Again, note any color changes. Repeat these procedures until no color change occurs when the closed flask is allowed to stand. This may require about 30 min. A white precipitate should have formed during this period.

Collect the white solid by vacuum filtration on a Hirsch funnel and wash it with a little ether. After allowing it to dry in air, take the melting point. The melting point of the white product obtained in this way by Gomberg was 185°.

EXERCISES

1. What is the purpose of the sodium hydroxide placed in the apparatus?
2. Identify the white precipitate formed when the mixture of triphenylmethyl bromide and zinc is exposed to the atmosphere.
3. What is the purpose of adding cyclohexene to the colored reaction mixture from the preparation of triphenylmethyl bromide?
4. Explain the chemistry involved in the color changes of the reaction mixture from triphenylmethyl bromide and zinc which you observed when the flask was alternately opened and closed.
5. Write resonance formulas illustrating delocalization of the unpaired electron of the triphenylmethyl radical.
6. On the basis of electron delocalization energy, predict the relative stability of the following radicals:

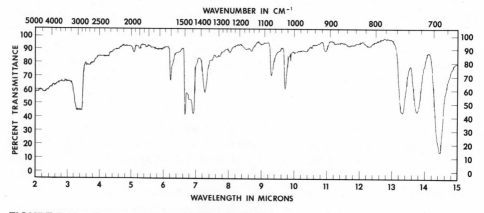

7. How might bulky groups in the *o*-positions of phenyl groups affect the stability of triphenylmethyl radicals?

REFERENCE

J. Workentin, *Journal of Chemical Education*, **43**, 331 (1966).

SPECTRA OF STARTING MATERIALS

FIGURE 5.16   IR spectrum of triphenylmethane.

FIGURE 5.17 NMR spectrum of triphenylmethane.

6

Alkenes
Preparation and reactions

6.1 Generation of Alkenes by Elimination Reactions

Elimination reactions are often used by organic chemists for synthetic purposes and to aid in determination of molecular structure. Fundamentally an elimination requires a *leaving group*, X, that may depart *with* its bonding electrons, and a group, usually hydrogen, on an adjacent carbon that may be abstracted *without* its bonding electrons. The product of the elimination may be either an alkene [equation (1)] or alkyne [equation (2)] depending on the reactant.

$$
\begin{array}{c}
\text{H--C--C--X} \longrightarrow \text{C=C} + HX
\end{array}
\tag{1}
$$

$$
\begin{array}{c}
\text{H--C=C--X} \longrightarrow \text{--C}\equiv\text{C--} + HX
\end{array}
\tag{2}
$$

An elimination may generally be classified according to one of three different mechanisms: (1) a two-step process in which X departs with its electrons leaving behind a carbonium ion which then in the second step loses a proton to form the double bond [E1 type, equation (3)]; (2) a one-step concerted process in which the proton

$$
\begin{array}{c}
\text{H--C--C--X} \xrightarrow{-X^{\ominus}} \text{H--C--C}_{\oplus} \xrightarrow{-H^{\oplus}} \text{C=C}
\end{array}
\tag{3}
$$

and X are lost simultaneously as the double bond forms [E2 type, equation (4)];[1] and

$$
\begin{array}{c}
\text{H--C--C--X} + B: (\text{Base}) \longrightarrow \left[\text{C--C} \atop B \cdots H \ X \right] \longrightarrow \text{C=C} + BH + X:
\end{array}
\tag{4}
$$

---

[1] Experiments involving eliminations following the E2 mechanism are found in Chapter 11.

(3) a two-step process in which base abstracts the proton to create a carbanion which then in the second step forms a double bond by ejection of the leaving group [equation (5)].

$$\overset{\displaystyle\;\;}{\underset{H}{>}C-C\overset{\displaystyle\;}{\underset{X}{<}} \longrightarrow \underset{\ominus}{>}C-C\overset{}{\underset{X}{<}} \longrightarrow >C=C< + X:^{\ominus} \tag{5}$$

The acid-catalyzed dehydration of an alcohol can usually be classified mechanistically as an elimination of the E1 type. However, due to the relative instability of primary carbonium ions, many primary alcohols undergo dehydration via the E2 pathway, with a molecule of water or alcohol functioning as base. The E1 dehydration [equation (6)] involves an initial protonation of the oxygen atom to form an oxonium

$$\underset{H}{>}C-C\overset{}{\underset{OH}{<}} \overset{H^{\oplus}}{\rightleftharpoons} -\underset{H}{>}C-C\overset{}{\underset{\overset{\oplus}{O}-H}{<}} \overset{-H_2O}{\rightleftharpoons} \underset{H}{>}C-\overset{\oplus}{C}< \overset{-H_3O^{\oplus}}{\rightleftharpoons} >C=C< \tag{6}$$

$$\underset{\mathbf{1}}{\qquad} \qquad\qquad H_2\ddot{O}: \qquad \underset{\mathbf{2}}{\qquad}$$

ion **(1)**. The next step, kinetically a first-order process, involves the decomposition of the oxonium ion to a carbonium ion **(2)** and water. The carbonium ion then loses a proton from a carbon adjacent to the charged atom, yielding the alkene. This final loss of the proton is probably aided by a molecule of water or alcohol in the medium acting as a base.

It should be noted that each of the steps along the reaction pathway is reversible: Under the conditions of the experiment, the alkene may be rehydrated to alcohol. In order to carry the elimination to completion (100% conversion), the alkene may be distilled from the reaction mixture as it is formed. This has the effect of shifting the equilibrium to the right and maximizing the yield of alkene. *The constant removal from a reaction mixture of products as they are formed is a technique often utilized to afford high yields of products from equilibrium reactions.*

An E1 elimination is usually accompanied to some extent by substitution reactions. These competing reactions involve combination of the intermediate carbonium ion with the anion of the acid used for catalysis or some other anion. For example, with hydrochloric acid, the chloride ion may react to give some chloroalkane, a product in which chlorine has replaced the hydroxy group. The extent to which substitution products will be formed depends on the amount and nature of the acid used. However, since the formation of substitution products is frequently reversible under the reaction conditions, these side reactions will not always constitute a practical problem in affecting the yield of an elimination, assuming the alkene is distilled as it is formed.

In an E1 reaction, it is not at all uncommon for two or more isomeric alkenes to be formed. If the charge of the carbonium ion is unsymmetrically located on the carbon skeleton, and hydrogens are attached to two of the adjacent carbon atoms, different products may be formed, depending on which of the hydrogens is lost. This possibility is illustrated in equation (7).

The proportion of the two alkenes (**3** and **4**) formed is dependent upon the relative activation energies of the two competing elimination processes. These activation

$$
\begin{array}{ccc}
 & H \quad OH \quad R'' \\
 & | \quad\;\; | \quad\;\; | \\
R- & C-CH-C-R' \\
 & | \qquad\;\; | \\
 & H_A \qquad H_B
\end{array}
$$

$$\Big\downarrow H^{\oplus},\ -H_2O \qquad\qquad\qquad (7)$$

$$
\underset{\textbf{3}}{\overset{\displaystyle R}{\underset{\displaystyle H}{{>}}}C=CH-\underset{H_B}{\overset{R''}{\underset{|}{\overset{|}{C}}}}H-R' \;\xleftarrow{-H_A^{\oplus}}\; R-\underset{H_A \quad H_B}{\overset{H \qquad R''}{\underset{|\qquad |}{\overset{|\qquad |}{C-\overset{\oplus}{C}H-C}}}}-R' \;\xrightarrow{-H_B^{\oplus}}\; \underset{\textbf{4}}{R-\underset{H_A}{\overset{H}{\underset{|}{\overset{|}{C}}}}-CH=C\overset{R''}{\underset{R'}{{<}}}}
$$

energies are directly related to the free energies of the two transition states, which may be represented by the general formula

$$
\underset{H \underset{\delta\oplus}{\cdots} \qquad \delta\oplus}{\overset{\diagdown}{C}\cdots\overset{\diagup}{C}}
$$

In these transition states the C—H bond is partially broken and the C—C double bond is partially formed. As each of the two transition states have partial double bond character, it is reasonable that their relative energies will parallel those of the corresponding alkenes, which of course have full double bonds. Therefore the more energetically favorable transition state will lead to the more stable alkene. Alkene stability usually increases as the number of alkyl substituents on the double bond is increased. Thus the more highly branched alkene (more highly substituted ethylene) is normally formed in higher yield when competing El eliminations are possible.

In many cases the number and kinds of alkenes formed as products in E1 reactions cannot be explained entirely by the type of reaction "choice" described above. In general, carbonium ions are highly susceptible to rearrangement if, by rearrangement, a more stable carbonium ion is produced. Indeed, some rearrangement may occur even when the new carbonium ion is less stable than the first, although the extent to which it occurs may be slight. Such rearrangements are accomplished by the migration of either an alkyl anion (R:⁻) or hydride ion (H:⁻) from an adjacent carbon atom to the carbon having the formal positive charge. Loss of a proton from the new carbonium ion intermediate may then lead to other isomeric alkenes.

Since carbonium ion stability increases with an increasing number of alkyl or aryl substituents, tertiary carbonium ions are more stable than secondary, which in turn are more stable than primary ions. Therefore, a migration that leads from a secondary or

$$
CH_3-\underset{\oplus}{\overset{CH_3}{\underset{|}{\overset{|}{C}}}}-CH_3 \;>\; CH_3-\underset{\oplus}{\overset{H}{\underset{|}{\overset{|}{C}}}}-CH_3 \;>\; CH_3-\underset{\oplus}{\overset{H}{\underset{|}{\overset{|}{C}}}}-H
$$

$$\xrightarrow{\text{decreasing stability}}$$
(increasing free energy)

primary ion to a tertiary carbonium ion will usually be quite favorable. For example, the neopentyl cation (**5**), a primary carbonium ion, rearranges by methyl migration to give the *t*-pentyl cation (**6**) as the predominant species. The same "driving force" (tendency to minimize free energy) will also cause the isobutyl cation (**7**), a primary carbonium ion, to rearrange to the *t*-butyl cation (**8**) by hydride migration. In this case

$$
\underset{\textbf{5}}{\underset{\underset{CH_3}{|}}{\overset{\overset{CH_3}{|\!\!\to\oplus}}{CH_3\!-\!C\!-\!CH_2}}}\xrightarrow{\;\sim CH_3:^{\ominus}\;}\underset{\textbf{6}}{\underset{\underset{CH_3}{|}}{\overset{\overset{CH_3}{\overset{\oplus}{|}}}{CH_3\!-\!C\!-\!CH_2}}}\tag{8}
$$

one should expect very little methyl migration, for this would lead to a secondary ion, whereas hydride migration leads to the even more stable tertiary ion.

$$
\underset{\textbf{7}}{\underset{\underset{H}{|}}{\overset{\overset{CH_3}{|}}{CH_3\!-\!C\!-\!CH_2}}}\xrightarrow{\;\sim H:^{\ominus}\;}\underset{\textbf{8}}{\underset{\underset{H}{|}}{\overset{\overset{CH_3}{|}}{CH_3\!-\!\overset{\oplus}{C}\!-\!\dot{C}H_2}}}\tag{9}
$$

Experimental procedures for the dehydration of 4-methyl-2-pentanol and cyclohexanol are presented below. 4-Methyl-2-pentanol yields a mixture of isomeric alkenes including 4-methyl-1-pentene (**9**), *trans*-4-methyl-2-pentene (**10**), *cis*-4-methyl-2-pentene (**11**), 2-methyl-2-pentene (**12**), and 2-methyl-1-pentene (**13**). Products **12**

and **13** are typically found to comprise approximately 45% of the product mixture showing that rearrangement is an important aspect of carbonium ion chemistry. These products arise by loss of a proton from the carbonium ion formed by two successive hydride migrations [equation (10)].

$$
\underset{\;\;\;H\;\;H}{\overset{\overset{CH_3}{|}}{CH_3\!-\!C\!-\!CH\!-\!\overset{\oplus}{CH}\!-\!CH_3}}\longrightarrow\underset{\;\;\;H\;\;\;\;\;H}{\overset{\overset{CH_3}{|}}{CH_3\!-\!C\!-\!\overset{\oplus}{CH}\!-\!CH\!-\!CH_3}}\longrightarrow
$$

$$
\underset{\qquad\qquad\qquad H\;\;\;H}{\overset{\overset{CH_3}{|}}{CH_3\!-\!\overset{\oplus}{C}\!-\!CH\!-\!CH\!-\!CH_3}}\tag{10}
$$

The rather complex reaction mixture may be analyzed by gas chromatography (see Chapter 3). By comparing retention times of authentic samples of the methylpentenes with those of the components of the mixture, the peaks in the gas chromatogram of the mixture can be assigned to the different methylpentenes. By measuring the relative areas under the peaks, the percentage of each component of the mixture may be obtained.

Cyclohexanol undergoes acid-catalyzed dehydration without rearrangement to yield a single product, cyclohexene. The reaction product may be identified, after purification, by comparing its infrared spectrum with that of pure cyclohexene.

$$\text{(cyclohexanol)} \xrightarrow{\text{H}^\oplus} \text{(cyclohexene)} + \text{H}_2\text{O} \qquad (11)$$

It is frequently desirable to determine whether or not a compound contains carbon-carbon multiple bonds. Two quick qualitative tests for such unsaturation are described below.[2]

Bromine will add to the carbon-carbon double bond of alkenes to produce dibromoalkanes [3] [equation (12)]. When this reaction occurs, molecular bromine is de-

$$\text{C=C} + \text{Br}_2 \longrightarrow \text{Br}\overset{|}{\underset{|}{\text{C}}}-\overset{|}{\underset{|}{\text{C}}}\text{Br} \qquad (12)$$

stroyed and its characteristic color disappears (if bromine is not present in excess). The rapid disappearance of the color of bromine is a positive test for unsaturation. However, the test is not unequivocal, for some alkenes do not react with bromine and some react very slowly. In case of a negative test, moreover, it is usually good practice to carry out the permanganate test (see below).

A second qualitative test for unsaturation, the Baeyer test, depends on the ability of potassium permanganate to oxidize the carbon-carbon double bond to give alkanediols [equation (13)] or the carbon-carbon triple bond to give carboxylic acids [equation (14)]. The permanganate ion is destroyed in the reaction and a brown precipitate of MnO_2 is produced. The disappearance of the characteristic color of the permanganate ion is a positive test for unsaturation. However, care must be taken, as compounds containing certain other types of functional groups (for example, aldehydes, containing the $-\text{CH}=\text{O}$ group) also decolorize permanganate ion.

$$\text{C=C} + \text{MnO}_4^\ominus \xrightarrow{\text{H}_2\text{O}} \overset{\text{HO}}{\underset{|}{-\text{C}}}-\overset{\text{OH}}{\underset{|}{\text{C}}}- + \text{MnO}_2 \qquad (13)$$

$$\text{R}-\text{C}\equiv\text{C}-\text{R}' + \text{MnO}_4^\ominus \xrightarrow{\text{H}_2\text{O}} \text{R}-\text{CO}_2\text{H} + \text{HO}_2\text{C}-\text{R}' \qquad (14)$$

---

[2] More detailed descriptions of these qualitative tests may be found in R. L. Shriner, R. C. Fuson and D. Y. Curtin, *The Systematic Identification of Organic Compounds*, 5th Ed., John Wiley & Sons, Inc., New York, 1967, Chapter 8.

[3] Addition reactions of alkenes are discussed in detail in Section 6.2.

EXPERIMENTAL PROCEDURE

A. DEHYDRATION OF 4-METHYL-2-PENTANOL

Assemble a fractional distillation apparatus as has been described earlier (Chapter 2). To minimize losses during distillation, the receiver, a test tube, should be cooled by an ice-water bath. Place 20 g (24.8 ml, 0.196 mole) of 4-methyl-2-pentanol in a 100-ml round-bottomed flask and to this add 5 ml of *concentrated* sulfuric acid.[4] Thoroughly mix the contents of the flask by swirling or stirring with a stirring rod. Add two or three boiling stones and connect the flask to the distillation apparatus. Distil the mixture until about 5 ml of the original mixture remains in the distillation flask. The rate of distillation should be such that the head temperature never exceeds 90°. *Caution:* The distillate and vapors are flammable. Transfer the distillate to a small separatory funnel, add 5–10 ml of 10% sodium hydroxide solution and shake well. Allow the layers to separate and drain off the aqueous layer. Transfer the organic layer to a dry 50-ml Erlenmeyer flask and add 1–2 g of *anhydrous* calcium chloride.★ Occasionally swirl the mixture during a period of 10 min and then decant the dried organic mixture into a 100-ml distilling flask. Add boiling stones and distil the mixture. Collect the fraction boiling between 53 and 69° in a pre-weighed, ice-cooled receiver. Weigh the product and calculate the yield obtained. If carried out properly, a yield of 75–85% may be expected.★ The expected products of the reaction and their boiling points are 4-methyl-1-pentene (53.9°), *trans*-4-methyl-2-pentene (58.6°), *cis*-4-methyl-2-pentene (56.4°), 2-methyl-1-pentene (61°), and 2-methyl-2-pentene (67.3°).

Test the distillate for unsaturation by each of the following methods: (1) To 1 or 2 ml of 2% bromine solution in carbon tetrachloride add 1 or 2 drops of the distillate. Rapid disappearance of the bromine color to give a colorless solution is a positive test for unsaturation. Perform this same experiment using cyclohexane instead of the distillate. (2) Perform the Baeyer test for unsaturation by dissolving 1 or 2 drops of the distillate in 2 ml of ethanol. Then add a 0.1 *M* potassium permanganate solution dropwise, observing the results. Count the number of drops added before the permanganate color persists. For a *blank determination*, count the number of drops that may be added to 2 ml of ethanol before the color persists. A significant difference in the number of drops required in the two cases is a positive test for unsaturation. Carry out the same experiment using cyclohexane in place of the distillate.

Submit a small sample of your product for glpc analysis. Figure 6.1 shows a typical glpc tracing of the products of this reaction.[5] Calculate the percentage composition of your reaction mixture. Also submit another small sample for infrared spectroscopic analysis. By comparing the absorption peaks from the infrared spectrum obtained with those from spectra of each of the pure authentic methylpentenes, identify as many of the components as you can.

B. DEHYDRATION OF CYCLOHEXANOL

Place 20 g (21.2 ml, 0.20 mole) of cyclohexanol and 5 ml of *concentrated* sulfuric

---

[4] 85% phosphoric acid is less satisfactory. The reaction is slower and the yields are lower. Use 5 ml of this acid should your instructor desire you to do so.

[5] The peaks in Figure 6.1 have been assigned as follows: (1) 4-methyl-1-pentene (14.4%); (2) *cis*- and *trans*-4-methyl-2-pentene (combined 52.1%); (3) 2-methyl-1-pentene (6.5%); (4) 2-methyl-2-pentene (36.9%); and (5) unidentified.

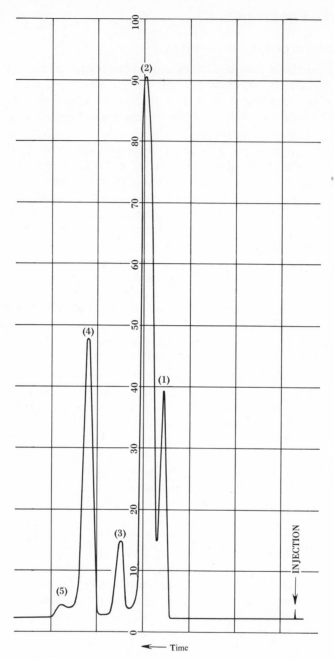

FIGURE 6.1 GLPC tracing of the product mixture from dehydration of 4-methyl-2-pentanol.[5,6]

acid [4] in a 100-ml round-bottomed flask and mix well. Add boiling stones and distil the mixture through a fractional distillation apparatus as described above for 4-methyl-2-pentanol until about 5 ml is left in the distillation flask. The receiver, a test tube, should be cooled in an ice-water bath. Transfer the distillate to a separatory funnel and

[6] This analysis was made with an 8-foot × ⅛-inch column packed with 5% SF-96 on 60/80 Chromosorb W at 35°.

shake well with 10 ml of 10% sodium hydroxide solution. Separate the aqueous layer, pour the organic layer into a 50-ml Erlenmeyer flask and add 1–2 g of *anhydrous* sodium sulfate.★ Occasionally swirl the flask during about 5 min and then decant the liquid into another dry 50-ml Erlenmeyer flask. Add another 1–2 g of *anhydrous* sodium sulfate, swirl for 5 min, and filter the liquid into a 100-ml round-bottomed flask. (The product *must be dry* at this stage in order to obtain pure cyclohexene since water and cyclohexene form a minimum-boiling azeotrope.) Fractionally distil the crude cyclohexene and collect the fraction boiling between 80 and 85° (12–14 g) in a pre-weighed receiver. Calculate the yield of the reaction.★ Perform each of the tests for un-saturation described under part A and submit a sample for infrared analysis. Compare your spectrum with that of authentic cyclohexene.

EXERCISES

1. Suggest the reason for washing the distillates in each of the above experiments with dilute sodium hydroxide.
2. Why, during the above dehydration experiments, is the distillation temperature kept below 90°?
3. Which of the following primary alcohols would be most likely to dehydrate via the E1 mechanism? the E2 mechanism? Explain.

$$CH_3CH_2OH \qquad (CH_3)_3CCH_2OH$$

4. Give a detailed mechanism explaining each of the products obtained from the dehydration of 4-methyl-2-pentanol.
5. Near the end of the dehydration of 4-methyl-2-pentanol, a white solid may usually be seen precipitating from the reaction mixture. What is this solid likely to be?
6. Give structures for the products of dehydration of each of the following al-cohols. For each, order the products with respect to preference of formation.

7. Suggest a reason for the necessity of a blank determination when performing the Baeyer test for unsaturation.

SPECTRA OF STARTING MATERIALS AND PRODUCTS

FIGURE 6.2 IR spectrum of 4-methyl–2-pentanol.

FIGURE 6.3 NMR spectrum of 4-methyl–2-pentanol.

FIGURE 6.4 IR spectrum of 4-methyl–1-pentene.

FIGURE 6.5   NMR spectrum of 4-methyl–1-pentene.

FIGURE 6.6   IR spectrum of 2-methyl–2-pentene.

FIGURE 6.7   NMR spectrum of 2-methyl–2-pentene.

FIGURE 6.8 IR spectrum of *trans*-4-methyl–2-pentene.

FIGURE 6.9 NMR spectrum of *trans*-4-methyl–2-pentene.

FIGURE 6.10 IR spectrum of *cis*-4-methyl–2-pentene.

FIGURE 6.11   NMR spectrum of *cis*-4-methyl–2-pentene.

FIGURE 6.12   IR spectrum of 2-methyl–1-pentene.

FIGURE 6.13   NMR spectrum of 2-methyl–1-pentene.

FIGURE 6.14   IR spectrum of cyclohexanol.

FIGURE 6.15   NMR spectrum of cyclohexanol.

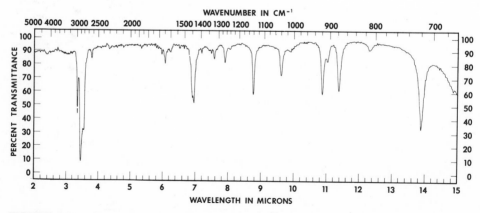

FIGURE 6.16   IR spectrum of cyclohexene.

FIGURE 6.17   NMR spectrum of cyclohexene.

6.2   Addition Reactions of Alkenes

As a class, alkenes are exceedingly useful starting materials for organic syntheses since these compounds undergo a large variety of reactions involving the carbon-carbon double bond. In general, much of the chemistry of the double bond of alkenes can be viewed as the addition of a reagent, $X-Y$, across the pi bond so that a saturated molecule results [equation (15)]. The mechanism through which the addition proceeds depends upon both the nature of $X-Y$ and the conditions under which the reaction is accomplished.

$$R_2C\!=\!CR_2 + X\!-\!Y \longrightarrow R_2\underset{\underset{X}{|}}{C}\!-\!\underset{\underset{Y}{|}}{C}R_2 \qquad (15)$$

One of the more common mechanisms of addition of a reagent to an alkene involves the attack of an electrophile, X^+, an electron-deficient species, on the double bond to give an intermediate 'onium ion **(14)** which reacts with a nucleophilic reagent, Y^-, to give the product **15** [equation (16)]. For unsymmetrical $X-Y$ compounds, Y is the more electronegative of the two atoms. The bromination of alkenes, a reaction used both as a qualitative test for unsaturation as described in Section 6.1 and as a quantitative measure of the amount of unsaturation, generally proceeds by this mechanism. The

$$R_2C\!=\!CR_2 + X^{\oplus} + Y^{\ominus} \Longleftrightarrow \left[R_2C\overset{\oplus}{\underset{\underset{X}{\diagdown\diagup}}{-\!-\!-\!-}}CR_2\right]^{\oplus} \overset{Y^{\ominus}}{\longrightarrow} R_2\underset{\underset{X}{|}}{C}\!-\!\underset{\underset{Y}{|}}{C}R_2 \qquad (16)$$

$$\mathbin{\Updownarrow}$$

$$X\!-\!Y \qquad\qquad \mathbf{14} \qquad\qquad\qquad \mathbf{15}$$

direction of addition of $X-Y$ to an unsymmetrical alkene is such as to result in bond formation between the electrophile, X^+, and the carbon atom bearing the larger number of hydrogen atoms [Markownikoff's Rule, *e.g.*, equation (17)]. The stereochemistry of the addition is such that X and Y are *trans* to one another, as shown by the observation that *trans*-1,2-dibromocyclohexane **(17)** is obtained from the addition of bromine to cyclohexene.

$$R_2C{=}CH_2 + HBr \xrightarrow{\text{CCl}_4} \underset{\substack{| \ \ | \\ Br \ H \\ \textbf{16}}}{R_2C{-}CH_2} \tag{17}$$

17

A related stepwise mechanism, which is observed less commonly, involves initial attack on the alkene by a nucleophile, Y^-, to produce the carbanion $\mathbf{\frac{13}{28}}$, which then reacts with X^+ to provide **19** [equation (18)]. In fact, addition by this mechanism occurs readily only if the double bond bears substituents such as cyano ($-C{\equiv}N$), nitro ($-NO_2$), or carbonyl ($-\underset{\|}{\underset{O}{C}}-$), which are capable of stabilizing a negative charge. The

$$R_2C{=}CR_2 + Y^{\ominus} \rightleftharpoons \underset{\substack{| \\ Y \\ \textbf{18}}}{R_2C{-}\overset{\ominus}{C}R_2} \xrightarrow[\text{(or X\,--Y)}]{X^{\oplus}} \underset{\substack{| \ \ | \\ Y \ X \\ \textbf{19}}}{R_2C{-}CR_2} \tag{18}$$

1,4-addition of aniline to benzalacetophenone [equation (19)], an experiment described in Chapter 15, follows a mechanism of this type.

$$C_6H_5CH{=}CHCOC_6H_5 + C_6H_5NH_2 \longrightarrow \underset{\substack{| \\ }}{C_6H_5\overset{\text{NHC}_6\text{H}_5}{CH}{-}CH_2COC_6H_5} \tag{19}$$

A reagent, $X-Y$, can often be added to an alkene by a free radical, stepwise process if free radical initiators are present or if the reaction mixture is exposed to light. In such reactions, the radical, $X\cdot$, adds to the double bond to yield an intermediate radical **(20)**. The subsequent reaction between **20** and $X-Y$ provides the product **(15)** and re-generates $X\cdot$ [equation (20)]. With unsymmetrical alkenes, X becomes bonded to that

$$R_2C{=}CR_2 + X\cdot \longrightarrow \underset{\substack{| \\ X \\ \textbf{20}}}{R_2C{-}\dot{C}R_2} \xrightarrow{X{-}Y} \underset{\substack{| \ \ | \\ X \ Y \\ \textbf{15}}}{R_2C{-}CR_2} + X\cdot \tag{20}$$

carbon bearing the larger number of hydrogen atoms to yield the more stable of the two possible radicals as an intermediate [equation (21)].

$$R_2C{=}CH_2 + HBr \xrightarrow[\text{(initiator)}]{\text{peroxides}} \underset{\substack{| \ \ | \\ H \ Br \\ \textbf{21}}}{R_2C{-}CH_2} \tag{21}$$

Thus, the products **16** and **21**, respectively, of polar and of free radical addition of hydrobromic acid to an alkene are different, since bromine atom, $Br\cdot$, is the chain carrier in the latter process. One must therefore carefully control reaction conditions in order to obtain pure products when free radical and polar addition can produce different products. Such control may still be important even in those instances in which the *direction* of addition is of no importance, as the *stereospecificity* of free radical addition is generally low, and products resulting from both *trans* and *cis* addition of $X-Y$ to the double bond can be obtained.

Reaction conditions which favor *electrophilic* addition are low temperatures, use of polar solvents such as water and alcohols, the presence of ionic salts like sodium or potassium bromide in the bromination reaction, and the absence of light and of peroxides. In contrast, *free radical* addition is favored by performing the reaction in the gas phase or in nonpolar solvents in the presence of strong light or peroxides.

A final point concerning additions to alkenes is noteworthy. We have now formulated three general mechanisms of addition of a reagent, $X-Y$, to an alkene, mechanisms which are distinguished by the type of species, electrophile, nucleophile or free radical, involved in the initial step of the reaction. This first step, at least in principle, could be *reversible*, and, in the case of appropriately substituted alkenes, could lead to geometric isomerization [equation (22)]. Thus, the intermediates **18** and **20**, which are generated by addition of a nucleophile and of a free radical, respectively, to an alkene,

$$A* + \underset{R}{\overset{R'}{C}}=\underset{R}{\overset{R'}{C}} \rightleftharpoons \left[\underset{R}{\overset{R'}{*C}}-\underset{R}{\overset{R'}{C}}-A \right] \rightleftharpoons \left[\underset{R'}{\overset{R}{*C}}-\underset{R}{\overset{R'}{C}}-A \right] \rightleftharpoons \underset{R'}{\overset{R}{C}}=\underset{R}{\overset{R'}{C}} + A* \quad (22)$$

$(* = \odot, \oplus, \ominus)$

are free to undergo *rotation* about the single bond *between* the carbons which were originally joined by a double bond. Regeneration of the alkene could produce a mixture of geometric isomers. Geometric isomerization might also be a consequence of reversible addition of an electrophile to an alkene if the intermediate so formed, **14**, exists in equilibrium with a carbonium ion [equation (23)] which would be free to rotate about the necessary carbon-carbon bond to produce isomerization [equation (22), $* = \oplus$].

$$\left[\underset{\underset{\oplus}{X}}{R_2C-CR_2} \right] \rightleftharpoons \left[\underset{X}{\overset{\oplus}{R_2C}-CR_2} \right] \quad (23)$$

$$\textbf{14}$$

As a specific example, consider the possible consequences of the addition of bromide ion to dimethyl maleate **(22)**, a *cis* isomer [equation (24)]. This process would give the stabilized anion 23 which could undergo rotation about the carbon-carbon bond to afford **24**. Loss of bromide ion from **24** provides the geometric isomer of **22**, dimethyl fumarate **(25)** [equation (24)]:

As described below, the conversion of dimethyl maleate to dimethyl fumarate is accomplished by treatment of the maleate with a dilute solution of bromine in carbon tetrachloride. A potentially important reaction, the attack of bromonium ion, Br^+, on the double bond of either the maleate or fumarate, is relatively unimportant because the presence of two electron-withdrawing carbomethoxy groups, $-CO_2CH_3$, on the double bond decreases the electron density of the bond sufficiently to preclude facile attack of an electrophile. Therefore the mechanism of the geometric isomerization may be either nucleophilic or free radical in character. Observations made in the experiment should provide a basis for a decision between the two.

A fourth type of mechanism for the reaction of the reagent $X-Y$ with an alkene is concerted addition [equation (25)]. As might be anticipated, the stereochemistry of the reaction is such as to provide *cis* addition of the reagent. Some reactions of this general mechanistic type are hydroxylation, hydroboration, ozonolysis, hydrogenation, and the Diels-Alder reaction. All of these reactions are of considerable synthetic utility and examples of some of them are described elsewhere in this manual (see Experimental Procedure, this section and Chapter 9).

$$R_2C{=}CR_2 + X{-}Y \longrightarrow \left[\begin{array}{c} R_2C{=\!=}CR_2 \\ | \quad\quad | \\ X{\text{-}{-}}Y \end{array} \right] \longrightarrow \begin{array}{c} R_2C{-}CR_2 \\ | \quad\quad | \\ X \quad\, Y \end{array} \qquad (25)$$

In an experiment described below, an alkene will be converted to an alkane by hydrogenation (X = Y = H). This reaction can be accomplished in several ways, but generally it involves use of hydrogen gas either at atmospheric pressure or greater and a noble metal such as palladium, platinum or nickel as the catalyst. For our purposes, it will be convenient to utilize sodium borohydride as the *in situ* source of hydrogen [equation (26)] and to generate the catalyst by reduction of platinic chloride with hydrogen in the presence of decolorizing carbon, which serves as a solid support for the finely divided metallic platinum produced [equation (27)].

$$NaBH_4 + HCl \xrightarrow{3\ H_2O} 4\ H_2 + B(OH)_3 + NaCl \qquad (26)$$

$$H_2PtCl_6 + 2\ H_2 \longrightarrow Pt° + 6\ HCl \qquad (27)$$

EXPERIMENTAL PROCEDURE

A. Geometric Isomerization of Dimethyl Maleate

Place 1.5 ml of dimethyl maleate in each of three 150-mm test tubes and with a dropper add enough of a 10% solution of bromine in carbon tetrachloride to *two* of the tubes so that an orange solution results. Stopper all the test tubes and place one of the tubes containing bromine in the dark and expose the other two tubes to strong light. If decoloration of a solution should occur, add an additional portion of the bromine solution. After 30 min cool all three solutions in ice water, observe in which test tube(s) crystals appear, and isolate the precipitate by vacuum filtration. Wash the crystals free of bromine with a little *cold* carbon tetrachloride and press them as dry as possible on the filter disk. Recrystallize the product from ethanol and determine its melting point and weight. The reported melting point for dimethyl fumarate is 101–102°.

B. Hydrogenation of 4-cyclohexene-*cis*-1,2-dicarboxylic acid

Caution: NO FLAMES! Hydrogen is highly flammable.

Prepare a reaction vessel for hydrogenation by tieing a heavy-walled balloon to the side-arm of a 125-ml filter flask. Also prepare a 1 M aqueous solution of sodium borohydride by dissolving 0.4 g (0.0001 mole) of sodium borohydride in 10 ml of water and adding 0.1 g of sodium hydroxide as a stabilizer. Place 10 ml of water, 1 ml of a 5% solution of chloroplatinic acid ($H_2PtCl_6 \cdot 6H_2O$) and 0.5 g of decolorizing carbon in the reaction flask and add, with swirling, 3 ml of the 1 M sodium borohydride solution. Allow the resulting slurry to stand for 5 min to permit formation of the catalyst. During this time, dissolve 1 g (0.007 mole) of 4-cyclohexene-*cis*-1,2-dicarboxylic acid (see Chapter 9) by heating it with 10 ml of water.

Pour 4 ml of *concentrated* hydrochloric acid into the catalyst-containing reaction flask, and then add the hot aqueous solution of the diacid to the flask. Seal the flask with a serum cap and wire the cap securely in place. Draw 1.5 ml of the 1 M sodium borohydride solution into a plastic syringe, push the needle of the syringe through the serum cap, and inject the solution dropwise while swirling the flask. If the balloon on the side arm of the flask becomes inflated, stop the addition of sodium borohydride solution until deflation occurs. When the syringe is empty, remove it from the flask and refill it with an additional 1.5 ml of solution. The dropwise addition of this further quantity of sodium borohydride should cause the balloon to inflate, and to remain inflated, indicating a positive pressure of hydrogen in the system. Remove the syringe from the serum cap and allow the flask to stand with occasional swirling for five minutes. Heat and swirl the flask on the steam cone until the balloon deflates to a constant size and then heat for an additional five minutes.

Release the pressure from the reaction flask by pushing the needle of a syringe from which the barrel has been removed through the serum cap, filter the hot reaction mixture by suction and place the filter paper in a container reserved for "recovered catalyst."★ Cool the filtrate and extract it with three 25-ml portions of technical ether. Combine the extracts, wash them with 10 ml of saturated sodium chloride solution and then filter the organic solution through a cone of sodium sulfate into a tared 50-ml Erlenmeyer flask.★ Evaporate the ether on a steam cone in the hood in order to isolate the saturated diacid. Determine the weight and melting point of the crude diacid.

Transfer the bulk of the crude diacid by scraping it into a 25-ml Erlenmeyer flask. Add about 2 ml of water to the larger flask, heat to dissolve any residual diacid, and pour the hot solution into the smaller flask. Bring all the diacid into solution at the boiling point by adding no more than one additional milliliter of water. After solution has been accomplished, add three drops of hydrochloric acid to decrease the solubility of the diacid in the solution, allow the mixture to cool to room temperature, and then place the flask in an ice-water bath to effect more complete crystallization of product. Isolate the product and determine its weight and melting point. The reported melting point is 192°. Ascertain that the product is saturated by performing the qualitative tests for unsaturation described in Section 6.1. Also determine whether hydrogenation has affected the acidic nature of the molecule by testing an aqueous solution of the diacid with litmus paper.

EXERCISES

A. GEOMETRIC ISOMERIZATION

1. Write a reasonable mechanism for the bromine-catalyzed geometrical isomerization of dimethyl maleate to dimethyl fumarate observed in this experiment.

2. What is the function of light in the isomerization reaction?
3. What is the purpose of exposing a sample of dimethyl maleate containing no bromine to light?
4. Suggest a reason why decoloration of the solution of bromine and dimethyl maleate is slow.
5. Offer an explanation for the observation that maleic anhydride (26) does not isomerize to 27 under the influence of bromine and light.

26   **27**

6. Figure 6.18 is an nmr spectrum of a mixture of dimethyl fumarate and dimethyl maleate. Calculate the percentage of each isomer in this mixture.

FIGURE 6.18 NMR spectrum of a mixture of dimethyl maleate and dimethyl fumarate.

B. HYDROGENATION
1. Draw the structure of the product of catalytic hydrogenation of
   a. 1,2-dimethylcyclohexene
   b. *cis*-1,2-dideuterio-2-butene
   c. *trans*-1,2-dideuterio-2-butene.
2. Is the same product obtained from (b) and (c) of question 1?
3. Why does the addition of hydrochloric acid to an aqueous solution of a carboxylic acid decrease the solubility of the carboxylic acid in water?

SPECTRA OF STARTING MATERIALS AND PRODUCTS

FIGURE 6.19 IR spectrum of dimethyl maleate.

FIGURE 6.20 IR spectrum of dimethyl fumarate.

FIGURE 6.21 NMR spectrum of dimethyl fumarate.

FIGURE 6.22 IR spectrum of 4-cyclohexene–*cis*-1,2-dicarboxylic acid.

FIGURE 6.23 IR spectrum of *cis*-cyclohexane–1,2-dicarboxylic acid.

7
Arenes
Electrophilic aromatic substitution

7.1 *Introduction to Electrophilic Aromatic Substitution*

Electrophilic aromatic substitution is a very important part of organic chemistry, for the introduction of many functional groups onto an aromatic ring is accomplished in this manner. The general form of this reaction may be represented by equation (1), in which Ar—H is an aromatic compound (an "arene") and E^+ is any one of a number

$$Ar\text{—}H + E^{\oplus} \rightarrow Ar\text{—}E + H^{\oplus} \tag{1}$$

of different electrophiles that may replace H. Although equation (1) represents the overall net reaction correctly, it is greatly oversimplified. For example, the electrophile must usually be generated from the starting materials during the course of the reaction.

The chemical kinetics of electrophilic aromatic substitution reactions have been studied extensively. For many, but not all,[1] cases the rate of substitution has been found to be first order in arene and first order in electrophile:

$$Rate = k_2[Ar\text{—}H]\,[E^+] \tag{2}$$

Because of its bimolecular nature, it is often termed an S_E2 reaction (S = substitution, E = electrophilic, and 2 = bimolecular). A general mechanism for the S_E2 reaction is given below. This mechanism has been substantiated by considerable evidence, especially that coming from isotope effect studies.

Step 1: Formation of Electrophile (an equilibrium reaction):

$$E\text{—}Y \xrightleftharpoons{\text{catalyst}} E^{\oplus} + Y^{\ominus} \tag{3}$$

---

[1] No attempt will be made here to differentiate specifically between the systems which follow the second-order rate expression and those which do not; in general, however, activated arenes do not follow this rate expression.

Step 2: Reaction of Electrophile with Arene (slow step):

$$\text{Ar—H} + \text{E}^{\oplus} \underset{slow}{\rightleftharpoons} \left[\overset{\oplus}{\text{Ar}}\overset{\nearrow \text{H}}{\underset{\searrow \text{E}}{}} \right] \qquad (4)$$

intermediate

Step 3: Loss of Proton to Give Product (fast step):

$$\left[\overset{\oplus}{\text{Ar}}\overset{\nearrow \text{H}}{\underset{\searrow \text{E}}{}} \right] \underset{fast}{\rightleftharpoons} \text{Ar—E} + \text{H}^{\oplus} \qquad (5)$$

The electrophile is most often produced by the reaction between a catalyst and a compound which contains a potential electrophile [equation (3)]. The second-order nature of the reaction may be seen to be related to equation (4), in which *two* molecules react to give the intermediate. If, indeed, the kinetics are second-order, then this must be the slow step in the reaction; the loss of a proton [equation (5)] which follows must be fast relative to the reaction of equation (4).

A listing of various common electrophiles, the conditions under which they are produced, and the reactions which they undergo is given in Table 7.1, along with references to other parts of this book where experiments utilizing them are given. The specific mechanisms, along with detailed information and side reactions, will accompany each of the experiments in the text.

TABLE 7.1   EXAMPLES OF ELECTROPHILIC AROMATIC SUBSTITUTION REACTIONS

Type of Reaction	Electrophile, E^+	Electrophile Precursor	Catalyst, if Any [a]	Structure of Product	Reference to this Reaction
Friedel-Crafts Alkylation	R^+	R—X	AlX_3	Ar—R	Chapter 7.2
Friedel-Crafts Acylation	$R—\overset{\oplus}{C}{=}O$	$R—\overset{O}{\overset{\|}{C}}—X$	AlX_3	$Ar—\overset{O}{\overset{\|}{C}}—R$	Chapter 17.1, 18
Chlorination	Cl^+	Cl—Cl	None	ArCl	Chapter 17.3
Bromination	Br^+	Br—Br	None	ArBr	Chapter 7.3, 17.3
Iodination	I^+	I—Cl	None	ArI	Chapter 17.3
Nitration	NO_2^+	HO—NO_2	H_2SO_4	$ArNO_2$	Chapter 17.2
Sulfonation	SO_3	Cl—SO_3—H	None	$ArSO_3H$	Chapter 17.2

[a] This refers to catalysts which are *actually used* in the experiments given; many of these types of reactions do require a catalyst for many substrates, though under the conditions of the experiments none is needed.

7.2   *Friedel-Crafts Alkylation of Benzene with Butyl Chlorides*

The Friedel-Crafts alkylation reaction is one of the classic types of electrophilic aromatic substitution and as such has been subjected to extensive mechanistic study.

It is also of great practical importance as the most versatile method for attaching alkyl side chains to aromatic rings. The two main limitations to this reaction as a synthetic tool are (1) the difficulty of preventing the introduction of more than one alkyl group onto an aromatic ring (owing to the activating effect of the first group introduced) and (2) the occurrence of rearrangements of the alkyl group. The first difficulty can often be resolved by using a large excess of the arene (note the proportion of benzene to butyl chloride used in the experiments below). There has been considerable confusion about the occurrence and extent of alkyl group rearrangement; it has sometimes been exaggerated and sometimes overlooked. The following experiments should give some indication of the types of alkylations which take place with and without accompanying rearrangements.

The generally accepted mechanism of alkylation [equations (6), (7), and (8)] involves a carbonium ion or partially ionized complex, shown for simplicity as R+, as the electrophile.

$$R{-}Cl + AlCl_3 \rightleftharpoons R^{\oplus} + AlCl_4^{\ominus} \tag{6}$$

this resonance-stabilized intermediate may be represented by the single symbol:

The reaction is catalyzed by a Lewis acid (AlCl₃) which assists in pulling the chlorine atom from the alkyl chloride to give the carbonium ion (R+) and $AlCl_4^-$. Observe that the AlCl₃ does indeed serve as a catalyst, for it is regenerated in the final step.

Since a carbonium ion intermediate is involved, rearrangement is to be expected in some but not necessarily all cases (for carbonium ion rearrangements accompanying elimination reactions, see Chapter 6). For example, rearrangement of an unstable (high energy) carbonium ion to a more stable (lower energy) carbonium ion is to be expected, but a rearrangement which involves little decrease in energy or requires a higher-energy intermediate is not necessarily to be expected. The extent of rearrangement is also dependent on other factors such as the nature of the arene (when other than benzene), the temperature, the solvent (if any), and the nature and concentration of the catalyst.

This type of rearrangement can be illustrated by the reaction between *n*-propyl chloride, aluminum chloride, and benzene. In this reaction, the *n*-propyl cation (a primary carbonium ion) can rearrange to the isopropyl cation (a secondary carbonium ion), which is more stable. The rearrangement is not complete, as is seen by examination of the products; both *n*-propyl- and isopropylbenzene are formed, in the ratio of 35:65. These facts are explained by the following formulation:

$$CH_3CH_2CH_2Cl \xrightarrow[\text{AlCl}_3]{-18 \text{ to } +80°} CH_3CH_2\overset{\oplus}{C}H_2 + AlCl_4^{\ominus} \xrightarrow[\substack{-\text{AlCl}_3 \\ -\text{HCl}}]{} CH_2CH_2CH_3$$

35%

~H:

$$CH_3\overset{\oplus}{C}HCH_3 \xrightarrow[\substack{-\text{AlCl}_3 \\ -\text{HCl}}]{} CH(CH_3)_2$$

65%

From the experiments which are given in this section, you will be able to demonstrate that this type of rearrangement is observed in alkylations with some isomeric butyl chlorides.

Rearrangements of the side chain may also occur after attachment to the aromatic ring. By this we do not mean the possible sequence of reversal of alkylation (dealkylation), rearrangement of carbonium ion, and realkylation, but rather a direct rearrangement of the intact alkyl benzene. For example, the rearrangement of t-pentylbenzene to 2-methyl-3-phenylbutane is explained by a carbonium ion chain mechanism, as follows:

$$C_6H_5-\underset{\underset{CH_3}{|}}{\overset{\overset{CH_3}{|}}{C}}-CH_2-CH_3 + R^{\oplus} \longrightarrow C_6H_5-\underset{\underset{CH_3}{|}}{\overset{\overset{CH_3}{|}}{C}}-\overset{\oplus}{C}H-CH_3 \xrightarrow{\sim CH_3:} C_6H_5-\underset{\underset{CH_3}{|}}{\overset{\overset{CH_3}{|}}{\overset{\oplus}{C}}}-CH-CH_3$$

$$+ RH$$

$$\downarrow RH$$

$$R^{\oplus} + C_6H_5-\underset{\underset{CH_3}{|}}{\overset{}{C}}H-\overset{\overset{CH_3}{|}}{C}H-CH_3$$

The R$^+$ may be produced initially from the reaction of RX or a trace of alkene with AlCl$_3$.

The following experiments with the butyl chlorides provide illustrations typical of the kinds of behavior described above: alkylation without rearrangement, alkylation with rearrangement at the carbonium ion stage, and rearrangement of the alkylbenzene after side chain attachment. Analysis of the products of the various reactions by means of gas chromatography (glpc) and nmr spectroscopy should allow you to decide what has occurred in each experiment.

The infrared spectra of the four butylbenzene isomers are given in Figure 7.1. Gas chromatography with several types of column materials (silicone gum rubber, apiezon, or polyesters are satisfactory) will provide good separation of n-butylbenzene and t-butylbenzene from sec-butylbenzene and isobutylbenzene (Figure 7.2). The latter two isomers cannot be separated readily by glpc, but they can easily be distinguished from one another by their infrared or nmr spectra, and the approximate composition of mixtures of these isomers can be estimated.

FIGURE 7.1 IR spectra of the butylbenzenes.

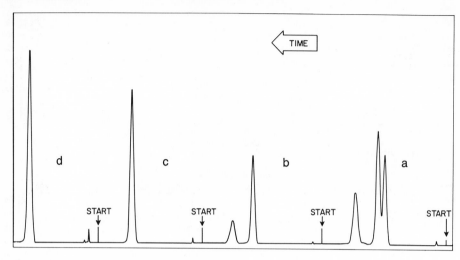

FIGURE 7.2 GLPC data on butylbenzenes.
(a) Synthetic mixture of four isometric butylbenzenes
(b) Fraction II from reaction of 2-butyl chloride, benzene, and AlCl₃ at 27°
(c) Fraction II from reaction of sec-butyl chloride, benzene, and AlCl₃ at 27°
(d) Fraction II from reaction of sec-butyl chloride, benzene, and AlCl₃ at 80°

EXPERIMENTAL PROCEDURE

Caution: If flames are used for heating in this experiment, be careful to keep them away from open flasks containing benzene or the butylbenzenes, as they are highly flammable.

A. *n*-BUTYL CHLORIDE AT ROOM TEMPERATURE

Provide a 250-ml round-bottomed flask with a reflux condenser equipped at the top with a one-hole rubber stopper carrying a tube leading to a simple gas-absorption trap such as shown in Figure 5.1. Clamp the flask high enough on the ring stand so that a cold water bath may be used to cool the reaction mixture as needed.

In the flask place 2.7 g (0.020 mole) of anhydrous powdered aluminum chloride and immediately cover it with 50 ml of dry benzene.[2] Measure 21 ml (18.5 g, 0.20 mole) of *n*-butyl chloride into a 100-ml graduated cylinder. Add about 3 ml of this through the top of the condenser, and quickly replace the stopper and tube leading to the gas trap. Swirl the reaction flask to mix the reactants; evolution of hydrogen chloride should start almost at once. If the evolution becomes too vigorous, raise the cooling bath momentarily around the flask. As soon as the reaction subsides, add another 3-ml portion of *n*-butyl chloride through the condenser. Continue the addition intermittently with swirling and cooling, if necessary, so that all of the *n*-butyl chloride is added within about 15 minutes. Swirl the reaction mixture frequently during an additional

[2] If dry benzene is not provided, you may dry ordinary benzene by azeotropic distillation. Take an amount about 25% in excess of the total amount of benzene you need for these experiments and distil it, using a dry flask. After the excess benzene has been distilled out, or whenever all signs of turbidity (due to water) in the drops of distillate have disappeared, the main part of the benzene (in the *distilling flask*, not the receiver) may be used in the alkylation reactions. Be sure it is cool (room temperature) before using it.

15 minutes, and then pour it into a 250-ml beaker containing 40 g of crushed ice. After stirring the mixture until all the ice has melted, pour it into a 125-ml separatory funnel and drain off the lower (aqueous) layer and discard it. Pour the organic layer into an Erlenmeyer flask containing about 3 g of *anhydrous* calcium chloride (4–6 mesh) and swirl it for about 5 min.* Filter the dried solution into a 100-ml round-bottomed flask and distil it through a short Vigreux column using a water-cooled condenser. Place a wire gauze beneath the flask and, after making sure that all connections are tight, use a burner flame for heating. Distil excess benzene and forerun into one receiver (I) until the distillation temperature reaches 167°, then extinguish the flame and allow the liquid in the Vigreux column to drain into the flask.* Replace the Vigreux column by a simple distilling head and resume distillation, collecting the fraction boiling between 167° and 185° in a second receiver (II). (If any forerun boiling below 167° is obtained in the second distillation, add it to the first receiver.) When the distillation temperature reaches 185°, stop the heating, allow the residue to cool, and pour it into running water in the sink.

Record the weights of fractions I and II. Submit samples of each for gas chromatographic analysis and infrared spectroscopic analysis. Using the weights of fractions I and II and the analytical results, calculate the yield of butylbenzene(s) and estimate the proportions of any isomers identified.

B. *sec*-BUTYL CHLORIDE AT ROOM TEMPERATURE

Carry out this experiment exactly as in A, except use *sec*-butyl chloride in place of *n*-butyl chloride.

C. *sec*-BUTYL CHLORIDE AT 80°

Carry out this experiment exactly as in B except for the following modifications. Attach a Claisen connecting tube at the top of the reflux condenser. Insert a separatory funnel in the opening of this tube directly above the reflux condenser, and connect the other opening to the gas trap. (If the Claisen connecting tube is not available, a two-hole stopper placed in the top of the reflux condenser may be used to hold the separatory funnel and a tube leading to the gas trap.) Heat the mixture of benzene and aluminum chloride to gentle reflux, and add the *sec*-butyl chloride dropwise from the separatory funnel during the course of 15 min. Continue heating an additional 30 min after the addition is complete. Allow the reaction mixture to cool to room temperature and then treat it as in A.

D. ISOBUTYL CHLORIDE AT ROOM TEMPERATURE

Carry out this experiment exactly as in A, except use isobutyl chloride in place of *n*-butyl chloride and use different distillation temperature ranges, as follows: fraction I, up to 162°, fraction II, 162–172°.

EXERCISES

1. Prepare in chart form a listing of the different products which are formed from each of the reactions above. Below each of these products, indicate the approximate amount which is produced. Account for the formation of these possible products using mechanisms. Pay particular attention to possible carbonium ion rearrangements of the types outlined in the discussion section, and indicate where you think these are important.

2. What method would you use to prepare pure *n*-butylbenzene, *i.e.*, without the necessity of separating it from isomers by chromatography?

3. What products would you expect from alkylation of benzene with aluminum chloride and each of the following alkyl chlorides at 25°?

    (a) *t*-butyl chloride;   (c) neopentyl chloride;

    (b) 2-chloropentane;   (d) 3-methyl-1-chlorobutane.

(a) reaction mixture from *n*-butyl chloride, benzene, and AlCl₃ at 27°

(b) reaction mixture from *sec*-butyl chloride, benzene, and AlCl₃ at 27°

(c) reaction mixture from *sec*-butyl chloride, benzene, and AlCl₃ at 80°

FIGURE 7.3 IR spectra of reaction mixtures from alkylations with butyl chlorides.

(a) reaction mixture from *sec*-butyl chloride, benzene, and AlCl₃ at 27°

(b) reaction mixture from *sec*-butyl chloride, benzene, and AlCl₃ at 80°

FIGURE 7.4 NMR spectra of reaction mixtures from alkylations with butyl chlorides.

4. Figure 7.2 shows gas chromatograms obtained in the authors' laboratories: (a) a synthetic mixture of the four isomeric butylbenzenes; (b) fraction II from reaction of *n*-butyl chloride, benzene, and AlCl₃ at 27°; (c) fraction II from reaction of *sec*-butyl chloride, benzene, and AlCl₃ at 27°; (d) fraction II from reaction of *sec*-butyl chloride, benzene, and AlCl₃ at 80°.

Figure 7.3 gives the infrared spectra of fraction II from (a) reaction of *n*-butyl chloride, benzene, and AlCl₃ at 27°; (b) reaction of *sec*-butyl chloride, benzene, and AlCl₃ at 27°; (c) reaction of *sec*-butyl chloride, benzene, and AlCl₃ at 80°.

Figure 7.4 gives the nmr spectra of fraction II from (a) reaction of *sec*-butyl chloride, benzene, and AlCl₃ at 27° and (b) reaction of *sec*-butyl chloride, benzene, and AlCl₃ at 80°.

From the data presented in Figures 7.2, 7.3, and 7.4, determine the identity and approximate proportions of the products from these alkylations with *n*-butyl chloride and *sec*-butyl chloride.

7.3 *Relative Rates of Electrophilic Aromatic Substitution*

Electrophilic aromatic substitution reactions are among the best understood of all organic reactions from the mechanistic standpoint. Most of the information which has been obtained about aromatic substitution is internally consistent. The qualitative principles of the reaction are usually discussed in organic chemistry textbooks. These include the effect of a substituent already present in an aromatic compound upon the reactivity of that compound towards a new, entering substituent and upon the orientation of the entering substituent (*ortho*, *meta*, or *para*). Normally, however, the subject of reactivity is discussed in terms of activation or deactivation of the molecule with respect to benzene, and little, if any, information is given about the *quantitative* differences in rates and reactivities of substituted aromatic compounds. The purpose of this discussion and accompanying experimental work is to give the student some feeling for the magnitude of the actual differences in reactivity.

TABLE 7.2 [a] **RELATIVE RATES OF REACTIVITY OF TOLUENE TOWARDS ELECTROPHILIC SUBSTITUTION**

Type of Reaction	Relative Rate [b]
Bromination	605
Chlorination	340
Acetylation	128
Nitration	24.5
Sulfonation	31.0
Methylation	2.9

[a] From L. M. Stock and H. C. Brown, *Advances in Physical Organic Chemistry*, V. Gold, editor, Academic Press, New York, 1963, Vol. I, pp. 50–52.

[b] The rates here are relative rates, where the rate of substitution on benzene has been taken as 1.0. Mathematically, the relative rate may be defined as:

$$\text{Relative Rate} = \frac{\text{Rate of Substitution on Toluene}}{\text{Rate of Substitution on Benzene}}$$

The values quoted were obtained at *ca.* 25° or have been corrected to 25° by extrapolation.

Table 7.2 lists some data which have been collected for some common types of electrophilic substitution reactions. Most of the reactions were carried out under identical conditions, so that a direct comparison of the numbers is reasonably valid. As mentioned in the footnote of Table 7.2 the rates are for toluene *relative* to benzene; for example, bromination is about 605 times more rapid for toluene than for benzene. Indeed, it can be seen that all electrophilic substitution reactions of toluene are faster than those for benzene, which supports the idea presented in most textbooks that the methyl group *activates* an aromatic ring and makes it more reactive toward an electrophile.

Table 7.3 gives some data for the bromination of different monosubstituted benzene derivatives. The rates are again relative to benzene (R = H in C_6H_5R), so that these results indicate the relative reactivities of the various substituted benzenes. Those exhibiting a rate greater than 1.0 have substituents already present which activate the aromatic ring towards further substitution, whereas those whose rate is less than 1.0 have substituents already present which deactivate the ring.

TABLE 7.3 [a]   **RATES OF BROMINATION FOR VARIOUS AROMATIC COMPOUNDS**

Substituent R in C_6H_5R	Relative Rate [b]
—OH	6.1×10^{11}
—OCH$_3$	1.8×10^9
—NHCOCH$_3$	2.1×10^8
—OC$_6$H$_5$	1.7×10^7
—C$_6$H$_5$	1.0×10^3
—CH$_3$	6.05×10^2
—C$_2$H$_5$	4.6×10^2
—CH(CH$_3$)$_2$	2.6×10^2
—C(CH$_3$)$_3$	1.38×10^2
—H	1.0 (standard)
—X (*e.g.*, Cl, Br)	0.5 (estimated value only)
—CO$_2$H	7.5×10^{-3}
—$\overset{+}{N}$(CH$_3$)$_3$	1.6×10^{-5}
—NO$_2$	1.6×10^{-5}

[a] From L. M. Stock and H. C. Brown, *Advances in Physical Organic Chemistry*, V. Gold, editor, Academic Press, New York, 1963, Vol. I, pp. 62–80.

[b] Rates are relative rates, where the rate of substitution on benzene has been taken as 1.0. The values were obtained at *ca.* 25° with bromine in aqueous acetic acid or have been corrected to 25° by extrapolation.

It can be seen from the data in Table 7.3 *how much* more activating one group is compared to another. In particular, it should be pointed out that phenol (R = OH) is *about 10^{16} times more reactive* towards bromine than is nitrobenzene (R = NO$_2$). Differences in reactivity can be seen only from data of this sort, and the student is advised to examine Table 7.3 carefully.

As has been stated before, the two most important steps in electrophilic aromatic substitution reactions are (1) the attack of the electrophile on the electron-rich aromatic ring and (2) loss of a proton [see equations (4) and (5) in Section 7.1]. The intimate details of the mechanism, particularly with regard to the rate expression, differ with various substrates and reaction conditions. Part of the problem lies in the role and concentration of the catalyst in these reactions, and some differences in mechanism are believed to be related to the speed with which the electrophile, E$^+$, is generated and consumed.

The measurement of reaction rates makes it necessary to review some of the factors which influence them (see Chapter 11.3 for a brief summary). Some of the variable factors are temperature, concentration of reactants, solvent composition and nature and concentration of catalysts. Since it is possible to change any of these variables while keeping all other factors constant, the effect of all of the variables can be determined.

In this experiment, it will be possible for the student to measure rates and rate differences for an electrophilic aromatic substitution reaction. The reaction chosen for study here is the bromination by elemental bromine of a series of typical aromatic compounds. The rate expression for this reaction is given by equation (9), in which Br_2 has

$$\text{Rate} = k_2[\text{ArH}][\text{Br}_2] \tag{9}$$

been inserted in place of E^+ of equation (2) in Section 7.1, since bromine is clearly the precursor to the actual electrophile. The rate expression could be more complex under certain conditions. Bromination of less reactive aromatic compounds requires the use of a Lewis acid catalyst for reaction at a reasonable rate. Water, in *trace* amounts, has been found to accelerate the rate of these electrophilic reactions. However, by avoiding the use of a catalyst and by carrying out the brominations in a mixed solvent composed of 90% acetic acid and 10% water,[3] both of these complications are excluded from the rate studies. The rate expression of equation (9) states that the rate of the reaction is proportional to the concentration of the aromatic substrate and to the concentration of bromine, and as written is a second-order rate expression. While a student could, in principle, work with this rate expression, the calculations are much more difficult and time-consuming than are those for first-order reactions. Thus, the experimental procedure is designed to provide a "pseudo-first-order" reaction to simplify the calculations. This is accomplished by maintaining the concentration of the aromatic substrate at a high level so that it does not change appreciably throughout the course of the reaction. Under these conditions the rate expression becomes

$$\text{Rate} = k_1[\text{Br}_2] \tag{10}$$

where k_1 is a new rate constant which will contain all of the factors of constancy for the reagents and solvents.[4]

The integrated form of equation (10) is

$$[\text{Br}_2]_t = [\text{Br}_2]_o\, e^{-k_1 t} \tag{11}$$

which can be rearranged to

$$\ln \frac{[\text{Br}_2]_o}{[\text{Br}_2]_t} = k_1 t \tag{12}$$

or

$$2.303 \, \log \frac{[\text{Br}_2]_o}{[\text{Br}_2]_t} = k_1 t \tag{13}$$

where $[\text{Br}_2]_o$ = initial concentration of bromine, and $[\text{Br}_2]_t$ = concentration of bromine at any time t.[5] From the experimental data, one can prepare a graph of $\log[\text{Br}_2]_o/[\text{Br}_2]_t$

[3] This much water in the solvent insures a concentration which remains essentially constant during the course of the reaction and hence does not enter into the rate expression.

[4] Rate constant k_1 is called a *pseudo*-first-order rate constant. Since all other factors are being held constant, the reaction rate will *appear* to depend only on the concentration of bromine, as the rate expression indicates. *Pseudo*-rate constants in general refer to reactions in which the concentration of one or more reactants which appear in the rate expression are chosen so that they do not appreciably change as the reaction proceeds. Thus they remain constant and are automatically incorporated in the pseudo-rate constant.

[5] See Chapter 11.3 for a more detailed discussion of these equations for first-order rate constants. The equations are very similar to those which are presented for the hydrolysis of tertiary halides.

vs. time, *t*. If the reaction is first-order, the graph should show a reasonably straight line with a slope of $k_1/2.303$ if common logarithms are used. It is important to note that $[Br_2]_t$ is the concentration of bromine *remaining* at time *t*, and not the amount of bromine which has reacted.

The reaction temperature and initial concentration of reactants must be controlled. The use of a large water bath in the experiment and careful preparation of the reagents will insure that these factors are kept constant.

One of the main reasons for choosing bromination reactions for carrying out rate studies is that it is easy to follow the rate of disappearance of the bromine color as the reagent reacts with the aromatic substrate. The virtue of choosing reaction conditions so that the rate of the reaction is proportional only to the concentration of the bromine should become quite apparent.

An important limitation must be placed on the types of aromatic compounds to be studied. As no Lewis acid catalyst is present, the study is by necessity limited to compounds which are more reactive than benzene. Benzene and compounds of similar or lower reactivity would react so slowly with bromine in the absence of a catalyst that it would not be possible to obtain good kinetic data in a single laboratory period. This limitation presents no serious problem, for it will still be possible to determine the relative reactivities of a series of compounds and to illustrate the principles which are involved in these sorts of measurements.

Experiments are described which allow one to obtain qualitative differences in reactivity. Depending on the facilities available, either semi-quantitative or quantitative measurements of absolute rates may also be determined.

The qualitative experiments involve preparing solutions of various substrates in 90% acetic acid containing bromine. From the length of time required for the bromine color to be discharged, one can deduce the relative reactivities.

The semi-quantitative experiments involve preparing a series of solutions of 90% acetic acid which contain known concentrations of bromine. If one then prepares a solution containing an excess of substrate and a known concentration of bromine, a rough estimate of the rate may be made by visual comparison of reaction mixture color with those of the known solutions.

The quantitative experiments require the use of an inexpensive visible/ultraviolet spectrophotometer,[6] which allows accurate measurement of the *rate of disappearance* of bromine. This instrument is non-recording, but the data can be read from a meter. If a compound strongly absorbs light at a given wavelength, then the ability of light to pass through a sample of that compound is diminished. The spectrophotometer measures the amount of light which has been absorbed, this value being called absorbance. Absorbance is defined as

$$\text{Absorbance } (A) = \log \frac{I_o}{I} \tag{14}$$

and is often called *optical density*. In equation (14), I_o is the intensity of incident light and I is the intensity of emergent light at a given wavelength. If a sample absorbs strongly, the ratio of I_o/I will be large and A will be large. The absorbance also depends on variables other than the nature of the absorbing species, these being the *concentration* of the absorbing species and the *length* of the cell which contains the solution. The

---

[6] Several relatively inexpensive instruments are available, such as the Bausch and Lomb "Spectronic 20" (about $250.00) which can be used in this experiment.

absorbance is *proportional* to the cell length and to the concentration, and this relationship, called the *Lambert-Beer Law*, may be written as

$$A = kcl \tag{15}$$

where A is the absorbance, k the proportionality constant, c the concentration of absorbing species, and l the path length of the solution. In these experiments you will use the same cell throughout, and the absorbing species (bromine) will remain the same, so that l and k will remain constant. Thus, if the absorptivities, A_1 and A_2, of two solutions of concentrations c_1 and c_2 are measured, equations (16a) and (16b) may be written and combined to give equation (17).

$$A_1 = kc_1l \tag{16a}$$

$$A_2 = kc_2l \tag{16b}$$

$$\frac{A_1}{A_2} = \frac{kc_1l}{kc_2l} = \frac{c_1}{c_2} \tag{17}$$

It may then be seen that if the concentration of one solution is known and the measurements of absorptivity made, the concentration of the other solution may be determined.

Additional instructions about treatment of data and use of equations will be given in the experimental procedure.

EXPERIMENTAL PROCEDURE

In the following experiments, it would be best to have two students working together.

A. QUALITATIVE RATE COMPARISONS

Prepare a water bath using a one-liter beaker, and adjust the temperature to about 35°. During the experiments which follow, keep the temperature at $35 \pm 2°$. Place 2.0 ml of 0.2 M solutions of each of the following substrates (contained in 90% acetic acid) in separate small test tubes: (a) phenol; (b) anisole; (c) diphenyl ether; (d) acetanilide; (e) p-bromophenol; (f) benzene; (g) chlorobenzene; (h) nitrobenzene; and (i) α-naphthol. If stock solutions are provided, use special care to insure that one solution is not contaminated by another. Suspend the carefully labeled test tubes partially in the water bath by looping a piece of copper wire around the neck of the test tube and over the rim of the beaker.

Prepare some droppers with a capacity of 2.0 ml. Calibrate these by comparison with known volumes, and mark the droppers by scratching the glass with a file. These will be used to introduce the bromine solution into the test tubes containing the substrates. It is essential that the addition be done as rapidly as possible and that the volumes be fairly accurate. Transfer about 45 ml of a solution containing 0.05 M bromine in 90% acetic acid to an Erlenmeyer flask and allow the solution to equilibrate in the water bath for a few minutes. Add 2.0 ml of the bromine solution to one of the test tubes containing a substrate. Make the addition rapidly, mix the solution quickly, and note the exact time of addition. Observe the reaction mixture and note how long it takes for the bromine color to become faint yellow or to disappear. Repeat this procedure with each of the substrates, making sure that you use the same end-point color for each one.

When the reaction is very slow (with which compounds would you expect this to arise?) go on to another compound while waiting for the end-point to be reached. Record the reaction times, and on the basis of these observations, arrange the compounds in order of increasing reactivity towards bromine.

If it is impossible to determine the relative rates accurately at 35°, repeat the experiment with the compounds in doubt at 0°, using an ice-water bath.

B. SEMI-QUANTITATIVE RATE MEASUREMENTS

Obtain stock solutions of each of the following contained in 90% acetic acid: 0.5 M bromine, 0.05 M p-nitrophenol, 0.05 M acetanilide, and 0.05 M diphenyl ether. Prepare standard color comparison solutions by diluting 2.0-ml portions of the 0.5 M bromine solution to each of the following volumes with 90% acetic acid solution: 4.0, 8.0, 16.0, 32.0, 64.0, 128, and 256 ml. Transfer 4.0 ml of each of these solutions to small test tubes; arrange the solutions in order of decreasing bromine concentration and thus decreasing intensity of color. (It would be helpful to place a white background behind these test tubes.) Prepare a reaction mixture comprised of 2.0 ml of one substrate solution and 2.0 ml of the bromine solution. After mixing the solutions, stir to insure homogeneity; note the time of mixing. With the aid of a partner, note the time (in seconds) that the color of the reaction mixture most closely matches each standard bromine sample. Be sure that the measurements are carried out during the same laboratory period in order to minimize the temperature effect. Repeat this procedure with the other substrates.

Treatment of data.

1. Determine the molar concentration of bromine in each of the standardized bromine solutions which you have prepared.
2. From the color comparison data and the bromine concentrations in each standard solution, determine the concentration of bromine which *remained* in the reaction mixture at the end of each of the time intervals.
3. Plot the molar concentration of bromine remaining (vertically) *vs.* time (seconds, horizontally) for each substrate.
4. Draw the best curve that you can through the points which you have drawn. Draw a line tangent to the curve where it crosses the $t = 0$ point. Determine the slope of this line. These slopes, when compared for the various substrates, give an estimate of the *initial relative rates* and thus of the relative reactivities.
5. The rate expression for bromination [equation (10)] may be rewritten as follows:

$$\text{Rate of Reaction} = \frac{\Delta[\text{Br}_2]}{\Delta t} = k_1[\text{Br}_2]$$

where $\Delta[\text{Br}_2]/\Delta t$ = the rate of disappearance of bromine.

When the reaction has just started, the percentage change in the bromine concentration is very small compared to the total amount present. Under these conditions, it is possible to substitute the *initial rate* determined in part 4 for $\Delta[\text{Br}_2]/\Delta t$. Since we know the initial rate and we know the initial concentration of bromine it is possible to substitute these values into the above rate equation and solve for k_1, the rate constant. Do this for each substrate.

C. Quantitative Measurements

Obtain three sample tubes for use with the spectrophotometer (these tubes resemble small test tubes, but the dimensions and glass thickness are carefully controlled; they are called "Cuvette" tubes). As all solutions absorb light to some extent, though they appear transparent to the naked eye, it is necessary to determine the absorbance of 90% acetic acid alone (without bromine or substrate) so that the absorbance of the solvent may be cancelled out in all future measurements. (The reason for doing this is that air would not be valid as a blank, for acetic acid will absorb slightly more than air.) Add 4 ml of 90% acetic acid to one Cuvette and stopper it. This is to be used as a blank for the experiment; your instructor will show you how to use the spectrophotometer and how to set it to zero absorbance for the blank. Determine the blank reading and all future readings at 520 mμ, the wavelength at which elemental bromine absorbs light.

Have at your disposal stock solutions (0.5 M) of acetylsalicylic acid (aspirin), diphenyl ether, and acetanilide in 90% acetic acid, as well as a 0.02 M solution of bromine in 90% acetic acid. It is necessary, in preparing for and executing the following experiments, that you be well organized so that you can work quickly. It may take several tries before you are able to work fast enough to get reasonable results. Before starting, equilibrate several 10-ml samples of each of the stock solutions at 35° (using a water bath).

Allow 2.0 ml of the stock solution of aspirin to equilibrate at 35° in a stoppered Cuvette. Now, rapidly and in the order listed, carry out the following operations: (a) open Cuvette; (b) add 2.0 ml of the equilibrated bromine solution; (c) stir quickly, one time only; (d) replace cork tightly; (e) remove Cuvette from water bath and dry it with paper towel; (f) place Cuvette in spectrophotometer, alongside the blank. A second person should note the exact time of mixing, so that this can be used as zero time.

Take readings at 30, 60, 90, 120, 150, and 180 sec. If an error is suspected, repeat the experiment since it will take only a little time. Remove the Cuvette and place it in the water bath. Take absorbance readings on the sample every 15 min for the remainder of the laboratory period.

Repeat this experiment for diphenyl ether and acetanilide. With these substrates, absorbance readings should be made every 10 sec for the first minute and every 30 sec for the next 4 min. Continue taking readings as long as there is a change in absorbance; however, this will normally not last more than 5 min. If possible, perform a duplicate run on the latter two substrates.

Treatment of data

1. Plot, for each substrate, the data which you have obtained, with absorbance (vertically) *vs.* time (seconds, horizontally). Extrapolate each curve to zero time; the point at which the curve crosses the zero time line is the A_o value (absorbance at time $t = 0$).

2. Divide the values of A for each datum point by A_o to obtain A/A_o ratios. Do this for each substrate, being careful not to intermix A and A_o values from the different runs.

3. Plot on a single sheet of graph paper A/A_o *vs.* time (seconds, from 0 to 500) for each substrate. Draw tangents to each curve at the point of zero time, and from

these tangents estimate the *relative initial rates*. (See a discussion of relative initial rates in part B.)

4. Calculate, for each substrate, A_o/A values. (These correspond to $[Br_2]_o/[Br_2]_t$.) Determine 1n A_o/A or log A_o/A, and plot these values *vs.* time. As discussed previously, the slope is the rate constant in the case of natural logarithms, but is 2.303 times the rate constant for common logarithms. The fact that a straight line is obtained here is further proof that the reaction is indeed first-order in bromine. Determine the rate constants for the bromination of each substrate studied and determine the relative reactivities from these.

Alkynes
Hydration

Alkynes are quite similar to alkenes (see Chapter 6) insofar as their reactions are concerned. In particular, alkynes undergo electrophilic *addition* reactions with reagents such as $X-Y$ [equation (1)]. Using carefully controlled reaction conditions, it is possible to stop the addition reaction at the alkene stage in some cases. As with alkenes, the

$$R-C\equiv C-R' \xrightarrow[\text{1 mole}]{X-Y} R-\underset{X}{\underset{|}{C}}=\underset{Y}{\underset{|}{C}}-R' \xrightarrow[\text{1 mole}]{X-Y} R-\overset{X}{\underset{X}{\overset{|}{\underset{|}{C}}}}-\overset{Y}{\underset{Y}{\overset{|}{\underset{|}{C}}}}-R' \tag{1}$$

electrophilic addition of unsymmetrical reagents to terminal alkynes follows Markownikoff's Rule, and the less electronegative atom of the reagent becomes attached to the carbon bearing the hydrogen [equation (2)]. Note that in the addition of the second mole of $X-Y$ to the alkene, the same orientation is followed as in the initial addition of $X-Y$ to the alkyne [equations (1) and (2)].

$$R-C\equiv C-H \xrightarrow[\substack{\text{1 mole} \\ \text{(X is less} \\ \text{electronegative} \\ \text{than Y)}}]{X-Y} R-\underset{Y}{\underset{|}{C}}=\underset{X}{\underset{|}{C}}-H \xrightarrow[\text{1 mole}]{X-Y} R-\overset{Y}{\underset{Y}{\overset{|}{\underset{|}{C}}}}-\overset{X}{\underset{X}{\overset{|}{\underset{|}{C}}}}-H \tag{2}$$

Acetylene, the simplest alkyne, is a widely used commercial organic compound, but because it is a gas at room temperature, it is not experimentally suitable for investigating the properties and reactions characteristic of the carbon-carbon triple bond. We shall study an alkyne having a terminal triple bond and therefore having chemical properties quite similar to those of acetylene. This compound, 2-methyl-3-butyn-2-ol (**1**) has the advantages of being inexpensive and readily available commercially, as well as easily handled.[1]

---

[1] Although this molecule contains a hydroxy group, its presence has little effect on the chemical properties of the carbon-carbon triple bond. The main effect of the hydroxyl group is on the physical properties of the molecule, such as melting and boiling point. Acetylenic *hydrocarbons* having the same molecular weight would melt and boil much lower.

$$CH_3-\underset{\underset{CH_3}{|}}{\overset{\overset{OH}{|}}{C}}-C\equiv C-H$$

1

As with alkenes, alkynes readily decolorize a solution of bromine in carbon tetrachloride. Either one or two moles of bromine may react per mole of alkyne, with two moles being required for complete reaction. Another common reaction which is used to detect the presence of unsaturation is the Baeyer test, in which a solution of potassium permanganate is decolorized.

The acetylenic hydrogen of a terminal alkyne ($-C\equiv C-H$) is relatively acidic compared to the hydrogens in alkanes and alkenes. The acidity of hydrocarbons has been studied extensively, and it has been found that the order of decreasing acidity is as follows:

$$H-C\equiv C- > \underset{H}{\overset{}{C}}=C < > H-\overset{|}{\underset{|}{C}}-\overset{|}{\underset{|}{C}}-$$

$$K_a \sim 10^{-20} \quad K_a \sim 10^{-40} \quad K_a < 10^{-40}$$
(most acidic) (least acidic)

This acidity is an important property of terminal alkynes, and a simple test for this group is the ready formation of a precipitate on addition of a solution containing silver ammonia complex (silver nitrate dissolved in ammonia). Terminal alkynes give a solid silver salt, which results when the terminal hydrogen is removed by ammonia acting as a base and replaced by silver ion. This reaction is shown in equation (3) in general terms, and compound **1** is a representative example of a compound which would so react.

$$R-C\equiv C-H + Ag(NH_3)_2^{\oplus} \longrightarrow R-C\equiv C-Ag + NH_3 + NH_4^{\oplus} \qquad (3)$$
a silver acetylide

This unique reaction of the terminal alkynes provides an easy method for separating them from nonterminal alkynes. The former form insoluble silver salts, whereas the latter do not and remain in solution. The silver acetylide salts can be reconverted to the terminal alkynes on treatment with hydrochloric acid, as shown in equation (4). *Dry* silver salts of this type are quite sensitive to shock and in this condition tend to decompose *explosively*. Care should be taken to insure that they never dry out before decomposition by hydrochloric acid.

$$R-C\equiv C-Ag + HCl \longrightarrow R-C\equiv C-H + AgCl \qquad (4)$$

A very useful reaction involving the carbon-carbon triple bond is its hydration to give an aldehyde or ketone. For example, the hydration of acetylene is a commercial method of preparation of acetaldehyde. The addition of water to acetylene is catalyzed by mercuric sulfate in the presence of sulfuric acid. The reaction proceeds via the initial formation of vinyl alcohol, which is unstable and tautomerizes to acetaldehyde, as shown by equation (5). With substituted acetylenes, hydration occurs in accordance

$$H-C{\equiv}C-H + H_2O \xrightarrow[\text{HgSO}_4]{\text{H}_2\text{SO}_4} \left[CH_2{=}C{\overset{OH}{\underset{H}{\Big\langle}}} \right] \rightleftharpoons CH_3-C{\overset{O}{\underset{H}{\Big\langle}}} \qquad (5)$$

vinyl alcohol
(unstable)

with Markownikoff's Rule (the proton going to the carbon atom already bearing the greater number of hydrogens), and the product is a ketone. 2-Methyl-3-butyn-2-ol **(1)** is a monosubstituted acetylene, which under typical hydration conditions, will give a ketone, 3-hydroxy-3-methyl-2-butanone **(2)**, as shown by equation (6).

$$\underset{\underset{\text{CH}_3}{|}}{\overset{\overset{\text{HO}}{|}}{CH_3-C}}-C{\equiv}C-H + H_2O \xrightarrow[\text{HgSO}_4]{\text{H}_2\text{SO}_4} \left[\underset{\underset{\text{CH}_3}{|}}{\overset{\overset{\text{HO}}{|}}{CH_3-C}}-\overset{\overset{\text{OH}}{|}}{C}{=}CH_2 \right] \rightleftharpoons \underset{\underset{\text{CH}_3}{|}}{\overset{\overset{\text{HO}}{|}}{CH_3-C}}-\overset{\overset{O}{\|}}{C}-CH_3 \quad (6)$$

1   (unstable)   **2**

The hydration reactions often require specific catalysts, some of which are cuprous, mercuric, and nickelic ions, but of these mercuric ion is used most often. The role these catalysts play in the reaction is not completely understood. It is felt that they may form a π-complex with the triple bond and thus render the alkynes more soluble in the polar aqueous solvent. It is also possible that these metal ions actually *add* directly to the triple bond and are then removed in a subsequent displacement reaction.

The hydration of **1** will be carried out using mercuric sulfate as the catalyst in the presence of sulfuric acid.[2] *Dilute* acid is used to eliminate the problem of having a highly exothermic reaction.

It has been asserted above that the addition of water (an unsymmetrical reagent) to **1** (an unsymmetrical alkyne) will occur according to Markownikoff's Rule to give **2** as the final product. This can readily be verified by identifying the product of the hydration experiment. If the water added in an anti-Markownikoff fashion, 3-methyl-3-hydroxybutanal **(3)** would be formed.

$$\underset{\underset{\text{CH}_3}{|}}{\overset{\overset{\text{OH}}{|}}{CH_3-C}}-CH_2-C{\overset{O}{\underset{H}{\Big\langle}}}$$

3

In order to distinguish between the two possible products, a solid derivative of the product can be prepared for identification purposes. (See Chapter 20 for additional information concerning identification of organic compounds through preparation of derivatives.) One suitable solid derivative for carbonyl-containing compounds is the semicarbazone (Chapter 15.1). Compounds **2** and **3** have been prepared by independent methods, and their semicarbazone derivatives have been made; the semicarbazone **(4)** of ketone **2** melts at 162.5°, whereas aldehyde **3** first dehydrates under the conditions of semicarbazone formation, and the resulting unsaturated aldehyde, $(CH_3)_2C{=}CHCHO$, then reacts to form a semicarbazone **(5)**, which melts at 222–223°. Thus, if the semi-

[2] The experimental procedure is a modification of that reported by N. C. Rose in *Journal of Chemical Education*, **43**, 324 (1966).

$$(CH_3)_2\underset{\underset{\textstyle HO}{|}}{C}-\underset{\underset{\textstyle CH_3}{|}}{C}=N NH\underset{\overset{\textstyle \|}{O}}{C}NH_2$$
4

$$(CH_3)_2C=\underset{\underset{\textstyle H}{|}}{CH}-\underset{\overset{\textstyle \|}{O}}{C}=NNH\underset{}{C}NH_2$$
5

carbazone of the product is prepared and its melting point determined, the principal mode of hydration may be deduced.

EXPERIMENTAL PROCEDURE

A. QUALITATIVE TESTS FOR THE TRIPLE BOND

Reaction with bromine in carbon tetrachloride. See the experimental procedure given in Chapter 6.1. Follow the same procedure for the qualitative test for the carbon-carbon triple bond in 2-methyl-3-butyn-2-ol.

Baeyer test for unsaturation. See the experimental procedure given in Chapter 6.1, and follow it for the reaction of 2-methyl-3-butyn-2-ol with potassium permanganate.

Reaction with silver ammonia complex; formation of a silver acetylide and its decomposition. Prepare a solution of silver ammonia complex from 5 ml of 0.1 *M* silver nitrate solution by adding ammonium hydroxide solution dropwise. Brown silver oxide forms first; add *just* enough ammonium hydroxide to dissolve the silver oxide. Dilute the solution by adding 3 ml of water. Add 3 ml of the diluted silver ammonia complex solution to about 0.1 ml of 2-methyl-3-butyn-2-ol.

Note the formation of the silver acetylide salt. (If you have at your disposal a disubstituted acetylene, treat it similarly and note any reactions which occur.) Filter the silver salt from the aqueous solution; be careful not to let it dry, for the dry salt is quite explosive. Treat the silver salt with a small amount of dilute hydrochloric acid, and observe what changes occur, especially in the color and form of the precipitate. Ultimately, destroy *all* solid salt by treatment with hydrochloric acid.

B. PREPARATION OF 3-METHYL-3-HYDROXY-2-BUTANONE;
THE HYDRATION OF 2-METHYL-3-BUTYN-2-OL

Add 18 ml of *concentrated* sulfuric acid *carefully* to 115 ml of water contained in a 500-ml round-bottomed flask. Dissolve 1 g of mercuric oxide in the resulting warm solution [3] and then cool the flask to about 50°. Attach a reflux condenser to the flask, and then add *in one portion* 12.6 g (0.15 mole) of 2-methyl-3-butyn-2-ol through the condenser. The precipitate, which forms immediately, is presumably the mercury complex of the alkyne. Shake the reaction flask to mix the contents. When this is done, an exothermic reaction ensues as the precipitate dissolves, and the solution turns light brown. Allow the reaction to proceed by itself for about two minutes and then heat the mixture to reflux.[4] As soon as the reflux begins, remove the heat and cool the reaction mixture to 50°. Add an additional 12.6 g of alkyne in one portion through the condenser. A precipitate again forms, and the contents of the flask should be mixed by shaking the flask. After several minutes, heat the mixture to reflux and continue heating

---

[3] It has been found that the quality (purity) of the mercuric oxide can greatly affect the yield of the hydration reaction. Use a good grade of mercuric oxide, or prepare fresh mercuric oxide by adding sodium hydroxide to mercuric chloride solution; ask your instructor for further details.

[4] It may be that the exothermic reaction will not begin on its own; if it does not, heat the mixture until it starts to reflux. Then proceed as directed.

for 15–20 min. Equip the flask for simple distillation, add 100 ml of water to the re-action mixture, and distil. Be especially careful during the collection of the first 75 ml of distillate, since foaming occurs. Continue distillation until a total of about 150 ml of distillate has been collected.* Transfer the distillate to a separatory funnel, add about 25 g of potassium carbonate sesquihydrate to the distillate, and then saturate the solu-tion with sodium chloride. (A second layer may form; if so, do not separate it but con-tinue with the extraction as described.) Extract the mixture with three 35-ml portions of benzene (technical grade) and combine the extracts. Dry the organic extracts over 8 g of *anhydrous* potassium carbonate, filter, and distil. After removal of the benzene, collect the product boiling between 138 and 141°. The average yield is between 10 and 15 g; the product should be a colorless liquid.

C. IDENTIFICATION OF THE PRODUCT; SPECTROSCOPIC METHODS

The infrared and nmr spectra of both the starting material (1) and the product (2) are provided at the end of this chapter. While it is possible to identify the product by converting it to a known solid derivative (see below), it is also quite easy to prove the structure using spectral methods. An examination of the infrared spectra provided will indicate the presence of a —C≡C—H bond in 1, and the absence of the C=O group, whereas the reverse of this will be found in 2. As an exercise and with the aid of Table 4.1, identify the absorptions attributed to these functional groups. The nmr spectra will also show a terminal acetylenic hydrogen for 1, whereas the hydrogens of the newly formed CH_3 group in $CH_3C=O$ will be observed for 2. Identify the peaks in the nmr spectra, using Table 4.2.

D. IDENTIFICATION OF PRODUCT; SEMICARBAZONE FORMATION

Add 1 ml of the product to a solution prepared from 1 g of semicarbazide hydro-chloride and 1.5 g of sodium acetate dissolved in 5 ml of water. Shake or stir vigorously. The solid which forms is the crude semicarbazone. Recrystallize about one-third of the crude solid as follows. Add to it about 5 ml of benzene, heat the mixture on a steam bath, and add 95% ethanol dropwise until all of the solid has dissolved. Cool and collect the crystals by filtration. Determine the melting point and use this information to identify the product obtained by hydration of 2-methyl-3-butyn-2-ol.

EXERCISES

1. It was stated that 2-methyl-3-butyn-2-ol is commercially available and is inex-pensive. Suggest a method of preparation of this compound from two simple in-expensive organic compounds.

2. How might you separate, and ultimately obtain in pure form, 1-octyne and 2-octyne from a mixture containing both of them?

3. Give the structures for the products which you would expect to obtain on hydra-tion of 1-octyne and of 2-octyne.

4. It is well known that alcohols can be oxidized to ketones by oxidizing agents such as potassium permanganate. An examination of alkyne 1 will show that it has a hydroxyl group as well as a triple bond. Therefore, a student might conclude that the positive permanganate test is due to the oxidation of the hydroxyl group and not to reaction of the triple bond. Could this be the case? If so, what addi-tional "control" experiments might you suggest to eliminate this possibility from further consideration?

5. Consider the compounds *n*-hexane, 1-hexene, and 1-hexyne. What similarities and what differences would you expect in the reactions of these compounds with: (1) bromine in carbon tetrachloride; (2) an aqueous solution of potassium permanganate; (3) an aqueous solution of sulfuric acid; (4) a solution of silver ammonia complex? Give the structures of the products, if any, which would be obtained from these reactions.

6. Suppose that you have three bottles each of which is known to contain one of the compounds listed in question 5 and that the labels have been removed. Based on your answers to question 5, how might you go about labeling each bottle correctly? Explain briefly.

7. Suggest a reason for adding the potassium carbonate and saturating the solution with sodium chloride in the work-up procedure of part B.

SPECTRA OF STARTING MATERIAL AND PRODUCT

FIGURE 8.1   IR spectrum of 2-methyl–3-butyn–2-ol.

FIGURE 8.2   NMR spectrum of 2-methyl–3-butyn–2-ol.

FIGURE 8.3 IR spectrum of 3-methyl–3-hydroxy–2-butanone.

FIGURE 8.4 NMR spectrum of 3-methyl–3-hydroxy–2-butanone.

9

Dienes:
The Diels–Alder reaction

One of the more useful synthetic methods available to the organic chemist is the reaction between a 1,3-diene and an alkene, sometimes referred to as the dienophile (Gr., *philos*, loving), to produce a derivative of cyclohexene [equation (1)]. If an alkyne is used as the dienophile, instead of an alkene, a derivative of 1,4-cyclohexadiene will be formed [equation (2)]. The reaction, which is termed the Diels-Alder reaction in honor of its primary developers, Otto Diels and Kurt Alder, results in the formation of new carbon-carbon bonds between the two π-bonded carbons of the dienophile and the 1- and 4-carbons of the diene. Thus, there is an overall 1,4-addition of the dienophile to the diene.

$$\tag{1}$$

$$\tag{2}$$

The Diels-Alder reaction is of quite general utility, and many different dienes and dienophiles have been employed to afford good yields of adducts. The presence of electron-releasing substituents such as alkyl and alkoxy (—OR) on the diene and/or electron-withdrawing groups like cyano (—CN) and carbonyl (—CO—) on the dienophile seems to enhance the yield obtained in the reaction.

Even though considerable research has been directed toward the elucidation of the mechanism of the Diels-Alder reaction, a detailed description of the reaction pathway is still not available. In deriving a mechanism for the reaction, the following facts must be considered. The reaction exhibits first-order dependency upon the concentration of both the diene and the dienophile so that the process is second-order overall. Kinetic measurements show that the rate of the reaction is not significantly changed by the addition of catalysts such as acids, bases, or free radicals, by irradiation with light of various wavelengths, or by the phase (gas or liquid) in which the reaction is performed. The evidence suggests that the transition state of the reaction is constructed from a single molecule of each of the reactants, and that neither highly polar intermediates, *e.g.*, carbonium ions, nor free radicals are involved in the mechanistic pathway linking starting materials and products. A mechanism consistent with these observations is one in which bond-breaking and bond-making occur simultaneously in the transition state, so that little polarity or free-radical character is developed [equation (3)].

Reactants   Transition State   Product

(3)

There is one important limitation to the Diels-Alder reaction which is a result of the concerted nature of the process, and it is the requirement that the diene must be capable of attaining an *s-cis* conformation ,[1] which is the geometry needed to give a *cis* double bond in the cyclic product. Reaction of a dienophile with a diene in an *s-trans* conformation would lead to a *trans* double bond which is not possible in a six-membered ring for geometric reasons (models are helpful here).

s-cis   *s-trans*

The strict orientation in the transition state of the dienophile relative to the diene might lead one to expect high stereoselectivity in the Diels-Alder reaction, and this expectation is fulfilled. As an example, the adduct resulting from reaction of *trans, trans*-2,4-hexadiene and a dienophile is exclusively *cis*-3,6-dimethylcyclohexene [equation (4)], whereas the product obtained when *cis, trans*-2,4-hexadiene is utilized is *trans*-3,6-dimethylcyclohexene.

(4)

A second type of stereoselectivity is observed for the reaction. The addition of a dienophile such as maleic anhydride **(1)** to a *cyclic* diene like cyclopentadiene **(2)** could, in principle, provide two products, **3** and **4** [equation (5)]. However, only a single adduct, **3**, is observed experimentally. The explanation for this stereoselectivity is not entirely clear, but one attractive theory is that stabilization, *i.e.*, a lowering in energy, of the transition state **5** leading to formation of **3**, is provided by interaction between the *p*-orbitals of the diene and those of the dienophile. Analogous stabilization is not possible in the transition state **6** required for production of **4**.

---

[1] The *s-cis* conformation of a 1,3-diene has the two conjugated double bonds on the same side of the single bond linking them. They are on opposite sides in the *s-trans* conformation.

$$\text{(5)}$$

1   2   3   4

5   6

It should be noted that all Diels-Alder reactions are not as stereoselective as that between cyclopentadiene and maleic anhydride so that mixtures of products are obtained. The product resulting from interaction of the diene and dienophile in a transition state analogous to **5** generally predominates over that arising from a transition state similar to **6**.

The Diels-Alder synthesis is remarkably free of complicating side reactions, and the yields of the desired product are often nearly quantitative. Probably the single most important side reaction observed is dimerization of the diene used, since one molecule of the diene can serve as the dienophile and another as the diene [equation (6)]. Such a reaction is usually important only in those instances in which a poor dienophile like ethylene is employed.

$$\text{(6)}$$

EXPERIMENTAL PROCEDURE

A. PREPARATION OF *endo*-NORBORNENE-*cis*-5,6-DICARBOXYLIC ANHYDRIDE

The cyclopentadiene required for this experiment is not commercially available as such because it readily dimerizes at room temperature to produce dicyclopentadiene [equation (7)]. Fortunately the equilibrium between the monomer and the dimer can be established at the boiling point of the dimer (170°) so that pure cyclopentadiene can be isolated by fractional distillation. The diene must be kept cold in order to prevent extensive dimerization before it can be used for a Diels-Alder reaction.

$$\text{(7)}$$

Place 20 ml of dicyclopentadiene in a 100-ml flask and attach the flask to an apparatus set for fractional distillation into an ice-cooled receiver. Gently heat the dimer until brisk refluxing occurs, and the monomer begins to distil in the range 40–42°. Distil the monomer as rapidly as possible, but do not permit the temperature of the vapor to exceed 43–45°. Approximately 6 g of monomer should be obtained after distillation for *ca.* 30 min. The distillation should be terminated when approximately 5 g of residue remains in the flask. If the distilled cyclopentadiene is cloudy due to condensation of moisture in the cold receiver, add about 1 g of *anhydrous* calcium chloride to dry the diene.

While the distillation is in progress, place 6 g (0.061 mole) of maleic anhydride in a 125-ml Erlenmeyer flask and dissolve it in 20 ml of ethyl acetate by heating on a steam bath. Add 20 ml of petroleum ether [2] (bp 60–80°) and cool the solution thoroughly in an ice-water bath. To this cooled solution, add 4.8 g (6 ml, 0.073 mole) of dry cyclopentadiene and swirl the resulting solution until the exothermic reaction is complete and the adduct separates as a white solid.★ Heat the mixture on the steam bath until the solid has dissolved and then allow the solution to cool slowly to room temperature.★ Filter the solution and determine the yield and melting point of the solid anhydride obtained. The reported melting point is 164–165°. The yield should be about 80% of theoretical. Test the product for unsaturation (Chapter 6.1).

The anhydride can be converted to the corresponding diacid in the following manner. Place 4 g of the anhydride and 50 ml of distilled water in a 125-ml Erlenmeyer flask, and swirl the flask over a Bunsen burner until boiling occurs and all the oil which initially forms has dissolved. Allow the solution to cool to room temperature and then scratch the inner wall of the flask at the air-liquid interface to induce crystallization.★ After crystallization has begun, cool the flask in ice to complete the process, and filter the solution. Determine the yield and melting point of the product. The reported melting point of the expected dicarboxylic acid is 180–182°. Perform appropriate tests to show whether the hydrolysis has destroyed the carbon-carbon double bond. Also test a saturated aqueous solution of the diacid with litmus paper and record the result.

B. Reaction of cyclopentadiene with *p*-benzoquinone

Benzoquinone (7) has two carbon-carbon double bonds which can serve as dienophiles. In this experiment, the Diels-Alder adduct resulting from reaction of two moles of cyclopentadiene with one mole of *p*-benzoquinone will be prepared [equation (8)].

$$(8)$$

Prepare dry cyclopentadiene as described in the preceding experiment. Add a solution of 3.2 g (4 ml, 0.049 mole) of cyclopentadiene and 10 ml of benzene to 2.7 g (0.025 mole) of *p*-benzoquinone [3] dissolved in 20 ml of benzene and swirl the resulting solution

[2] See footnote 15, p. 39, Chapter 2.

[3] Commercial *p*-benzoquinone is generally rather impure and may require purification before being used in this experiment. However, a grade of this chemical having a melting point of 113–115° can be used successfully without further purification. *p*-Benzoquinone can be purified, if necessary, by sublimation in an apparatus such as that pictured in Fig. 2.16. Consult your instructor for details of this method of purification.

until the mildly exothermic reaction has subsided.★ After one hour, cool the reaction mixture in an ice-water bath and filter to isolate the solid adduct.

In order to collect a second crop of product, concentrate the mother liquor to one-third of its original volume by heating on a steam bath, cool the solution, collect the precipitate and wash it with a few milliliters of ice-cold benzene. Combine the two crops of the diadduct and recrystallize by dissolving the product in a minimum amount of boiling acetone (*Caution:* No flames!) and then adding water until turbidity begins to appear in the boiling mixture. Clarify the solution by addition of a *small* amount of acetone and allow the solution to cool slowly to room temperature. Complete the recrystallization by cooling in an ice-water bath.

The pure adduct forms beautiful irridescent white needles or platelets and has a re-ported melting point of 157–158°. The yield should be 60–70%.

C. REACTION OF 1,3-BUTADIENE AND MALEIC ANHYDRIDE

1,3-Butadiene is a gas at room temperature (bp −4.4°). Diels-Alder reactions in which this diene is utilized are normally performed in closed steel pressure vessels (autoclaves) into which the diene is introduced under pressure. However, it has been discovered that 1,3-butadiene may be generated conveniently by the thermal decompo-sition of 3-sulfolene [equation (9)] and that the diene will then react with a dienophile present in the reaction vessel. Such an *in situ* preparation of 1,3-butadiene will be used in this experiment to prepare 4-cyclohexene-*cis*-1,2-dicarboxylic anhydride [equation (10)].

$$
\overset{}{\underset{}{\bigcirc}}SO_2 \xrightarrow{\text{heat}} \; \Big\Vert \;\; + \; SO_2 \tag{9}
$$

$$
\text{diene} \; + \; \text{maleic anhydride} \longrightarrow \text{adduct} \tag{10}
$$

Place 10 g (0.085 mole) of 3-sulfolene, 6 g (0.061 mole) of finely pulverized maleic anhydride and 4 ml of dry xylene in a 100-ml round-bottomed flask equipped with a water-cooled condenser, which is connected by way of a 1-hole stopper and rubber tubing to the gas trap described in Chapter 5 (Figure 5.1). Warm the flask gently while swirling to effect solution, then heat the solution to a gentle reflux with a Bunsen burner; continue the heating for about 30 min. Cool the solution, add 60 ml of ben-zene and about 1 g of decolorizing carbon, and heat on a steam cone with constant swirling for a few minutes. Filter the hot mixture (*Caution:* no flames!) through a short-stemmed funnel, preferably using a fluted filter, and carefully add petroleum ether to the filtrate until cloudiness develops. Set the solution aside to cool to room tempera-ture.★ Collect the crystals by suction filtration and dry them thoroughly. Record their weight and determine the melting point of the product. The reported melting point is 103–104°. Perform qualitative tests for unsaturation and hydrolyze 4 g of the anhydride to the diacid according to the procedure described in part A above. Determine the yield and melting point of the diacid and test it for unsaturation. The reported melting point of 4-cyclohexene-*cis*-1,2-dicarboxylic acid is 164−166°.

EXERCISES

1. Why should 3-sulfolene and maleic anhydride be completely dissolved in the xylene before decomposition of the sulfolene is attempted?

2. Suggest a reason for the observation that cyclopentadiene dimerizes much more readily than do acyclic dienes.

3. Explain why cyclopentadiene reacts more rapidly with *p*-benzoquinone than with another molecule of itself.

4. Write the structure of the product resulting from the reaction of *cis, trans*-2,4-hexadiene and tetramethylethylene.

5. Write the structures of the products expected in the following reactions. If no reaction is to be anticipated, write N.R.

(a) [structure] $+ CH_2{=}CH_2$

(e) [structure] $+ CH_3O_2C{-}C{\equiv}C{-}CO_2CH_3$

(b) [structure] $+ CH_3O_2C{-}C{\equiv}C{-}CO_2CH_3$

(f) [structure] $\longrightarrow C_8H_{10}$

(c) [structure] $+$ [structure]

(g) [structure] $+$ [structure] (2 moles)

(d) [structure] $+ CH_2{=}CHCHO$

6. The "cracking" of dicyclopentadiene to two moles of cyclopentadiene [equation (7)] is an example of a reverse Diels-Alder reaction. Predict the products to be anticipated from an analogous reaction with the compounds shown below.

(a) [structure] CO_2CH_3 CO_2CH_3

(c) [structure]

(b) [structure] H H

(d) [structure] CH_3 CH_2CH_3

(e) [structure] CH_3 CH_3

REFERENCES

1. J. Sauer, *Angewandte Chemie, International Edition,* **5,** 211 (1966). A recent general review.

2. M. Kloetzel, in *Organic Reactions*, John Wiley & Sons, Inc., New York, 1948, Vol. IV, Chapter 1.

3. H. L. Holmes, *ibid.*, Chapter 2.

4. L. W. Butz and A. W. Rytina, *ibid.*, Vol. V, 1949, Chapter 3.

5. T. E. Sample, Jr. and L. F. Hatch, *Journal of Chemical Education*, **45**, 55 (1968).

SPECTRA OF STARTING MATERIALS AND PRODUCTS

FIGURE 9.1   IR spectrum of 1,3-cyclopentadiene.

FIGURE 9.2   NMR spectrum of 1,3-cyclopentadiene.

FIGURE 9.3 IR spectrum of maleic anhydride.

FIGURE 9.4 IR spectrum of *p*-benzoquinone.

FIGURE 9.5 IR spectrum of 3-sulfolene.

FIGURE 9.6 NMR spectrum of 3-sulfolene.

FIGURE 9.7 IR of 4-cyclohexene–*cis*-1,2-dicarboxylic anhydride.

FIGURE 9.8 NMR of 4-cyclohexene–*cis*-1,2-dicarboxylic anhydride.

10

Kinetic and equilibrium control of a reaction

In predicting which of two competing reactions will predominate, a useful "rule of thumb" is to choose the reaction which is more exothermic. This rule is based on the fact that *most commonly* the more exothermic reaction will have the smaller heat of activation, and thus will have the greater rate of reaction. For example, Figure 10.1 depicts the

FIGURE 10.1

potential energy changes involved in the conversion of compound X into the products Y and Z by two competing reactions [equations (1) and (2)]. The more exothermic

$$X \rightleftharpoons Y \tag{1}$$
$$X \rightleftharpoons Z \tag{2}$$

reaction [equation (2)] produces the more stable product, Z (the one lying at the lower energy level). Note that the energy of the transition state leading to Z (TS_{XZ}) is lower than the energy of the transition state leading to Y (TS_{XY}); *i.e.*, the relative energies of the transition states are in the same order as the relative energies of the products to

152

which they lead. Hence, the heat of activation required for production of Z ($\Delta H_Z^{\ddagger}$) is lower than the heat of activation required for production of Y ($\Delta H_Y^{\ddagger}$), so that the rate of the reaction producing Z is greater than the rate of the reaction producing Y.

Most organic reactions are either practically irreversible or they are carried out under conditions such that equilibrium between products and starting materials is not attained, so that the yields of products are determined by the relative rates of competing reactions as described above. The major product of such a reaction is said to be the product of *kinetic control*. However, when the experimental conditions are favorable for equilibrium to be reached between starting materials and products, the product that predominates initially because of kinetic control *may not* be the major product after equilibrium has been attained. In some reactions, the product of kinetic control is less stable than another product that is formed at a lower reaction rate. An example of such a system [equations (3) and (4)] is represented by Figure 10.2, in which product

$$X \rightleftarrows A \tag{3}$$

$$X \rightleftarrows B \tag{4}$$

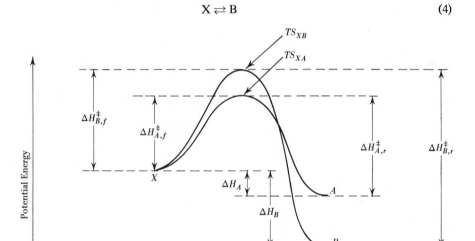

FIGURE 10.2

B is shown to be more thermodynamically stable than product A. It should be noted that, in contrast to the more usual relationships depicted in Figure 10.1, the relative energies of the transition states are not in the same order as the relative energies of the products to which they lead. Thus, although product A will predominate initially in the reaction mixture, [1] if the reactions are allowed to come to equilibrium, the more

---

[1] It is interesting to note that although A is produced initially more rapidly than B because of the relative heats of activation for the competing forward reactions ($\Delta H_{A,f}^{\ddagger} < \Delta H_{B,f}^{\ddagger}$), A is also reconverted to starting material (X) more rapidly than B because the heats of activation for the reverse reactions have the same relationship ($\Delta H_{A,r}^{\ddagger} < \Delta H_{B,r}^{\ddagger}$), owing to the greater stability of B than A. This is not true for competing reactions that have the more usual energy relationships represented by Fig. 10.1, so that even when equilibrium conditions are attained in these systems, the same product (Z) will predominate; *i.e.*, the product of kinetic control is also the product of equilibrium control, although the proportion of products is not necessarily the same under all experimental conditions.

stable product B will be found to be the major product. For this reason it is called the product of *equilibrium control* (or thermodynamic control).

There are many well known reactions of various mechanistic types in which kinetic and equilibrium control lead to different major products under different experimental conditions. For example, the addition of hydrogen bromide to 1,3-butadiene at low temperatures gives the 1,2-adduct, 3-bromo-1-butene, as the main product [equation (5a)], whereas at room temperature the main product is the 1,4-adduct, 1-bromo-2-

$$CH_2\!\!=\!\!CH\!-\!\underset{\underset{Br}{|}}{CH}\!-\!CH_3 \qquad (80\%)$$

$$+$$

$$BrCH_2\!-\!CH\!\!=\!\!CH\!-\!CH_3$$
$$(\textit{cis and trans}) \qquad (20\%)$$

$$(5a)$$

$$CH_2\!\!=\!\!CH\!-\!CH\!\!=\!\!CH_2$$
$$+$$
$$HBr$$

$$CH_2\!\!=\!\!CH\!-\!\underset{\underset{Br}{|}}{CH}\!-\!CH_3 \qquad (20\%)$$

$$+$$

$$BrCH_2\!-\!CH\!\!=\!\!CH\!-\!CH_3$$
$$(\textit{cis and trans}) \qquad (80\%)$$

$$(5b)$$

butene [equation (5b)]. The 1,2-adduct is the product of kinetic control, since equilibrium is not attained at low temperatures. The 1,4-adduct, although formed more slowly, is the more stable product, so it predominates after equilibrium has been established at higher temperatures.

Sulfonation of naphthalene at 120° gives mainly 1-naphthalenesulfonic acid [equation (6a)]. However, at temperatures of 160° or above, or with prolonged heating at

$$+ \text{ HOH} \qquad (6a)$$

major product

$$+ \text{ H}_2\text{SO}_4$$

$$+ \text{ HOH} \qquad (6b).$$

major product

lower temperatures, the major product is 2-naphthalenesulfonic acid, the more stable isomer [equation (6b)]. A similar behavior is observed in the sulfonation of phenol, which gives mainly the *ortho*-isomer at lower temperatures [equation (7a)] and mainly the *para*-isomer at higher temperatures [equation (7b)].

Another aromatic substitution reaction which is reversible, and hence subject to either kinetic or equilibrium control, is Friedel-Crafts alkylation. Reaction of toluene with methyl chloride and aluminum chloride at low temperatures gives chiefly *o*- and *p*-xylene [equation (8a)]; at higher temperatures the chief product is *m*-xylene [equation (8b)].

$$\text{(7a)}$$

OH + H$_2$SO$_4$ $\underset{100°}{\overset{15\text{-}20°}{\rightleftharpoons}}$

major product

$$\text{(7b)}$$

major product

$$\text{(8a)}$$

CH$_3$ + CH$_3$Cl + AlCl$_3$ $\underset{80°}{\overset{0°}{\rightleftharpoons}}$

major products

$$\text{(8b)}$$

major product

The elimination reaction that occurs when 2-phenyl-2-butyl acetate is heated in acetic acid gives a very different proportion of products when a small amount of a strong acid is added to the reaction mixture [equations (9a) and (9b)]. Apparently the

$$\text{(9a)}$$

C$_6$H$_5$C—CH$_2$CH$_3$ $\underset{50°}{\overset{CH_3CO_2H}{\rightleftharpoons}}$ C$_6$H$_5$C=CHCH$_3$ + C$_6$H$_5$C—CH$_2$CH$_3$

OCOCH$_3$

CH$_3$

CH$_3$ CH$_2$

cis (53%), *trans* (2%)   (45%)

$$\text{(9b)}$$

$\underset{+ C_7H_7SO_3H}{\overset{50°}{\underset{CH_3CO_2H}{\rightleftharpoons}}}$ C$_6$H$_5$C=CHCH$_3$ + C$_6$H$_5$C—CH$_2$CH$_3$

CH$_3$ CH$_2$

cis (81%), *trans* (16%)   (3%)

strong acid catalyzes the reaction so that equilibrium is attained, and the equilibrium product mixture contains a much higher proportion of the *cis*- and *trans*-isomers of the internal alkene, which are more thermodynamically stable than the terminal alkene.

The particular example chosen to illustrate the principles of kinetic and equilibrium control of products involves the competing reactions of two carbonyl compounds, cyclohexanone [**2**, equation (10)] and 2-furaldehyde [**4**, equation (11)], with semi-carbazide (**1**). The organic products of these reactions (**3** and **5**), which are known as

$$\text{H}_2\text{N}\overset{\overset{\text{O}}{\|}}{\text{C}}\text{NHNH}_2 + \text{O}{=}\hexagon \rightleftharpoons \text{H}_2\text{N}\overset{\overset{\text{O}}{\|}}{\text{C}}\text{NHN}{=}\hexagon + \text{H}_2\text{O} \qquad (10)$$

$$\mathbf{1}\mathbf{2}\mathbf{3}$$

$$\text{H}_2\text{N}\overset{\overset{\text{O}}{\|}}{\text{C}}\text{NHNH}_2 + \text{O}{=}\text{CH}{-}\!\!\furan \rightleftharpoons \text{H}_2\text{N}\overset{\overset{\text{O}}{\|}}{\text{C}}\text{NHN}{=}\text{CH}{-}\!\!\furan + \text{H}_2\text{O} \qquad (11)$$

$$\mathbf{1}\mathbf{4}\mathbf{5}$$

the semicarbazones of the respective carbonyl compounds, are crystalline solids and have distinctive melting points by which they may be identified easily. For the purposes of the experiment, semicarbazide may be taken to be X of either Figure 10.1 or Figure 10.2, and the problem to be solved is identification of the semicarbazones of cyclohexanone and 2-furaldehyde with either Y and Z or A and B; that is, the experimental determination of which compound is the kinetically controlled product and which is the equilibrium controlled product (they might be the same as in Figure 10.1 or different as in Figure 10.2).

The reactions of carbonyl compounds with nucleophiles such as semicarbazide are subject to significant effects of pH on rates and equilibrium constants.[2] This is because of the different way in which the reactants are affected by reaction with hydrogen ions. The addition of a hydrogen ion to a carbonyl compound [equation (12)] makes the carbonyl carbon atom *more* electrophilic because of the partial positive charge, as shown in the resonance hybrid of the conjugate acid **6**. On the other hand, addition of a hydrogen ion to the nucleophilic semicarbazide [**1**, equation (13)] converts it to its

$$\text{R}_2\text{C}{=}\text{O} + \text{H}^{\oplus} \rightleftharpoons \left[\text{R}_2\overset{\oplus}{\text{C}}{-}\text{OH} \longleftrightarrow \text{R}_2\text{C}{=}\overset{\oplus}{\text{O}}\text{H} \right] \qquad (12)$$

$$\mathbf{6}$$

$$\text{H}_2\text{N}\overset{\overset{\text{O}}{\|}}{\text{C}}\text{NHNH}_2 + \text{H}^{\oplus} \rightleftharpoons \text{H}_2\text{N}\overset{\overset{\text{O}}{\|}}{\text{C}}\text{NHN}\overset{\oplus}{\text{H}}\text{H}_3 \qquad (13)$$

$$\mathbf{1}\mathbf{7}$$

conjugate acid (**7**) which is *not* nucleophilic. At very low pH (high H+ concentration) the concentration of the nucleophile (**1**), is reduced and at high pH (low H+ concentration) the concentration of the activated electrophile (**6**) is reduced. Accordingly, for the reaction of each carbonyl compound with each nucleophilic reagent there is an optimum pH at which the product of the concentrations of the conjugate acid of the carbonyl compound and the nucleophilic reagent is maximized. In order to produce and maintain this optimum pH for reactions of aldehydes and ketones with reagents such as semicarbazide, phenylhydrazine, and hydroxylamine, these reactions are carried out in *buffered* solutions.

In parts A–C of the experimental procedure, the use of a phosphate buffer, which provides reaction rates that are most satisfactory for demonstrating the principles of kinetic and equilibrium control, is recommended. In part F a bicarbonate buffer is em-

[2] Actually, these reactions are affected not only by the concentration of hydrogen ions (of which pH is a measure) but also by the concentration (and nature) of any weak acids that may be present. A discussion of these effects is beyond the scope of this book; further information may be found in textbooks on physical-organic chemistry, in connection with specific and general acid catalysis.

ployed, providing a higher pH than the phosphate buffer. The results from the experiments of part F should be compared with those obtained in the experiments of part C.

EXPERIMENTAL PROCEDURE

A. In a 25-ml Erlenmeyer flask, dissolve 1.0 g of semicarbazide hydrochloride and 2.0 g of dibasic potassium phosphate (K_2HPO_4) in 25 ml of water. This should give a solution having a pH of about 6.1–6.2.[3] Using a 1-ml graduated pipet, deliver 1.0 ml of cyclohexanone into a test tube and dissolve it in 5 ml of 95% ethanol. Pour the ethanolic solution into the aqueous semicarbazide solution and swirl or stir the mixture immediately. Allow 5 or 10 min for crystallization of the semicarbazone to reach completion, then collect the crystals by suction filtration and wash them on the filter with a little cold water. Dry the crystals in air and determine their weight and melting point. The reported melting point of cyclohexanone semicarbazone is 166°.★

B. Prepare the semicarbazone of 2-furaldehyde by following the procedure of part A exactly, except use 0.8 ml of 2-furaldehyde [4] instead of 1.0 ml of cyclohexanone. The reported melting point of 2-furaldehyde semicarbazone is 202°.

C. Dissolve 3.0 g of semicarbazide hydrochloride and 6.0 g of dibasic potassium phosphate in 75 ml of water.[3] This solution will be referred to below as *solution W*.

Prepare a solution of 3.0 ml of cyclohexanone and 2.5 ml of 2-furaldehyde in 15 ml of 95% ethanol. This solution will be referred to below as *solution E*.

(1) Cool a 25-ml portion of solution W and a 5-ml portion of solution E *separately* in an ice bath to 0–2°, then add solution E to solution W and swirl them together; crystals should form almost immediately. Place the mixture in an ice bath for about 3–5 min, and then collect the crystals by suction filtration and wash them on the filter with a little cold water. Dry the crystals and determine their weight and melting point.

(2) Add a 5-ml portion of solution E to a 25-ml portion of solution W at room temperature; crystals should be observed in about 1–2 min. Allow the mixture to stand at room temperature for 5 min, cool it in an ice bath for about 5 min, and then collect the crystals by suction filtration and wash them on the filter with a little cold water. Dry the crystals and determine their weight and melting point.

(3) Warm a 25-ml portion of solution W and a 5-ml portion of solution E *separately* on a steam bath or in a water bath to 80–85°, then add solution E to solution W and swirl them together. Continue to heat the mixture for 10–15 min, cool it to room temperature, and then place it in an ice bath for about 5–10 min. Collect the crystals by suction filtration and wash them on the filter with a little cold water. Dry the crystals and determine their weight and melting point.★

D. In a 25-ml Erlenmeyer flask, place 0.3 g of cyclohexanone semicarbazone (prepared in part A), 0.3 ml of 2-furaldehyde, 2 ml of 95% ethanol, and 10 ml of water. Warm the mixture until a homogeneous solution is obtained (about 1 or 2 min should

---

[3] Sodium acetate (2.0 g) may be substituted for dibasic potassium phosphate in parts A and B of this experiment, and may be used for the general preparation of semicarbazones of aldehydes and ketones as needed for qualitative organic analysis (Chapters 15 and 20). The solution prepared from semicarbazide hydrochloride and sodium acetate will have a pH of about 4.9–5.0. Sodium acetate *must not* be substituted for dibasic potassium phosphate in part C of this experiment.

[4] For best results the 2-furaldehyde should be redistilled just before use.

suffice) and then an additional 3 min. Cool the mixture to room temperature and then in an ice bath. Collect the crystals on a filter and wash them with a little cold water. Dry the crystals and determine their melting point.★

E. Repeat the experiment in part D, but use 0.3 g of 2-furaldehyde semicarbazone (prepared in part B) and 0.3 ml of cyclohexanone in place of the cyclohexanone semi-carbazone and 2-furaldehyde.★

F. Dissolve 2.0 g of semicarbazide hydrochloride and 4.0 g of sodium bicarbonate in 50 ml of water. (The pH of this solution should be about 7.1–7.2.) Prepare a solution of 2.0 ml of cyclohexanone and 1.6 ml of 2-furaldehyde in 10 ml of 95% ethanol. Divide each of these solutions into two equal portions.

Mix one half of the aqueous solution and one half of the ethanolic solution at room temperature. Allow the mixture to stand at room temperature for 5 min, cool it in an ice bath for 5 min, and then collect the crystals by suction filtration and wash them on the filter with a little cold water. Dry the crystals and determine their weight and melting point.

Warm the other portions of the aqueous and ethanolic solutions *separately* on a steam bath or in a water bath to 80–85°, then combine them and continue heating the mixture for 10–15 min. Cool the solution to room temperature and then place it in an ice bath for 5 to 10 min. Collect the crystals by suction filtration and wash them on the filter with a little cold water. Dry the crystals and determine their weight and melting point.

EXERCISES

1. On the basis of the results from the experiments in parts C, D, and E state which semicarbazone is the product of kinetic control and which is the product of equilibrium control. Are they the same or different?
2. On the basis of the results from the experiments of parts C, D, and E draw a diagram similar to either Figure 10.1 or 10.2, whichever is appropriate, and clearly label the products corresponding to Y and Z or A and B.
3. Explain the differences, if any, in the results of the experiments of parts D and E.
4. On the basis of the results from the experiments of part F, explain the effect of the higher pH on the reactions between semicarbazide and the two carbonyl compounds.
5. What results might be expected if sodium acetate buffer (which provides a pH of *ca.* 5) were used in experiments analogous to those of part C? Explain.

SPECTRA OF STARTING MATERIALS

FIGURE 10.3   IR spectrum of 2-furaldehyde.

FIGURE 10.4   NMR spectrum of 2-furaldehyde.

FIGURE 10.5 IR spectrum of cyclohexanone.

FIGURE 10.6 NMR spectrum of cyclohexanone.

Elimination and substitution reactions

11.1   Base-Induced Elimination Reactions

The elimination reaction was introduced in Chapter 6.1. It was noted that all elimination reactions do not proceed via the same mechanism. In that chapter, the E1 mechanism was discussed and experiments were performed which demonstrated both the techniques of performing a dehydration (an elimination which is often of the E1 variety) and the principles of carbonium ion rearrangements. The following discussion and experiments are concerned with E2 elimination reactions.

E2 eliminations are *base-induced* and proceed by a direct attack of the base, B:, on the substrate molecule, with concurrent formation of the double bond and loss of the leaving group, L, as shown in equation (1). The dotted lines in the transition state

$$ \text{B:} + \overset{\text{H}}{\underset{}{\text{C}}}-\text{C}\overset{\text{L}}{\underset{}{}} \longrightarrow \left[\overset{}{\underset{\text{H}}{\text{C}}}\cdots\text{C}\overset{\text{L}}{\underset{}{}} \atop \text{B} \right] \longrightarrow \text{C}=\text{C} + \text{BH} + \text{L:} \qquad (1) $$

transition state

represent bonds partially formed or broken. It is important that the atoms HCCL are pictured in a *trans*-coplanar arrangement, for E2 eliminations occur most readily when this geometry can be attained. Generally, *strong* bases are employed, such as alkoxide ion (RO⁻) in alcohol (ROH), amide ion (NH₂⁻) in either diethyl ether or benzene, or potassium hydroxide in ethanol solution. Types of leaving groups frequently encountered are the halogens (—Cl, —Br, and —I), the sulfonate ester groups (1), the dimethylsulfonium group (2), and the trimethylammonium group (3). All of these, after

$$ \underset{\textbf{1}}{-O-\overset{O}{\underset{O}{\overset{\|}{\underset{\|}{S}}}}-R} \qquad \underset{\textbf{2}}{-\overset{\oplus}{S}\overset{CH_3}{\underset{CH_3}{}}} \qquad \underset{\textbf{3}}{-\overset{CH_3}{\underset{CH_3}{\overset{\oplus}{N}}}-CH_3} $$

departing, becomes species that are relatively *weak* bases. The ability of a group to serve as a leaving group is dependent on its ability to accept the electron pair with which it is bonded to the molecule, and this ability varies inversely with the base strength of the species formed.

Substitution reactions may sometimes compete with elimination reactions [equation (2)], decreasing the yield of the desired alkene. The degree to which this side reac-

$$\text{(2)}$$

tion is important depends on the nature of the substrate and the base employed. The competition is most important when the carbon bearing the leaving group is primary, becomes less so when it is secondary, and is usually unimportant when it is tertiary. With strong bases such as those listed above, substitution reactions are seldom competitive. The competition between substitution and elimination is discussed in more detail in Section 11.2.

In cases where the leaving group is unsymmetrically located on the alkyl carbon skeleton, elimination may occur in two different directions. Thus, for example, *t*-pentyl chloride yields both 2-methyl-2-butene (4) and 2-methyl-1-butene (5).

$$\underset{H\ \ \ Cl\ H}{\overset{CH_3}{CH_2-C-CH-CH_3}} \xrightarrow{\text{base}} \underset{H}{\overset{CH_3}{CH_2-C=CH-CH_3}} + \underset{H}{\overset{CH_3}{CH_2=C-CH-CH_3}} \qquad (3)$$

$$\mathbf{4} \qquad\qquad\qquad \mathbf{5}$$

As these reactions are *irreversible* under the experimental conditions, the alkenes **4** and **5** may be considered to be the products of two competing eliminations, and their proportions thus subject to kinetic control (Chapter 10). Their proportions are therefore determined by the relative free energies of their respective transition states. The transition states for elimination in each direction for *t*-pentyl chloride are shown below (**4‡** and **5‡**).[1] When there are no complicating factors (primarily steric, see below),

4‡   **5‡**

the predominant product in an E2-type elimination is found to be the more highly substituted ethylene. [Recall that increasing the number of alkyl substituents on the double bond increases the stability of alkenes (decreases free energy).] This is interpreted to mean that the free energy of the partial double bond is the major contributor to the total free energy of the transition state. Thus, the free energy of activation for the for-

[1] The symbol ‡ is frequently used by organic chemists to refer to a transition state.

mation of **4** is less than that for the formation of **5**, and **4** should be the preferred product in the t-pentyl chloride example. Prediction of the major product is not always this simple, however, for there may be other complicating factors.

The relative free energies of the transition states of the competing reactions may also be influenced by steric crowding. This crowding will tend to increase the free energy of the transition state and thus the energy requirements of the reaction. If steric effects are more important along the pathway to one of the products than along the pathway to the other, the proportion of products formed may be quite dramatically affected. This is particularly so when the pathway to the more highly substituted ethylene suffers from the greater steric crowding, in which case it may well be the *minor* product.

Unfavorable steric interactions in the transition state may be caused by steric interference of the substituents on the HCCL grouping [*cf.* equation (1)] to the approach of the base (B:), particularly when the base and/or substituents are bulky. Substituents attached to the carbon from which the hydrogen is removed will cause greater hindrance to the base than will substituents on the carbon bearing the leaving group. Therefore, because of the methyl group attached to carbon 3, the transition state **4‡** should be more adversely affected than **5‡** if a more bulky base is used to effect the elimination. If, then, the base were large enough, the lesser substituted ethylene should become the major product.

Similar trends toward favoring the lesser substituted ethylene would also be observed by using a larger leaving group L,[2] or by changing the substrate so that the size of the group(s) crowding the base is increased.

The following series of experiments is designed to demonstrate the steric effect as a molecular parameter of the E2 reaction and the general techniques of performing base-induced elimination reactions.

EXPERIMENTAL PROCEDURE [3]

Elimination with alcoholic potassium hydroxide.   Place 0.2 mole of potassium hydroxide and 100 ml of *absolute* ethanol in a dry 200-ml round-bottomed flask. Connect the flask to an efficient reflux condenser which is fitted with a cold trap bearing a calcium chloride drying tube (Figure 11.1).[4] The cold trap, constructed from a side-arm test tube, is cooled with an ice bath. A Hempel column, filled with pieces of broken glass tubing to increase its efficiency, will serve conveniently as a reflux condenser. Heat the mixture with a burner until the potassium hydroxide dissolves. Cool the solution with a water bath, remove the flask and add to it 0.05 mole of 2-chloro-2-methyl-butane (t-pentyl chloride).[5] Reattach the flask to the reflux condenser and heat at

[2] Steric effects caused by increasing the size of the leaving group make it more difficult to attain the *trans*-coplanar arrangement of HCCL leading to the more substituted ethylene. In examples in which the dimethylsulfonium or trimethylammonium group functions as the leaving group, the lesser substituted ethylene is usually the major product, regardless of the nature of the base.

[3] It is suggested that each student do the experiment with one or two different bases that are assigned at random. The results may then be collected, averaged for each base, and these results presented to the class in order to observe the trend of the effect of base size.

[4] If you are using glassware equipped with ground glass joints, be sure to lubricate the joint connecting the condenser to the flask with a hydrocarbon or silicone grease. Lubrication of ground glass joints is good laboratory practice at all times; however it is particularly important in this experiment because the strong bases being used may cause the joints to "freeze."

[5] A four-fold excess of base is used in order to minimize complications which might arise in the interpretation of the results of this experiment due to competing elimination by the E1 mechanism.

Cold-Trap (Side-Arm Test Tube)

Hempel Column
(with Packing of
Broken Glass)

Drying Tube

Water

Ice Bath

Reaction Flask

FIGURE 11.1 Apparatus for elimination reactions (with cold-trap to help contain low-boiling liquids).

gentle reflux for two hours. By the end of this time, some solid should have precipitated, and it may cause some bumping.★

At the end of the reflux period, cool the reaction mixture with an ice bath and arrange the flask with its contents for fractional distillation.[6] If a Hempel column has been used, drain the water from the cooling jacket and use it as a fractional distillation column. The condenser should be fitted with a vacuum adapter bearing a calcium chloride drying tube and connected tightly to a tared receiving flask, cooled in ice water. If any liquid has condensed in the cold trap during the reflux period, add this to the reaction mixture before distillation. Fractionally distil the mixture, collecting all distillate boiling below 50° (2-methyl-1-butene, bp 31°; 2-methyl-2-butene, bp 38°). Determine the yield of the reaction and perform qualitative tests which will demonstrate the presence of alkenes in the distillate. Submit a sample of your product for glpc analysis and, after receiving the results, calculate the ratio of the two isomeric alkenes formed. Typical glpc tracings of the products from this elimination and from one in which potassium *t*-butoxide was used as base are shown in Figure 11.2. Figure 11.9 shows the nmr spectrum of the product mixture from a typical experiment.

Elimination with sodium methoxide. Prepare a solution of sodium methoxide in *anhydrous* methanol by either of the following methods. (1) If solid sodium methoxide is available, prepare the base solution by dissolving 0.2 mole of the solid in 100 ml of *anhydrous* methanol contained in a dry 200-ml round-bottomed flask. Use a clean *dry* spatula for transferring the methoxide, taking care to perform the weighing operation

---

[6] An alternative method of obtaining the products would be to pour the mixture into 200 ml of ice-cold water contained in a separatory funnel and collect the organic product which forms a second layer. This procedure is satisfactory when ethanol is used as the solvent, but appears not to work well when other higher molecular weight alcohols are used, as below.

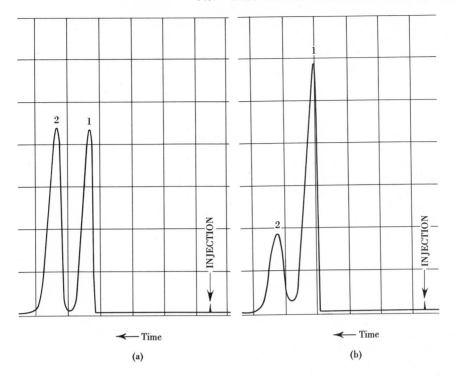

FIGURE 11.2 Typical glpc traces of the products of elimination of 2-chloro–2-methylbutane.[8] Assignments: peak (1) 2-methyl–1-butene; peak (2) 2-methyl–2-butene:
(a) Elimination with KOH, showing approximately 44 percent 2-methyl–1-butene.
(b) Elimination with KOC (CH₃)₃, showing approximately 76% 2-methyl–1-butene.

as rapidly as possible to minimize the reaction between the methoxide and atmospheric water vapor [equation (4)]. The bottle from which the sodium methoxide is taken

$$NaOCH_3 + H_2O \rightleftarrows HOCH_3 + NaOH \tag{4}$$

should be kept *tightly* closed. (2) The sodium methoxide solution may also be prepared by dissolving 0.2 mole of sodium metal in 100 ml of *anhydrous* methanol (*Caution: No flames; hydrogen is evolved!*). In order to weigh the sodium metal, place a 50-ml beaker containing 25 ml of *dry* benzene on a triple beam balance and determine its weight. Add an amount equal to the weight of sodium desired to this weight on the beam. Sodium metal is normally stored under mineral oil for protection from the air (water and oxygen). Use a small knife or spatula (*dry*) to cut a small piece of sodium metal about the size of a pea. Impale this piece of sodium on the tip of the knife and "swish" it in a second beaker containing benzene to remove the mineral oil. Remove the piece of sodium, blot it with a *dry* paper towel and transfer it to the beaker on the balance. Continue until the correct amount of sodium has been obtained. Place 100 ml of *anhydrous* methanol in a 200-ml round-bottomed flask and, with forceps, remove a piece of sodium from the beaker or benzene, again blotting it to remove benzene, and add it to the methanol. Wait until the reaction subsides, then continue adding the sodium one piece at a time until it is all added. Attach a calcium chloride drying tube to the flask and allow time for the sodium to go completely into solution.

If the base solution prepared by either of the above methods is warm, cool it to room temperature in a water bath, and add 0.05 mole of 2-chloro-2-methylbutane to the flask. Add a few boiling stones, connect the flask to the reflux apparatus described in part A (Figure 11.1) and continue from this point following the directions of part A.

Elimination with sodium ethoxide.   Prepare a solution of sodium ethoxide in *absolute* ethanol by dissolving 0.2 mole of sodium metal in 100 ml of *absolute* ethanol contained in a 200-ml round-bottomed flask. The procedure for handling the sodium is described in part B. Cool the solution to room temperature and add 0.05 mole of 2-chloro-2-methylbutane to the flask. Add a few boiling stones, connect the flask to the reflux apparatus described in part A (Figure 11.1) and continue from this point following the directions of part A.

Elimination with sodium isopropoxide.   Obtain 100 ml of 1.8 M sodium isopropoxide solution in *anhydrous* isopropyl alcohol.[7] Place this solution in a *dry* 200-ml round-bottomed flask and add 0.045 mole of 2-chloro-2-methylbutane to the flask. Add a few boiling stones, connect the flask to the reflux apparatus described in part A (Figure 11.1) and continue from this point following the directions of part A.

Elimination with potassium t-butoxide.   Place 0.16 mole of solid potassium *t*-butoxide and 100 ml of *anhydrous t*-butyl alcohol in a *dry* 200-ml round-bottomed flask. Use the same precautions for handling potassium *t*-butoxide as used for solid sodium methoxide described in part B. Attach the flask to the reflux apparatus described in part A (Figure 11.1) and heat the mixture at gentle reflux with a water bath for about 10 min in order to dissolve most of the solid. Not all of the base will dissolve. Cool the mixture to room temperature and add 0.04 mole of 2-chloro-2-methylbutane to the flask along with a few boiling stones. Reattach the flask to the reflux apparatus and continue from this point following the directions of part A with the following exception: Use a water bath instead of a burner for heating during the reflux period. A typical glpc tracing [8] and a typical nmr spectrum of the products of this elimination are shown in Figures 11.2 and 11.10, respectively.

EXERCISES

1. What would be the expected results of the alkoxide-induced eliminations if the alcohol solvents employed were wet?
2. What is the solid material that precipitates as the eliminations proceed?
3. Why does the excess of base used in these eliminations favor the E2 elimination as opposed to the E1 elimination? (Hint: consider the rate law expressions for the E1 and E2 processes.)
4. If all of the elimination reactions in the experimental section had proceeded by the E1 mechanism, would the results have been different from those actually obtained? Why?

[7] A maximum concentration of 1.8 M sodium isopropoxide in isopropyl alcohol may be obtained by heating sodium and isopropyl alcohol at reflux for 6 hr.

[8] These analyses were performed at 45° on a 10 ft column of 30% Silicone Gum Rubber supported on Chromosorb P.

5. From the results of the above experiments and/or from the data in Figure 11.2, what conclusions can be drawn concerning the relative sizes of the bases employed?

6. What differences might you expect in product distributions for the eliminations of 2-chloro-2-methylbutane and 2-chloro-2,3-dimethylbutane with excess sodium methoxide in methanol solution?

7. If the leaving group in the 2-methyl-2-butyl system were *larger* than a methyl group, why would 2-methyl-1-butene be expected to be formed in greater amounts than if the leaving group were smaller than methyl, regardless of the base used. Use "sawhorse" structural formulas in your explanation.

8. What is the reason for packing the Hempel column with pieces of broken glass tubing in these experiments?

9. What is the purpose served by the cold trap during the reflux period in these experiments? What problems might arise if the cold trap were not used and if the mixture were heated too vigorously at reflux?

10. By reference to Figures 11.6, 11.8, 11.9 and 11.10, calculate the percentage compositions of the mixtures of isomeric methylbutenes obtained from reaction of 2-chloro-2-methylbutane with potassium hydroxide and potassium *t*-butoxide.

SPECTRA OF STARTING MATERIALS AND PRODUCTS

FIGURE 11.3 IR spectrum of 2-chloro–2-methylbutane.

FIGURE 11.4   NMR spectrum of 2-chloro–2-methylbutane.

FIGURE 11.5   IR spectrum of 2-methyl–2-butene.

FIGURE 11.6   NMR spectrum of 2-methyl–2-butene.

FIGURE 11.7 IR spectrum of 2-methyl–1-butene.

FIGURE 11.8 NMR spectrum of 2-methyl–1-butene.

FIGURE 11.9 NMR spectrum of the product mixture from the elimination of 2-chloro–2-methylbutane with potassium hydroxide.

FIGURE 11.10   NMR spectrum of the product mixture from the elimination of 2-chloro–2-methylbutane with potassium *t*-butoxide.

11.2   *Nucleophilic Substitution at Saturated Carbon*

Many organic reactions involve the *substitution* of one group for another at a saturated carbon atom [equation (5)]. Substitutions are also known to occur at unsaturated carbon atoms (*e.g.*, benzene derivatives, alkenes, etc.); these reactions are discussed elsewhere (see Chapters 7 and 17). In equation (5), L represents the *leaving group* and

$$R\!-\!L + Nu: \rightarrow R\!-\!Nu + L: \tag{5}$$

Nu: represents a molecule or ion which has *nucleophilic* character (*i.e.*, Cl^-, Br^-, HO^-, CN^-, H_2O, NH_3, etc.). These substitution reactions are discussed in the following paragraphs with regard to application and mechanism.

Types of nucleophilic substitution.   Although nucleophilic substitution is a very general reaction for aliphatic compounds ($R\!-\!L$), the mechanism followed in the conversion is highly dependent upon the structure of the alkyl group and upon the nature of the leaving group. It has been found that two distinctly *different* mechanisms of substitution may operate, depending on the structure of the starting material. These are designated as the S_N1 (**S**ubstitution, **N**ucleophilic, and **1** for **uni**molecular) and S_N2 (**S**ubstitution, **N**ucleophilic, and **2** for **bi**molecular) mechanisms [equations (6) and (7), respectively].

S_N1 MECHANISM: $\qquad\qquad R_3C\overset{\frown}{-}L \overset{\text{slow}}{\rightleftarrows} R_3C^{\oplus} + L:^{\ominus} \tag{6a}$

$$R_3C^{\oplus} + Nu:^{\ominus} \xrightarrow{\text{fast}} R_3C\!-\!Nu \tag{6b}$$

S_N2 MECHANISM: $Nu: \longrightarrow \underset{H}{\overset{R}{H\!-\!C\!-\!L}} \longrightarrow \left[Nu\cdots\underset{H\ H}{\overset{R}{C}}\cdots L \right] \longrightarrow Nu\!-\!\underset{H}{\overset{R}{C}}H + L: \tag{7}$

transition state

It is seen that the S_N1 mechanism occurs in two steps. The first step of the reaction is the departure of the leaving group taking the pair of bonding electrons with it; the carbon from which L departs becomes a carbonium ion. The second step involves a nucleophilic attack on the carbonium ion; this is an exceedingly fast process, as all ionic reactions are, in general. Normally the concentration of the nucleophile is large compared to that of $L:^-$, so that the reverse of reaction (6a) is relatively unimportant in most cases. It is important to note, for purposes of classifying the mechanism as S_N1, that the reaction rate is dependent only on the concentration of R_3C—L, and independent of the nature and concentration of the nucleophile

The S_N2 mechanism, on the other hand, is represented as a direct attack of the nucleophile on the backside (with respect to L) of the carbon atom. The C—L bond is broken concurrently with the formation of the C—Nu bond. As the substrate (RCH_2—L) and nucleophile are both involved in the transition state, this is a bimolecular reaction. Therefore, the rate of an S_N2 reaction is dependent on the natures and concentrations of both substrate and nucleophile.

In light of the two basic mechanisms, S_N1 and S_N2, outlined for nucleophilic substitution, it is not surprising to find that two factors play an important role in dictating the preferred mode of reaction for a particular substrate: (1) In going from CH_3—L to R_3C—L the number of alkyl groups attached to the central carbon atom is increased, making it more difficult for the backside attack by the nucleophile because of steric hindrance; this serves to decrease the ease of S_N2 reaction; (2) as alkyl groups are added to the central carbon, the incipient carbonium ion of the S_N1 reaction becomes more stable, facilitating the reaction along the S_N1 pathway. (The relative stability of substituted carbonium ions has been discussed earlier in Chapter 6.1.) These two effects reinforce one another in producing the general trends shown in the following diagram:

Increasing ease of S_N2 reaction

$\longleftarrow$

CH_3—L RCH_2—L R_2CH—L R_3C—L

$\longrightarrow$

Increasing ease of S_N1 reaction

Competition between substitution and elimination. In all reactions in which nucleophilic substitution occurs, there is the possibility of competing elimination reactions which produce alkenes. For the most part, bimolecular elimination reactions (E2) compete with S_N2 substitution, and unimolecular elimination reactions (E1) compete with S_N1 substitution. E1 reactions have been discussed in detail in Chapter 6.1 and E2 reactions in Section 11.1.

The competition in the two cases may be represented by the following equations. Generally, elimination is favored when strongly basic and only slightly polarizable

S_N2 *vs.* E2:

Elimination: $\diagup C=C \diagdown$ + L: + HNu

Substitution: $-\overset{|}{\underset{Nu}{C}}-\overset{|}{\underset{H}{C}}-$ + L:

(8)

Substitution (S_N2)   Nu: Elimination (E2)

S_N1 *vs.* E1 :

$$
\text{Elimination:} \quad \underset{}{>}C{=}C\underset{}{<} + HNu
$$

(9)

$$
\text{Substitution:} \quad -\underset{H}{\underset{|}{C}}-\underset{|}{C}-
$$

Substitution
(S_N1)

Nu:

Elimination
(E1)

species are employed (*i.e.*, RO^-, NH_2^-, H^-, HO^-, etc.), and substitution is favored when weakly basic and highly polarizable species are used (*i.e.*, I^-, Br^-, Cl^-, H_2O, $CH_3CO_2^-$, etc.). Polarizability is a measure of the ease of distortion of the electron cloud of the species by a nearby charged center (positively charged in this case).

The preparation of *n*-butyl bromide; an S_N2 reaction. A common method of converting a primary alcohol to an alkyl halide is shown in equation (10). This method

$$
HX + R{-}OH \underset{\text{heat}}{\overset{H_2SO_4}{\rightleftharpoons}} R{-}X + H_2O
$$

(10)

involves the establishment of equilibrium between an alcohol and a mineral acid ($HX = HCl, HBr, HI$). The best yield of alkyl halide is of course obtained when the position of equilibrium lies well to the right (see below).

The preparation of *n*-butyl bromide may be accomplished by heating *n*-butyl alcohol with concentrated hydrobromic acid in the presence of concentrated sulfuric acid [equation (11)]. The mechanism is best written in two steps, which are given in equa-

$$
CH_3CH_2CH_2CH_2{-}OH + HBr \underset{\text{heat}}{\overset{H_2SO_4}{\rightleftharpoons}} CH_3CH_2CH_2CH_2{-}Br + H_2O
$$

(11)

tion (12). The first step involves the protonation of the alcohol to give oxonium ion **6**,

$$
n\text{-}C_4H_9{-}\overset{..}{\underset{..}{O}}H + H^{\oplus} \rightleftharpoons n\text{-}C_4H_9{-}\overset{\oplus}{O}\overset{H}{\underset{H}{<}}
$$

(12a)

6

$$
Br^{\ominus} + n\text{-}C_3H_7{-}CH_2{-}\overset{\oplus}{O}\overset{H}{\underset{H}{<}} \rightleftharpoons n\text{-}C_4H_9{-}Br + H_2O
$$

(12b)

6

an example of Lewis acid-Lewis base complex formation. Once formed the oxonium ion undergoes displacement by the bromide ion to form the alkyl bromide and water [equation (12b)]. In this S_N2 reaction, *water is the leaving group* and *bromide ion is the nucleophile*. The sulfuric acid serves two distinct and important purposes: (1) as a dehydrating agent to reduce the activity of water and shift the position of equilibrium to the right, and (2) as an added source of hydrogen ions to increase the concentration of oxonium ion **6**. The use of *concentrated* hydrobromic acid also helps to establish a favorable equilibrium.

The mixture of hydrobromic acid and sulfuric acid may be prepared in two ways. One method is to add concentrated sulfuric acid to concentrated hydrobromic acid. The second way is to generate the hydrobromic acid *in situ* by adding *concentrated* sulfuric acid to sodium bromide [equation (13)]. These methods work well and give

$$NaBr + H_2SO_4 \rightleftharpoons HBr + NaHSO_4 \tag{13}$$

high yields of the bromide when low-molecular-weight alcohols are used. With higher-molecular-weight alcohols, however, it has been found that the sulfuric acid is best omitted; this is because of the decreased solubility of the alcohol when the reaction is carried out in concentrated salt solutions. With very high-molecular-weight alcohols, the addition of hydrogen bromide gas to the alcohol at elevated temperatures appears to give the best results.

Although the added sulfuric acid is desirable as discussed above, it may also give rise to two important side reactions. The alcohol may react with sulfuric acid to form a hydrogen sulfate ester [7, equation (14)]. This ester formation is reversible and the

$$ROH + H_2SO_4 \rightleftharpoons RO\text{—}SO_3H + H_2O \tag{14}$$

7

alcohol is regenerated (position of equilibrium shifted to the left) as its concentration is reduced in the production of alkyl bromide. The formation of the hydrogen sulfate ester would not directly decrease the yield, were it not able to undergo reaction in two other ways to give undesired products. (1) On heating, it can undergo elimination to

$$RO\text{—}SO_3H \xrightarrow{\text{heat}} \text{alkenes} + H_2SO_4 \tag{15a}$$

$$RO\text{—}SO_3H + ROH \xrightarrow{\text{heat}} R_2O + H_2SO_4 \tag{15b}$$

give alkenes [equation (15a)], and (2) it can react with another molecule of alcohol to give a dialkyl ether by an S_N2 reaction in which the nucleophile is ROH [equation (15b)]. Both of these side reactions consume alcohol and result in a decreased yield of alkyl bromide. For primary alcohols, the temperatures which are used for the substitution reaction are generally not high enough to cause these side reactions to be of great importance.

In the preparation of secondary bromides from secondary alcohols, the method involving the use of concentrated sulfuric acid cannot be employed. This is because secondary alcohols are far more easily dehydrated by sulfuric acid to give the corresponding alkenes [see Chapter 6 and equation (15a)] than are primary alcohols. One means of circumventing this problem is to use 48% hydrobromic acid without the sulfuric acid. However, the secondary halides are generally best prepared by use of phosphorus trihalides [PX_3, X = Cl or Br, equation (16)] or thionyl halides [SOX_2, X = Cl or Br, equation (17)]. The latter reagents are required when the alcohol is susceptible to *carbonium ion rearrangements* (Chapter 6.1) so that the presence of acid is to be avoided. In these cases, pyridine is used as the solvent, since this base reacts with the acid (HX) as it is formed in the reaction and serves to neutralize it [equation (17)].

$$3\,ROH + PX_3 \longrightarrow 3\,RX + H_3PO_3 \tag{16}$$

$$\text{ROH} + \text{SOX}_2 \longrightarrow \text{RX} + \text{SO}_2 + \text{HX} \longrightarrow \qquad \qquad X^\ominus \qquad (17)$$

The preparation of tertiary alkyl chlorides; an S_N1 reaction. Although the same general approaches as described above can be used for conversion of tertiary alcohols to their corresponding halides, only one method is found to give acceptable yields. This method involves the use of a mineral acid with the alcohol in a reaction which is typically S_N1. Phosphorus trihalides or thionyl halides are not applicable for the preparation of t-alkyl halides from the alcohols because elimination reactions predominate.

The reaction of t-butyl alcohol with hydrochloric acid [equation (18)] is illustrative

$$\begin{array}{ccc} \text{CH}_3 & & \text{CH}_3 \\ | & & | \\ \text{CH}_3{-}\text{C}{-}\text{OH} + \text{HCl} \rightleftharpoons \text{CH}_3{-}\text{C}{-}\text{Cl} + \text{H}_2\text{O} \\ | & & | \\ \text{CH}_3 & & \text{CH}_3 \end{array} \qquad (18)$$

of the S_N1 reaction. The mechanism, shown in equation (19), involves three distinct steps. The second step [equation (19b)] is the slowest step in this sequence although it is faster than the corresponding S_N2 attack of Cl^- on oxonium ion **8**. The reverse is

$$\begin{array}{cc} \text{CH}_3 & \text{CH}_3 \quad \text{H} \\ | & | \; \oplus \\ \text{CH}_3{-}\text{C}{-}\ddot{\text{O}}\text{H} + \text{H}^\oplus \rightleftharpoons \text{CH}_3{-}\text{C}{-}\text{O} \\ | & | \qquad \text{H} \\ \text{CH}_3 & \text{CH}_3 \end{array} \qquad (19a)$$

$$\textbf{8}$$

$$\begin{array}{cc} \text{CH}_3 \quad \text{H} & \oplus \; \text{CH}_3 \\ | \; \oplus & \\ \text{CH}_3{-}\text{C}{-}\text{O} \xrightarrow{\text{slow}} \text{CH}_3{-}\text{C} \qquad + \text{H}_2\text{O} \\ | \qquad \text{H} & \text{CH}_3 \\ \text{CH}_3 & \end{array} \qquad (19b)$$

relatively stable
tertiary carbonium ion

$$\begin{array}{cc} \oplus \; \text{CH}_3 & \text{CH}_3 \\ & | \\ \text{CH}_3{-}\text{C} \qquad + \text{Cl}^\ominus \xrightarrow{\text{fast}} \text{CH}_3{-}\text{C}{-}\text{Cl} \\ \text{CH}_3 & | \\ & \text{CH}_3 \end{array} \qquad (19c)$$

true for the oxonium ions of primary and secondary alcohols. This reactivity difference between tertiary oxonium ions and primary and secondary oxonium ions is a reflection of the relative stabilities of the three types of carbonium ions which result from the loss of a water molecule. Although *relatively* slow compared to the other steps in the sequence, reaction (19b) is quite a rapid one from an *absolute* viewpoint. Thus, with concentrated hydrochloric acid a favorable equilibrium giving high yields of t-butyl chloride is established in a few minutes at room temperature. Concentrated hydrochloric acid is necessary to establish a favorable equilibrium [equation (18)].

The principal side reaction is the E1 elimination, resulting from the loss of a proton from the incipient carbonium ion to give 2-methylpropene [isobutylene, equation (20)].

Under the reaction conditions, however, the reaction is readily reversible through the addition of HCl to the alkene to give *t*-butyl chloride [according to Markownikoff's Rule, equation (21)]. Therefore, this is not a serious complication.

$$CH_3-\overset{\oplus}{\underset{CH_3}{C}}\overset{CH_2-H}{\diagdown} \longrightarrow CH_3-\underset{CH_3}{\overset{CH_2}{C}} + H^{\oplus} \qquad (20)$$

$$CH_3-\underset{CH_3}{\overset{CH_2}{C}} + H^{\oplus} \longrightarrow CH_3-\overset{\oplus}{\underset{CH_3}{C}}\overset{CH_3}{\diagup} \xrightarrow{Cl^{\ominus}} CH_3-\underset{CH_3}{\overset{CH_3}{\underset{|}{\overset{|}{C}}}}-Cl \qquad (21)$$

EXPERIMENTAL PROCEDURE

n-**Butyl bromide.**   Place 0.30 mole of sodium bromide in a 250-ml round-bottomed flask and add about 35 ml of water and 0.40 mole of *n*-butyl alcohol. Mix thoroughly and cool the flask in an ice bath. Slowly add 35 ml of *concentrated* sulfuric acid to the cold mixture; add the acid slowly, with swirling and continuous cooling. Remove the flask from the ice bath. Fit it with a reflux condenser and warm the flask gently until most of the salts have dissolved. Add boiling stones and heat the mixture to gentle reflux. A noticeable reaction occurs, and two layers form, the upper layer being the alkyl bromide; the inorganic layer has a high concentration of salts and thus is much more dense than *n*-butyl bromide. Continue heating at reflux for 45 min.[9]★

Equip the flask for simple distillation. Distil the mixture rapidly and collect the distillate in an ice-cooled receiver. Codistillation of *n*-butyl bromide and water occurs. Continue the distillation until the distillate is clear. The head temperature should be around 115° at this point. (The increased boiling point is caused by codistillation of sulfuric acid and hydrobromic acid with water.)★

Transfer the distillate to a separatory funnel. Add about 30 ml of water and shake gently, with venting. Note that two layers are formed; decide which of these is the organic layer by adding a couple of drops of water to the heterogeneous mixture and seeing in which layer it dissolves.[10] Remove the organic layer and transfer it to a separatory funnel. Add an equal volume of *cold, concentrated* sulfuric acid and shake the mixture gently. Remove the organic layer. If it is difficult to determine where the interface of the two layers is, view the mixture from several different angles. Wash the organic layer, first with about 20 ml of water, then with 10 ml of 5% sodium hydroxide solution, and finally with 20 ml of water.

Transfer the cloudy *n*-butyl bromide layer to a small Erlenmeyer flask, and dry it over granular *anhydrous* calcium chloride.★ The drying process may be hastened by warming slightly on a steam bath and by swirling gently. Decant the dry liquid into a

[9] A well-organized student will be doing other laboratory work during the 45-min reflux period. Reactions go no faster when watched, and it would be well for students to assemble the needed equipment for the rest of the experiment or for some other experiment during the reflux time.

[10] The drop of water will be readily soluble in the inorganic layer, but will be insoluble in the organic layer. If the organic layer is on top, the water will fall through it and dissolve in the inorganic layer; if the inorganic layer is on top, the water will dissolve directly in it. A common misfortune that students encounter is in finding that the organic layer has been thrown away and the inorganic layer has been kept. A good, general practice in the laboratory is to keep *both* layers, until you are sure which is the one you are to keep. It is recommended that the student do this throughout the experiments.

round-bottomed flask, add boiling stones, and equip for simple distillation. Distil and collect the material boiling at 99–103°. Weigh the product and compute the yield. The yield should be 70–82%.

t-Butyl chloride. Place 0.5 mole of t-butyl alcohol and 1.5 moles of reagent grade *concentrated* hydrochloric acid (specific gravity 1.19) in a 250-ml separatory funnel. After mixing, swirl the contents of the separatory funnel gently, without the stopper on the funnel. After about a minute of swirling, stopper and invert the funnel; after inverting, vent to release excess pressure by carefully opening the stopcock (be careful to avoid shaking until this is done). Shake the funnel for several minutes, with intermittent venting. Allow the contents to stand until the mixture has separated into two distinct layers; these should be completely clear.

Separate the layers and wash the organic layer with 50 ml of *saturated* sodium chloride solution and then with 50 ml of *saturated* sodium bicarbonate solution. On initial addition of the bicarbonate solution, vigorous gas evolution will ensue; gently swirl the *unstoppered* separatory funnel until the vigorous effervescence ceases. Stopper the funnel and invert gently; vent immediately to release gas pressure. Shake the separatory funnel gently, with *frequent* venting; then shake vigorously, again with frequent venting. Separate the organic layer (the density of a saturated sodium bicarbonate solution is about 1.1), wash it with about 40 ml of water, and then with 40 ml of *saturated* sodium chloride solution. Carefully remove the aqueous layer.

Transfer the t-butyl chloride to a small Erlenmeyer flask and dry with *anhydrous* calcium chloride; shaking expedites the drying process.* When the organic layer is clear, transfer the product by decantation to a small round-bottom flask and distil, collecting the distillate in an ice-cooled receiver. Make sure that the distilling flask and condenser used are *dry*. Collect the fraction boiling from 49–52°. If any lower boiling material is obtained, it should be redistilled and any material boiling from 49–52° retained. The yield of t-butyl chloride should be 78–88%.

t-Pentyl chloride. t-Pentyl chloride may be prepared from t-pentyl alcohol using the same general procedure as that described for t-butyl chloride, using 0.5 mole of t-pentyl alcohol. The t-pentyl chloride will be obtained as the product on final distillation; material boiling between 83–85° should be collected. The yield will be about 75%.

EXERCISES

1. Observe that some water is added to the initial reaction mixture. How might the yield of n-butyl bromide be affected by the failure on the part of the student to add the water? What products would be obtained? How might the yield of n-butyl bromide be affected by adding twice as much water as is called for?
2. In the purification process, the impure n-butyl bromide is "washed" with concentrated sulfuric acid. What impurities would a wash of this sort remove? Why?
3. Suppose that a student added 0.4 mole of trimethylene glycol, HOCH₂CH₂CH₂OH, instead of 0.4 mole of n-butyl alcohol. In principle, what product would you expect to be produced from the glycol? Write an equation for its formation. Do you think that it would be produced in good yield under the conditions of this experiment (and using the same quantities of reagents which are called for herein)? Why or why not?

4. Draw the structures of the other alcohols which are isomeric with *t*-butyl alcohol. Arrange these alcohols in order of *increasing* reactivity toward concentrated hydrochloric acid. Which, if any, of these alcohols would you expect to give a reasonable yield of the corresponding alkyl chloride under such conditions?

5. The work-up procedure in the *t*-butyl chloride preparation calls for using sodium bicarbonate. From your experience in the laboratory, you know that this work-up is accompanied by vigorous gas evolution which increases the difficulty of handling and requires considerable caution. Alternatively, you might consider using a dilute solution of sodium hydroxide instead of the sodium bicarbonate. Comment on the relative advantages and disadvantages of using these two basic solutions in the work-up. Based on these considerations, why were you instructed to use sodium bicarbonate, even though it is more difficult to handle?

SPECTRA OF STARTING MATERIALS AND PRODUCTS

The ir and nmr spectra of *n*-butyl chloride (Figures 5.5 and 5.6, respectively) are quite similar to those of *n*-butyl bromide, so that the spectra of the latter compound are not included. For ir and nmr of *t*-pentyl chloride (2-chloro-2-methylbutane), see Figures 11.3 and 11.4, respectively.

FIGURE 11.11   IR spectrum of *n*-butyl alcohol.

FIGURE 11.12   NMR spectrum of *n*-butyl alcohol.

FIGURE 11.13 IR spectrum of *t*-pentyl alcohol.

FIGURE 11.14 NMR spectrum of *t*-pentyl alcohol.

11.3 Chemical Kinetics: Evidence for Nucleophilic Substitution Mechanisms

Some of the most important criteria in support of the S_N1 and S_N2 mechanisms have been obtained from kinetic studies on various types of substrates. For primary and most secondary compounds, it is found that the *rate* of substitution is proportional to the concentrations of the substrate *and* the nucleophile. This situation is represented by the following rate law:

$$\text{Rate} = k_2[\text{R–L}][\text{Nu:}] \qquad (22)$$

The proportionality constant, k_2, relating rate and concentrations is known as the *rate constant*. This equation indicates that the transition state involves both the substrate and the nucleophile, and that the reaction can be represented mechanistically as shown in equation (7), *i.e.*, as an S_N2 reaction.

On the other hand, the rates of substitution of tertiary and some secondary compounds are found to be proportional to the concentration of the substrate but *independent* of the concentration of the nucleophile. Therefore the transition state of the slow step in these cases involves only the substrate and not the nucleophile.[11] This is interpreted to indicate an initial slow ionization of the substrate to yield a carbonium ion as represented in the S_N1 mechanism [equation (6)]. For this reaction, then, the rate law is as given in equation (23).

$$\text{Rate} = k_1[\text{R–L}] \tag{23}$$

In addition to the concentration dependence, rates of substitution are also dependent on such factors as the temperature and the nature of the reactants and solvent. The dependency of the rate on these factors may be considered, for the sake of simplicity, as being reflected in the value of the rate constant; *e.g.*, if a temperature change results in a doubling of the rate (at the same concentration of reactants) then the rate constant will be larger by a factor of two.

The S_N1 reaction and its rate law. S_N1 reactions are first-order reactions in that the rate of reaction is proportional to the first power of the concentration of the substrate [equation (23)].[12] It should be evident that a graphical plot of *rate vs. concentration* in such a case would yield a straight line of slope k_1, and that doubling the concentration would double the rate. The rate constant is independent of the concentration.

When a substance is being consumed in a first-order reaction, its concentration decreases exponentially with time. If C_o is the *initial* concentration of the substrate (at time $t = 0$, or t_o), and C_t is its concentration at any elapsed time t (where t is measured in any unit of time, *e.g.*, seconds, minutes, hours, etc.) after the reaction is started, the relationship between these variables is

$$C_t = C_o\, e^{-k_1 t} \tag{24}$$

The rate constant, k_1, has the units of $(\text{time})^{-1}$, *e.g.*, $(\text{sec})^{-1}$. Equation (24) may be rewritten as

$$\ln (C_o/C_t) = k_1 t \tag{25}$$

or

$$2.303 \log (C_o/C_t) = k_1 t \tag{26}$$

If the initial concentration of the reactant (C_o) is known and if one measures the concentration of the reactant (C_t) at a series of measured time intervals while the reaction is proceeding, then the rate constant may be determined in either of two ways:

[11] Molecules of solvent are involved as well, since the reaction is performed in solution. For simplicity, the involvement of the solvent (often called "solvation") will not be considered since any *changes* due to solvation are insignificant when a large excess of solvent is present.

[12] In a *hypothetical* reaction, if the rate were proportional to the second power of the concentration of the substrate, then the rate law would contain the concentration of the substrate *squared:*

$$\text{Rate} = k_2[\text{R–L}]^2$$

(1) The values of C_t and t measured at each point during the reaction may be substituted into equation (26), which is then solved for k_1. Thus several values of k_1 are determined and these values are then averaged. The rate constant (the *correct* one) is not easily obtained by this method, for the average value will be affected without bias by any points (determinations) that may be "bad" (incorrect due to experimental error). (2) An alternative and better method of calculation of k_1 from the experimental data consists of preparing a plot of log (C_o/C_t) *vs.* the time t. Assuming the reaction to be first-order and the data to be fairly accurate, a reasonably straight line should be obtained. This line is drawn on the graph in such a way as to lie closest to the largest number of points. The line may be drawn *with bias*, purposely giving more "weight" to the majority of points that will lie close to the line and ignoring or giving less credence to those points that appear to be in error. The slope of this line ("the best straight line") [13] is the rate constant k_1 if equation (25) is used (natural logrithms) or $k_1/2.303$ if equation (26) is used. In the latter case the slope must be multiplied by 2.303 to obtain k_1.

It should be noted that equations (25) and (26) contain only the *ratio* of the concentrations of the reactants, so that the *units* of concentration cancel. Hence, one does not need to express the concentrations in the customary units, but may instead use any units so long as both C_o and C_t are given in the same units. In fact, one may use *total amounts*, *i.e.*, grams or moles, of reactants at time zero and at time t (A_o and A_t, respectively) to give the following equation:

$$2.303 \log (A_o/A_t) = k_1 t \tag{27}$$

The usefulness of this equation will be demonstrated later.

Both S_N1 and S_N2 reactions can be studied from a kinetic standpoint, but bimolecular reactions are generally more difficult to examine experimentally and the calculations for them are more tedious. Therefore, we shall limit our studies to a detailed examination of the kinetics of S_N1 reactions, the rate law for which we have just discussed. The experiments presented here are designed to illustrate methods of studying chemical kinetics and measuring the effects of structure on reactivity, as exemplified by the solvolysis of tertiary alkyl halides.

Solvolysis refers to a substitution reaction in which the solvent (SOH) functions as the nucleophile [equation (28)]. In principle, solvolysis reactions can be carried out in

$$\text{R—X} + \text{SOH} \rightarrow \text{R—OS} + \text{H}^{\oplus} + \text{X}^{\ominus} \tag{28}$$

any nucleophilic solvent, such as water (hydrolysis), alcohol (alcoholysis), carboxylic acids (acetolysis with acetic acid and formolysis with formic acid) or liquid ammonia (ammonolysis). A major limitation in choosing a solvent is dictated by the solubility of the substrate in the solvent. It is important that the reaction mixture be homogeneous, for if it is not, surface effects at the interface of the phases will make the kinetic results difficult to interpret and, probably nonreproducible, as well. In the experiments described here, the solvolyses will be carried out in mixed solvents consisting of isopropyl alcohol and water.

[13] An alternative method of obtaining the best straight line is the application of a *least-squares treatment* to the data. This mathematical treatment will also yield a "biased" line, but will do so without the subjective wishful-thinking that may contaminate human judgment. The student who wishes to apply this method may consult the following reference for details: G. H. Brown and C. M. Sallee, *Quantitative Chemistry*, Prentice-Hall, Inc., Englewood Cliffs, N. J., 1963, p. 123 ff.

In order to determine the rate constants (k_1) for these solvolysis reactions [equation (28)], one must make use of the rate expression given in equation (26). Recall that the quantities which must be obtained in order to use equation (26) are C_o and C_t. Of these, C_t is the more difficult to obtain. One possible method would be to withdraw an aliquot (a measured sample of the reaction mixture, *e.g.*, 5 ml) at time t and to isolate the remaining alkyl halide contained therein. This could be a very inaccurate procedure, for it would be difficult to isolate remaining starting material quantitatively from such a mixture. The best procedure is based upon the fact that for every molecule of alkyl halide that reacts, one molecule of HX is produced [equation (28)]. Thus the progress of the reaction may be followed by determining the concentration of hydrogen ion, [H^+], produced as a function of time. Note, however, that [H^+] gives the amount of alkyl halide which has *reacted* and not the amount which remains, *i.e.*, [H^+] = C_o − C_t. C_t is then determined by subtracting [H^+] at time t from C_o [equation (29)]. Experi-

$$C_t = C_o - [H^+] \tag{29}$$

mentally, [H^+] at time t is determined by withdrawing an aliquot from the reaction mixture with a pipet (the volume must be accurately measured) and adding it to a sufficient quantity of 98% isopropyl alcohol sufficient to "quench" the reaction so that it no longer proceeds at a measureable rate. The elapsed time is noted, and the quenched sample is titrated with standardized dilute sodium hydroxide solution.

The reaction is started by dissolving a weighed quantity of alkyl halide in an appropriate amount of the solvent to be used. If a temperature other than room temperature is to be used, the solvent and halide should be separately adjusted to this temperature before mixing. If a volumetric flask is available, the solution should be prepared in this flask and enough additional solvent added to bring the total volume to the mark on the flask. Thus the exact volume is known, as well as the weight of sample, so that the initial concentration, C_o, may be calculated. The time, t_o, is taken as that time when the final adjustment of volume is made.

An easier and more reliable method of determining C_o is to allow the reaction to go to completion (the "infinity point," the point at which essentially all of the alkyl halide has been converted into product) and to titrate an aliquot of this final reaction mixture. The [H^+] at this point should be equal to the initial concentration of alkyl halide. The solvolysis of *t*-butyl chloride will reach 99.5% completion in about 12 hr with 60% isopropyl alcohol, or in about 4 hr with 50% isopropyl alcohol.

If the latter method is used to obtain C_o, equation (27) may be used to determine k_1, where A_o = (ml of NaOH from final titration), and A_t = (ml of NaOH from final titration) − (ml of NaOH from titration at time t). As noted before, the concentration units in equation (26) cancel giving rise to equation (27). To apply equation (27) in this way, the aliquot size must be the same for each titration, including the final titration.

Factors which influence the rate of S_N1 reactions. Some of the particular factors which influence the rate of S_N1 reactions are described in the following paragraphs; experiments are provided which will allow some of them to be investigated.

1. Solvent composition. The nature of the solvent influences greatly the rate of S_N1 reactions. The solvent effect may be considered as a dependence of the rate on the *polarity* of the solvent. The polarity influences the rate of the initial ionization [equation (6a)], the slow step of the S_N1 reaction. The more polar the solvent, the more rapid is the ionization. This is because of the greater ability of the more polar solvent to stabilize charged species through solvation.

As mentioned before, the experiments make use of mixed solvents consisting of isopropyl alcohol and water. The effects of solvent composition on rate can be studied using mixtures containing 60:40, 55:45, and 50:50 isopropyl alcohol:water.

2. Effect of concentration. Equation (23) indicates that the rate of solvolysis is proportional to the concentration of alkyl halide. This equation assumes that the rate constant is independent of the concentration of alkyl halide. To show that these statements are true, experiments are presented in which the initial concentration of alkyl halide is changed.

3. Effect of temperature. Temperature changes also influence the rate of a chemical reaction. A *rough* "rule of thumb" is that the rate doubles for each 10° rise in temperature. Thus, in going from 25° to 45° (a 20° rise in temperature), the rate of reaction increases by a factor of *about four*. To examine the effects of temperature on solvolysis reactions, there are two temperatures which can be utilized conveniently in the laboratory; one of these is room temperature, and the other is 0°, which can be obtained with the use of an ice-water bath.

4. Effect of alkyl group. In solvolysis reactions, tertiary halides may be used with complete confidence that the substitutions are proceeding by the S_N1 mechanism. Two compounds which are suitable for the study are *t*-butyl chloride and *t*-pentyl chloride, whose preparations are described in Section 11.2. A determination of the rate of solvolysis of these compounds will provide data which can be used to illustrate the reactivity difference between two tertiary alkyl groups, the *t*-butyl group, $(CH_3)_3C—$, and the *t*-pentyl group, $(CH_3)_2(C_2H_5)C—$.

5. Effect of leaving group. Two common leaving groups are chloride and bromide; although iodide and fluoride could be used, compounds containing them are generally difficult to obtain and expensive. Also, tertiary alkyl iodides are notably unstable. *t*-Butyl chloride and *t*-butyl bromide are used to illustrate the effect of the leaving group on reactivity. If all other factors are kept constant, then the relative abilities of chloride and bromide to leave can be measured by comparing the rates of solvolysis of these two compounds.

Products from S_N1 solvolysis reactions.   Equation (28) implies that the product of a solvolysis reaction is only R—OS, but the reaction is not as simple as it might appear. Recall that the intermediate in an S_N1 reaction is a carbonium ion [equation (6)], which may undergo several reactions. Equation (30) indicates the major products that occur when *t*-butyl chloride is solvolyzed in an isopropyl alcohol-water solvent mixture. It should be noted that the carbonium ion **9** [equation (30a)] undergoes reactions which are typical of most carbonium ions: (1) loss of a proton to give an alkene [equation (30b)], (2) reaction with water to give an alcohol [equation (30c)], and (3) reaction with isopropyl alcohol to give an ether [equation (30d)].

The products which are obtained in solvolysis reactions are highly dependent upon the nature and composition of the solvent. In the present experiment, the relative amounts of alcohol, ether and alkene produced are different for the different solvent mixtures and conditions used. The polarity of the solvent plays an important role in dictating the product distribution, and there are no generalities which can be used to predict this distribution. The reaction temperature and nature of the leaving group also play a role in determining the product ratio.

$$CH_3-\underset{\underset{CH_3}{|}}{\overset{\overset{CH_3}{|}}{C}}-Cl \xrightarrow{\text{slow}} CH_3-\overset{\oplus}{\underset{CH_3}{\overset{CH_3}{C}}} + Cl^{\ominus} \qquad (30a)$$

$$\underset{\mathbf{9}}{}$$

$$CH_3-\overset{\oplus}{\underset{CH_3}{\overset{CH_3}{C}}} \xrightarrow{\text{fast}} CH_3-\underset{CH_3}{\overset{CH_2}{C}} + H^{\oplus} \qquad (30b)$$

$$CH_3-\overset{\oplus}{\underset{CH_3}{\overset{CH_3}{C}}} + H-OH \xrightarrow{\text{fast}} CH_3-\underset{\underset{CH_3}{|}}{\overset{\overset{CH_3}{|}}{C}}-OH + H^{\oplus} \qquad (30c)$$

$$CH_3-\overset{\oplus}{\underset{CH_3}{\overset{CH_3}{C}}} + (CH_3)_2CHOH \xrightarrow{\text{fast}} CH_3-\underset{\underset{CH_3}{|}}{\overset{\overset{CH_3}{|}}{C}}-OCH(CH_3)_2 + H^{\oplus} \qquad (30d)$$

Although no directions are given here for determining the composition of the mixture obtained from the solvolysis reactions, gas-liquid partition chromatography (glpc) is frequently used.

EXPERIMENTAL PROCEDURE

Successful results in the present experiment depend greatly upon being systematic and organized. The best way to insure being systematic is to prepare a neat table in the notebook beforehand, and to check regularly during the course of the kinetic run to see that all necessary entries have been recorded. Provision should be made for recording the following data: (1) the volume of solvent used in the reaction mixture; (2) the exact weight, or volume, of alkyl halide added; (3) the time, t_o, at which the kinetic run is initiated; (4) the temperature of the reaction solution; (5) the results of a "blank" titrimetric determination; (6) a series of times at which aliquots are withdrawn; and (7) the titrimetric data for each aliquot.

This experiment involves a series of quantitative measurements carried out in a relatively short time. Therefore the student should be prepared to work *rapidly*, although *carefully* so that a high standard of accuracy can be maintained. Buret readings should be made to the nearest 0.02 ml if possible, although precision within 0.05 ml will normally be satisfactory. If a rate measurement is to be made at 0° or if *t*-butyl bromide has been assigned, read the "Special Kinetic Procedures" section completely *before* beginning the experiment.

A. General kinetic procedure

Throughout this experiment, use the same pipet and buret as well as standardized base solution.

Measure out 100 ml of the appropriate solvent accurately with a 50-ml graduated cylinder, transferring the solvent into a 250-ml Erlenmeyer flask equipped with a well-fitting rubber stopper. Obtain 80 ml of 98% [14] isopropyl alcohol to use for quenching purposes.

---

[14] Commercial isopropyl alcohol is 98% pure and is satisfactory for quenching the solvolysis reactions. The term "98% isopropyl alcohol" is used throughout to indicate the commercial grade.

Obtain 125–150 ml of approximately 0.04 *N standardized* sodium hydroxide solution. Stopper the flask with a well-fitting solid rubber stopper. *Note and record the concentration of the standard base solution.* Set up a buret, rinse it with a small amount of the base, fill it, see that all air bubbles are out of the tip, and cover the top of the buret with a test tube or small beaker to minimize absorption of carbon dioxide from the air.

Put about 2 ml of phenolphthalein indicator solution in a test tube and have it available, with a dropper, for use in the different titrations. Also connect a short length of rubber tubing to the nearest aspirator or vacuum line for use in drawing air through the pipet for a minute or two after each sampling in order to dry it before taking the next sample. Have available a watch with a minute hand for use. *Record the laboratory temperature.*

To initiate a kinetic run, add the alkyl halide to the solvent mixture. (A sample size of approximately 1 g is satisfactory; the sample must be measured *accurately* by means of a pipet, or by weighing.) Swirl the solution gently to obtain homogeneity. Note the time of addition to the nearest minute. Keep the flask well stoppered to avoid evaporation.

While waiting to make the first measurement, determine a "blank" correction for the solvent. Using a graduated cylinder, measure a separate 10-ml portion of the isopropyl alcohol-water mixture being used into a 125-ml Erlenmeyer flask. (*Note:* Do *not* use the mixture containing the alkyl halide.) Next, add 10 ml of 98% isopropyl alcohol and 4–5 drops of phenolphthalein to the "blank," and titrate with base to a faint pink color that persists for 30 sec. In all titrations, use a white background (paper or a towel) below the titration flask, and accentuate the lower edge of the meniscus in the buret by holding dark paper or some other dark object just below it to make it easier to read. The blank correction will probably be no more than 0.05–0.15 ml.

At regular intervals, take a 10-ml sample from the reaction mixture with a 10-ml pipet and add it to 10 ml (measured with a graduated cylinder) of 98% isopropyl alcohol contained in a 125-ml flask. Be sure to note the time of addition of the aliquot, probably best taken as the time at which half of the aliquot has been added to the alcohol used to "quench" the reaction. Titrate with base to the phenolphtholein end-point as in the blank determination above.

The suggested *approximate* times of taking aliquots using various solvents under various conditions are listed below:

a. 50% isopropyl alcohol-water and *t*-butyl chloride at room temperature: 10, 20, 35, 50, 75, and 100 min.

b. 60% isopropyl alcohol-water and *t*-butyl chloride at room temperature: 20, 40, 70, 100, 130, and 170 min.

c. 55% isopropyl alcohol-water and *t*-butyl chloride at room temperature: 15, 30, 50, 75, 100, and 135 min.

d. 50% isopropyl alcohol-water and *t*-butyl chloride at 0°: 20, 40, 70, 100, 130, and 170 min.

e. 60% isopropyl alcohol-water and *t*-butyl chloride at 0°: 20, 40, 80, 120, 160, and 200 min.

f. 55% isopropyl alcohol-water and *t*-butyl chloride at 0°: 20, 40, 75, 110, 145, and 185 min.

g. 50% isopropyl alcohol-water and *t*-pentyl chloride at room temperature: 10, 20, 30, 40, 50, and 60 min.

h. 55% isopropyl alcohol-water and *t*-pentyl chloride at room temperature: 20, 40, 60, 80, 100, and 120 min.

i.   60% isopropyl alcohol-water and *t*-pentyl chloride at room temperature: 15, 30, 45, 60, 80, 110, and 140 min.

j.[15] 50% isopropyl alcohol-water and *t*-butyl bromide at room temperature: 1, 2, 3, 4, 5, 6, 8, and 10 min.

k.[15] 55% isopropyl alcohol-water and *t*-butyl bromide at room temperature: 2, 4, 6, 8, 10, 12 and 15 min.

l.[15] 60% isopropyl alcohol-water and *t*-butyl bromide at room temperature: 2, 5, 7, 10, 15, 20, and 25 min.

The "infinity" titrations (after the reaction is *complete*) for each of these sets of conditions should be done during the next laboratory period. Be sure to leave the flask containing the reaction mixture *tightly* stoppered.

B. SPECIAL KINETIC PROCEDURES

1. Kinetics using *t*-butyl bromide.   The kinetics of the solvolysis of *t*-butyl bromide may be done at room temperature. The reaction is very fast, and a special technique must be used. At the end of the suggested time intervals, 10-ml portions of the reaction mixture are removed and placed in an Erlenmeyer flask which already contains 20 ml of 98% isopropyl alcohol and which has been cooled previously to 0° in an ice-water bath. Carry out the titration at 0°. If this is not done, the titrations are very inaccurate, because of the speed with which the solvolysis proceeds at room temperature. Since six titration samples will be taken over a maximum period of 25 min, it is best to collect all six samples, quench and store them at 0° in stoppered flasks, and to run all titrations at the end of the experiment.

2. Kinetics at 0°.   If you are assigned to carry out a kinetic determination at 0°, proceed as follows. Add the designated solvent mixture to a dry 250-ml Erlenmeyer flask, and place this flask in a pan. Add enough ice and water so that the level of the water in the pan is about the same as that in the flask. Be sure that the flask is fitted with a rubber stopper. Note and record the temperature of the bath, and with a *dry* thermometer, determine the temperature of the contents *within* the flask after it has come to equilibrium with the ice water. Proceed as directed previously. More ice will have to be added from time to time and some water will have to be removed periodically to maintain a constant level in the pan. It may be helpful to hold the flask with a clamp; however, be sure to shake the contents of the flask well after adding the alkyl halide.

C. TREATMENT OF DATA

1. If an infinity titration was made, proceed as follows. From the titration volumes, determine C_o and C_t for each time t. If the same pipet, buret, and standardized base solution were used for each titration, "ml of NaOH" may be used as A_o and A_t. For each kinetic point, compute the logarithm of C_o/C_t *or* A_o/A_t, and then multiply this value by 2.303 [see equation (26) or (27)]. Plot the resulting number vertically *vs.* t (in hr) for each kinetic "point." Draw the best straight line through the points which were plotted. Determine the slope of the line; the slope is the rate constant, k_1. It should be about 0.2 hr^{-1} for 60% isopropyl alcohol and *t*-butyl chloride at room temperature and about 0.8 hr^{-1} for 50% isopropyl alcohol and *t*-butyl chloride at room temperature.

[15] The kinetic measurements with *t*-butyl bromide at room temperature involve rather closely spaced sampling intervals (the reaction is relatively rapid). It is suggested that these be attempted only by those students performing a second kinetic experiment, and then as the second experiment (see section on "Special Kinetic Procedures").

When this method is used, it is not necessary to apply the blank correction since it is a constant which does not affect the slope of the line (see Exercise 3).

2. If an infinity titration was *not* made, proceed as follows. Using the known weight of the alkyl halide, determine the initial concentration (C_o) of the alkyl halide in the original reaction mixture. (If only the volume of the starting alkyl halide was known, its weight may be determined from its density.) Using the normality of the standard base solution and its volume, as well as the volume of each sample which was titrated, determine C_t for each kinetic point. With this method, it will be necessary to apply the blank correction to C_o and to each value of C_t. Determine the values of $2.303 \log (C_o/C_t)$ for each time t, and plot these data as described in 1 (above). Determine the value of k_1 from the slope of the line.

3. Take the same data and calculate the rate constant separately for *each* kinetic point by putting the data directly into equation (26) and solving for k_1 (in hours^{-1}).

4. Compare each of the values obtained in 3 (above) with the average of those values and with the value obtained from the graph in 1 or 2 (above). Which method allows one to spot an error more easily?

5. From equation (26), it can be seen that the half-life ($t_{1/2}$, that time necessary for half of the original alkyl chloride to react) is given by the relation:

$$t_{1/2} = \frac{0.69}{k_1} \tag{31}$$

Calculate the half-life using the value of k_1 obtained from the graph. After the half-life has been calculated, go back and examine the experimental data. If about one-half of the total volume of NaOH used in the final ("infinity") titration would not have been consumed in a titration at this time, an error has been made in the calculations. Recheck them carefully.

6. Consider the magnitudes and types of errors involved in the various measurements made in this experiment. (Most general chemistry laboratory manuals include a discussion of errors.) Using the unavoidable errors in measurement which would have been expected in volume, time, and titration measurements, calculate the *maximum* error in k_1 which can be expected. (*i.e.*, $k_1 = 0.80$ hour$^{-1} \pm 0.05$ hour^{-1}.)

EXERCISES

1. Give as many advantages as possible for allowing the reaction to proceed essentially to completion, allowing for a final titration, as opposed to adding a known volume of alkyl halide and *not* making a final titration.
2. Explain clearly why "ml of NaOH" can be used [equation (27)] instead of using C_o and C_t in moles/liter in equation (26). Is the normality of the NaOH needed? Why or why not?
3. In the plot of $2.303 \log (C_o/C_t)$ *vs.* t, why doesn't the line go through $2.303 \log (C_o/C_t) = 0$ when $t = 0$, *i.e.*, why is C_o not equal to C_t at $t = 0$?
4. Show how equation (31) can be derived from equation (26).
5. Suppose that the flask had not been stoppered until the next laboratory period when the final titration was taken. Would the rate constant be expected to be larger or smaller than it would have been with the flask stoppered?
6. Why does the titration endpoint fade after 30–60 seconds?

7. Using the chemical equations involved for the hydrolysis and for the titration of the acid formed with base, show that in order to get C_t one must subtract $(C_o - C_t)$ from C_o. Explain the reasoning. (C_t is the concentration of alkyl halide which remains at any time t.)

8. In a titration done at $0°$, the reaction mixture is stored at room temperature until the next laboratory period before carrying out the final titration. Since the rest of the titrations were done at $0°$, why is the value of the final titration valid?

9. List the errors involved in the determinations of rate constants by the procedure used. State the relative importance of each.

12

Carbenes and arynes:
Highly reactive intermediates

For a more complete understanding of any organic reaction, recognition and characterization of possible intermediates formed during the conversion of starting materials to products is required. The types of intermediates encountered in organic chemistry are varied, both in their relative reactivities and in their electronic structures. Charged intermediates, such as carbonium ions and carbanions, are usually very reactive, and their preparation and isolation commonly requires special experimental precautions such as the use of inert atmospheres and the rigorous exclusion of moisture from the reaction. For instance, the preparation of a Grignard reagent, RMgX (X = halogen), an organometallic substance having the properties anticipated for a carbanion, is successful only if the reagents and apparatus used are dry. Some charged intermediates, however, are sufficiently stable so that they can be isolated as salts without the use of special experimental techniques, as is the case in the preparation of triphenylmethyl fluoborate, $(C_6H_5)_3C^+BF_4^-$, a carbonium ion salt (Chapter 18).

Uncharged intermediates quite often are stable molecules which can be readily isolated by use of appropriate reaction conditions. As an example, benzophenone (1), a stable ketone, undoubtedly intervenes as an intermediate in the preparation of triphenylmethanol by reaction of methyl benzoate with two moles of phenylmagnesium bromide [equation (1)]. Although benzophenone is not isolated in the procedure

$$C_6H_5CO_2CH_3 + C_6H_5MgBr \longrightarrow \underset{1}{C_6H_5COC_6H_5} \xrightarrow[2)\ H_3O^\oplus]{1)\ C_6H_5MgBr} (C_6H_5)_3COH \qquad (1)$$

described in Chapter 13, it could be if the conditions of the reaction were modified. Many neutral intermediates cannot be easily isolated, however, because of their high reactivity. Typical of these are free radicals such as those generated in the free radical-chain halogenation of alkanes (Chapter 5).

In this chapter, we shall consider the generation and reactions of two very reactive neutral intermediates which owe their high reactivity to two quite different molecular properties. The first of these substances has the general formula 2 (R = H, alkyl, aryl, halogen, cyano, etc.), and is called a *carbene*. The presence of only six rather than eight electrons in the outer valence shell imparts high reactivity to such intermediates. The experiment described in Section 12.1 will involve the generation and chemistry of *dichlorocarbene* (**2**, R = Cl).

The other intermediate to be considered in this chapter is *benzyne* (**3**), a member of a class of compounds designated as *arynes*. The high reactivity of benzyne, or dehydrobenzene as this molecule is sometimes called, is considered to be due to the large amount of strain resulting from introduction of a triple bond into a six-membered ring.

$$R—\overset{..}{C}—R$$

2   **3**

12.1 Carbenes

Carbenes can be generated in a variety of ways, two of which are given below. One method involves decomposition of a diazo compound (**4**) by irradiation or heat, or with a metallic catalyst [equation (2)]. This method is generally much too hazardous to

$$\left[R_2C{=}N{=}N \longleftrightarrow R_2\overset{\ominus}{C}{-}\overset{\oplus}{N}{\equiv}N : \right] \xrightarrow[\text{or metallic catalyst}]{\text{light or heat}} R_2C : + :N{\equiv}N : \qquad (2)$$

$$\mathbf{4}\mathbf{2}$$

be attempted in the introductory organic laboratory because the diazo compounds are often explosive.

An alternative approach to producing carbenes is that termed α-elimination. This technique fundamentally involves initial formation of a carbanion of the type **5**, followed by decomposition of the carbanion to a carbene by loss of an anion such as halide ion [equation (3), M = metal, X = halogen]. This route to carbenes is quite safe

$$R{-}\overset{\overset{\displaystyle R}{|}}{\underset{\underset{\displaystyle X}{|}}{C}}{:}\overset{\ominus}{}M^{\oplus} \longrightarrow R_2C: + X^{\ominus} + M^{\oplus} \qquad (3)$$

$$\mathbf{5}\mathbf{2}$$

experimentally but does have the disadvantage that yields of products are lower than when diazo compounds are utilized as sources of carbenes.[1]

Two types of reactions which are characteristic of carbenes are their *insertion* into carbon-hydrogen bonds [equation (4)] and their *addition* to carbon-carbon double and triple bonds [equation (5)]. The insertion reaction is of considerable interest mechanistically, but has little synthetic utility. The addition reaction, however, has received wide application in the synthesis of molecules containing three-membered rings. Of importance in terms of the mechanism of the reaction is the observation that the addition of carbenes usually occurs in a stereospecifically *cis* fashion. As an example, the reac-

---

[1] It should be noted here that current investigations indicate that α-elimination may not lead to a "free" carbene, as does photochemical or thermal decomposition of a diazo compound, but rather that the reactive intermediate in the former procedure is either **5** or **2** complexed with the salt MX. Whatever the true nature of the active intermediate resulting from α-elimination, it is observed to react in much the same way as a carbene generated from a diazo compound.

$$R_2C: + -\overset{|}{\underset{|}{C}}-H \longrightarrow -\overset{R}{\underset{R}{\overset{|}{\underset{|}{C}}}}-\overset{|}{\underset{|}{C}}-H \tag{4}$$

$$R_2C: \xrightarrow{\quad \diagdown C=C \diagup \quad} R_2C \diagup \overset{C}{\underset{C}{\bigtriangleup}} \diagdown \tag{5a}$$

$$R_2C: \xrightarrow{\quad -C\equiv C- \quad} R_2C \diagup \overset{C}{\underset{C}{\parallel}} \diagdown \tag{5b}$$

tion of *trans*- and *cis*-2-butene with a carbene, $R_2C:$, produces the two different adducts **6** and **7**, respectively.

6   **7**   **8**

Precursors **5**, required for generation of carbenes by the method of α-elimination are commonly prepared by the reaction of a strong base with a halogenated substrate. Thus, treatment of chloroform with potassium *t*-butoxide yields **8** (M = K), which can subsequently decompose, with loss of potassium chloride, to dichlorocarbene, (R = Cl). Alternatively, the salt **8** (M = Na), can be produced by the thermal decomposition of sodium trichloroacetate [**9**, equation (6)], which is the procedure to be followed in this experiment.

$$CCl_3CO_2^{\ominus}Na^{\oplus} \xrightarrow{-CO_2} Cl_3C^{\ominus}Na^{\oplus} \xrightarrow{-NaCl} :CCl_2 \tag{6}$$

$$\qquad\quad \mathbf{9} \qquad\qquad\qquad \mathbf{8}$$

The decomposition of sodium trichloroacetate will be performed in a mixture of tetrachloroethylene and diglyme, $(CH_3OCH_2CH_2)_2O$, as solvent. This mixture has the advantages that its boiling point, approximately 110°, is high enough to produce a suitable rate of decomposition of the sodium trichloroacetate, and that the acetate is only slightly soluble in the medium so that its concentration in solution is rather low. This low concentration minimizes the importance of side reactions, such as that shown in equation (7), which act to decrease the yield of the desired addition product.

$$CCl_3^{\ominus}Na^{\oplus} + :CCl_2 \longrightarrow Cl_2\overset{Cl\quad Na^{\oplus}}{\underset{}{C-C}}{}^{\ominus}Cl_2 \longrightarrow \overset{Cl}{\underset{Cl}{\diagdown}}C=C\overset{Cl}{\underset{Cl}{\diagup}} + NaCl \tag{7}$$

Indene **(9)** will be used as the unsaturated substrate with which the dichlorocarbene is to react. In principle a number of products resulting from both the insertion and addition reactions is possible. Fortunately, the reaction mixture is not as complex as it might be since carbenes formed by α-elimination preferentially give products resulting from the addition rather than from the insertion of the carbene. However, even if the reaction is limited to addition of dichlorocarbene to indene, there is still possible a variety of products, some of which are shown in Chart 12.1.

CHART 12.1 Possible addition products of dichlorocarbene and indene.

A prediction concerning which of the possible products might be favored can be made on the basis that a fundamental difference exists between compound **10**, which bears the imposing name, 1,1-dichloro-1*a*,6*a*-dihydrocycloprop[a]indene, and all the other potential products of addition. The formation of **10**, by attack of dichlorocarbene on the double bond between carbon atoms 2 and 3 of indene, permits preservation of the aromaticity of the benzene ring, whereas the production of any of the other adducts results in the loss of the aromaticity in the molecule. Therefore a preference for addition of dichlorocarbene across the 2,3-double bond might be anticipated. In fact, this preference is realized experimentally, since generation of dichlorocarbene in the presence of indene, followed by work-up of the reaction mixture under mild conditions, affords **10** in 70% yield as the only isolable product.

Although the attack of dichlorocarbene on indene turns out to be selective, a possible further complication should be noted. Three-membered rings bearing *geminal* halo-

$$(8)$$

gens, such as that in **10**, are relatively unstable and rearrange in polar media and/or with heating to ring-opened isomers [equation (8)]. In the case of **10**, rearrangement is known to occur readily at temperatures greater than 50°. Compound **10**, by analogy to the general reaction given in equation (8), would be expected to rearrange to a mixture of **11** and **12** [equation (9)], but it has been observed experimentally that upon treatment with an ethanolic solution of potassium hydroxide, **10** yields 2-chloronaphthalene **(13)** in high yield. Compounds **11** and **12** may serve as intermediates in the formation of **13**, but they readily lose the elements of hydrogen chloride to produce the more highly stable naphthalene system.

In the experimental procedure described below, the addition product **(10)** is produced by heating indene and sodium trichloroacetate at a temperature greater than 100° in a polar medium. As a consequence **13**, the rearrangement product from **10**, is isolated in this experiment.[2] The overall equation for the reaction is therefore

It is important to note that this experiment demonstrates the marked effect that reaction conditions and purification techniques may sometimes have on the structures of products ultimately isolated from organic reactions. Minor modifications in an experimental procedure may be the difference between obtaining a desired product in high yield or producing major amounts of undesired compounds.

EXPERIMENTAL PROCEDURE

Equip a 250-ml round-bottomed flask with a reflux condenser fitted with a thermometer adapter, or 1-hole rubber stopper, through which is passed a glass tube. Attach to this tube a length of rubber tubing bearing a short piece of glass tubing. Place 0.081

---

[2] Students interested in preparing compound **10** should consult reference 2 given at the end of this section.

mole of sodium trichloroacetate,[3,4] which has previously been dried for at least one hour at 90°, 20 ml of tetrachloroethylene, 5 ml of diglyme, and 0.086 mole of indene [5] in the flask. Bring this mixture to a gentle reflux with the aid of a microburner, swirling the flask occasionally to insure good mixing. Continue to heat until the evolution of carbon dioxide has ended, as indicated by the absence of bubbling when the tube leading from the top of the condenser is inserted into a few milliliters of tetrachloroethylene contained in a test tube. At this time, the granular sodium trichloroacetate should have disappeared, and finely divided sodium chloride should have separated.★

Before removing the burner from beneath the flask, disconnect the bubbler in order to prevent suction of tetrachloroethylene into the reaction flask. Detach the flask from the condenser, add 25 ml of water to the reaction mixture, and attach the flask to an apparatus set for steam distillation using an external source of steam (see Figure 2.9).

Steam-distil the reaction mixture as rapidly as the ability of the condenser to cool the distillate will permit, collecting the distillate in a graduated cylinder. When approximately 26 ml of an organic layer has been obtained, which will require that a total volume of 60 to 65 ml of distillate be collected, replace the graduated cylinder with a 250-ml Erlenmeyer flask and discard the liquid in the graduated cylinder.★ Continue the distillation until no more oily material can be detected in the distillate. Collection of about 200 ml of distillate will be required to obtain all of the crude 2-chloronaphthalene.★ During the course of the distillation, check to make sure that solid does not collect in the condenser and block it. If solid does begin to form, shut off the flow of water through the condenser and allow the uncondensed steam to sweep the solid out of the condenser.

With the aid of a water bath, cool the distillate to room temperature and then transfer it to a separatory funnel. Some crystalline 2-chloronaphthalene may remain in the Erlenmeyer flask, but it will be recovered later. Extract the aqueous solution in the funnel three times with separate 25-ml portions of either dichloromethane or technical ether and combine the extracts in the 250-ml Erlenmeyer flask used to collect the steam distillate.★

Remove the solvent from the crude product by heating the solution on a steam bath (preferably in the hood).★ Alternatively, the solvent can be removed by performing a simple distillation (*No flames* if ether was used!). Transfer the crude 2-chloronaphthalene to a smaller Erlenmeyer flask, rinse the larger flask with 15 ml of 95% ethanol and add this rinse along with some decolorizing carbon to the product. Boil this solu-

---

[3] If commercial sodium trichloroacetate (Columbia Organic Chemicals) is used, take into account the fact that the product is of 95% purity when making the calculation of the number of grams required.

[4] If desired, sodium trichloroacetate can be prepared in the following manner. Prepare a solution of 0.090 mole of sodium hydroxide in 10 ml of water and cool this solution thoroughly in an ice-water bath. Place 0.086 mole of trichloroacetic acid in a 125-ml suction flask and, while swirling this flask in the ice-water bath, add to it about 9 ml of the cold solution of base. Now add one drop of 0.04% Bromocresol Green solution so that a yellow color is visible. Complete the titration of trichloroacetic acid by adding the hydroxide solution dropwise to a point where a single drop produces a color change from yellow to blue. If the end-point is overshot, add some additional acid and repeat the titration.★ Stopper the flask and attach it to a water aspirator by means of suction tubing. Place the flask *within* the rings of a steam cone and concentrate the solution by evacuating the flask and heating with steam. After the solution has been evaporated to dryness,★ which will require about 20 minutes, transfer the solid to a watch glass, break up any large lumps, and dry the sodium trichloroacetate in an oven at 90°.

[5] Indene is not stable for long periods of time and therefore should be purified immediately before use by means of a simple distillation.

tion for a few minutes on the steam cone and then filter to remove the carbon.★ Adjust the temperature of the filtrate to about 45° and, while maintaining this temperature, add water dropwise until the solution becomes turbid. At this point, allow the solution to stand undisturbed until it cools to room temperature.★ If the 2-chloronaphthalene fails to crystallize but rather begins to oil out, a problem which is characteristic of many low-melting solids, scratch the walls of the flask at the interface between the solution and the air in an attempt to induce crystallization. In some cases, it may be necessary to evaporate a small sample of the solution to obtain the oily crude product which, upon scratching, may provide a few seed crystals that can be added to the solution to encourage crystallization.

When no more solid crystallizes at room temperature, cool the flask in an ice-water bath to complete the crystallization, and collect the product. A second crop of product can be obtained if the filtrate is reheated to 45° and water added dropwise until the solution again becomes turbid. Cooling and scratching as before should provide additional 2-chloronaphthalene of somewhat lower purity than the first crop.

Pure 2-chloronaphthalene is an irridescent white solid [6] (mp 58–59°). The yield of product in this experiment should be 20–30%.

EXERCISES

1. Why is tetrachloroethylene rather than some other liquid such as water used to test for the evolution of CO_2?

2. Predict which double bond in phenanthrene would be most reactive toward a carbene and give an explanation for your choice.

Phenanthrene

3. What properties of the diglyme-tetrachloroethylene mixture make it suitable as a solvent in the reaction?

4. What organic compounds distil in the first stage of the steam distillation?

5. When attempting a mixed-solvent recrystallization of the type performed in this experiment, the solvent in which the solute is sparingly soluble is normally added to a *boiling* solution of the solute and the other solvent. What special circumstances prohibit the addition of water to a boiling solution of 2-chloronaphthalene in 95% ethanol in order to effect saturation? Why is it more efficient to saturate the ethanolic solution at about 45° than at, for example, room temperature?

6. The 2-chloronaphthalene isolated by the procedure described in this experiment sometimes has a bluish tinge caused by the presence of a small amount of a chloroazulene. The chlorine substituent is in the seven- rather than the five-membered ring but its exact position is unknown. Propose a mechanism for the formation of the chloroazulene from indene and dichlorocarbene. Suggest a

[6] For a discussion of the reason for the blue color often observed in the crude 2-chloronaphthalene and even in the recrystallized product obtained in this experiment, see Exercise 6.

reason why it is difficult to remove the azulene from 2-chloronaphthalene even after repeated recrystallization.

7. Write out the expected products of the reaction between dichlorocarbene and the following alkenes under mild conditions.

(a) $CH_2{=}C\overset{\displaystyle CH_3}{\underset{\displaystyle CH_3}{\big\langle}}$ + :CCl_2 (1 mole)

(b)   + :CCl_2 (1 mole)

(c)   + :CCl_2 (2 moles)

8. Suggest why sodium trichloroacetate decarboxylates much more readily than sodium acetate.

REFERENCES

1. W. Kirmse, *Carbene Chemistry*, Academic Press, New York, 1964, Chapter 8.
2. W. E. Parham, H. E. Rieff and P. Swartzentruber, *Journal of the American Chemical Society*, **78**, 1437 (1956).

SPECTRA OF STARTING MATERIAL AND PRODUCT

FIGURE 12.1   IR spectrum of indene.

FIGURE 12.2 NMR spectrum of indene.

FIGURE 12.3 IR spectrum of 2-chloronaphthalene.

12.2 Arynes

Arynes, which can be considered to be cyclohexadienynes, have been recognized as reaction intermediates only since the early 1950s. Evidence which prompts postulation of their existence includes the observations that treatment of **14** with amide ion produces an equimolar mixture of **15** and **16** [equation (10)], that reaction of **17** with hydroxide ion generates **18** and **19** [equation (11)] and that pyrolysis of **20** in the presence of anthracene yields triptycene [**22**, equation (12)]. Equations (10) and (11)

$$\underset{\substack{\mathbf{14}\\ \text{* denotes }^{14}\text{C}}}{\text{Cl}} \quad \xrightarrow[\text{heat}]{\text{H}_2\text{N}^{\ominus}} \quad \left[\right] \quad \xrightarrow{\text{NH}_3} \quad \underset{\mathbf{15}}{\text{NH}_2} \quad + \quad \underset{\mathbf{16}}{\text{NH}_2} \tag{10}$$

$$(11)$$

$$(12)$$

show the production of an aryne by an elimination reaction followed by reaction of the aryne with a nucleophile. Equation (12) exemplifies the generation of the aryne, benzyne (**21**), by thermal decomposition of an unstable "zwitterionic" species, **20**, with subsequent reaction of the benzyne, acting as a dienophile, with a diene.

Although benzyne itself is often generated by reaction of an aryl halide with a strong base [equations (13) and (14)], a more convenient procedure involving synthesis and decomposition of a benzyne precursor closely related to **20** will be utilized here. Thus reaction of isoamyl nitrite (**23**) and hydrochloric acid with anthranilic acid (**24**) yields benzenediazonium-2-carboxylate hydrochloride (**25**), possibly by the pathway outlined in Chart 12.2. The salt **25** is considerably more stable than benzenediazonium-2-carboxylate (**20**), a compound which is highly sensitive to shock, particularly when dry.

CHART 12.2   The preparation of benzenediazonium–2-carboxylate hydrochloride.

Benzenediazonium-2-carboxylate hydrochloride decomposes to benzyne upon gentle heating. If the decomposition is carried out in the presence of anthracene and an *aprotic* solvent, a Diels-Alder adduct, triptycene (**22**), is isolated [equation (13)]. Anthracene, an aromatic compound, functions as the diene in the reaction. An aprotic solvent is used in order to minimize side reactions which would result from the nucleophilic attack of solvent on benzyne.

$$\longrightarrow 22 + HCl + CO_2 + N_2 \qquad (13)$$

The formation of triptycene indicates that benzyne adds preferentially across the 9- and 10- positions of anthracene. The fundamental reason for this selectivity is that **22** has the greatest amount of aromatic stabilization of all the possible products.[7]

In this experiment, benzenediazonium-2-carboxylate hydrochloride is decomposed in the presence of a slight excess of anthracene to produce triptycene and a mole of hydrochloric acid [equation (13)]. The presence of the acid in the reaction mixture is undesirable, since it will react with benzyne and thereby lower the yield of desired product. The acid can be efficiently removed as it is formed by trapping it with propylene oxide (**26**), which reacts rapidly and irreversibly with the acid as shown in equation (14). The product, propylene chlorohydrin **27**, can readily be removed from the reaction mixture by distillation.

$$CH_3CH-CH_2 + HCl \longrightarrow CH_3CH-CH_2 \qquad (14)$$

An important technical problem that is encountered in the purification of the reaction mixture is the removal of unreacted anthracene which, as might be anticipated on structural grounds, has solubility characteristics similar to those of triptycene. This problem is solved by preparing a derivative of anthracene which is soluble in aqueous base. Thus, treatment of the crude reaction mixture with maleic anhydride will result in the destruction of excess anthracene by the formation of the Diels-Alder adduct, **28**, which, on hydrolysis with aqueous potassium hydroxide, affords the water-soluble salt **29** [equation (15)].

$$(15)$$

EXPERIMENTAL PROCEDURE

To a solution of 0.050 mole of anthranilic acid and 75 ml of 95% ethanol contained in a 250-ml Erlenmeyer flask, add 5 ml of concentrated hydrochloric acid. Cool this solution to 0–5° in an ice-water bath and then add, with intermittent swirling, 0.093 mole of isoamyl nitrite over a period of *ca.* 5–10 min. Maintain the temperature of the

---

[7] A similar explanation was used to rationalize the preferred attack of dichlorocarbene on the 2,3-carbon-carbon double bond of indene (Section 12.1).

reaction mixture at 0–5° throughout the course of the addition and for an additional ten minutes afterwards. Then add about 75 ml of technical ether to the reaction mixture, swirl and cool the solution for a few minutes and finally collect the crystalline benzene-diazonium-2-carboxylate hydrochloride by suction filtration. Wash the filter cake with two 10-ml portions of ether and then air-dry the product for 5–10 min. Determine the weight and percentage yield of the diazonium hydrochloride.

Place 40 ml of 1,2-dichloroethane in a 100-ml round-bottomed flask and add 0.022 mole of benzenediazonium-2-carboxylate hydrochloride, 0.022 mole of anthracene and 0.043 mole of propylene oxide.[8] Attach a reflux condenser to the flask and heat the mixture at reflux on the steam cone for two hours. Allow the reaction mixture to cool for a few minutes and then add 30 ml of diethylene glycol diethyl ether, a high-boiling solvent (bp 162°).★ Attach the flask to an apparatus arranged for simple distillation, and distil until a head temperature of 150° is reached. Now add 0.025 mole of maleic anhydride [8] to the reaction mixture and heat this mixture at reflux for five minutes.★ Cool the flask with the aid of a water bath and then slowly add, with stirring, a solution of 10 g of potassium hydroxide in 30 ml of water and 20 ml of methanol.★ Cool the resulting suspension in an ice-water bath, collect the precipitated triptycene by suction filtration and wash it with 5 ml of a cold 4:1 methanol-water solution. Recrystallize the product by first dissolving the triptycene in butanone (25 ml per gram of triptycene), by heating on a steam cone, decolorizing the butanone solution, and then filtering to remove the decolorizing carbon.★ Heat the filtrate to boiling on a steam cone (in the hood, if possible) and gradually introduce an equal volume of methanol. Concentrate the solution to one-third of its original volume, allow it to cool to room temperature, and collect the product by suction filtration. Determine the yield and melting point of the product.

EXERCISES

1. What substances are removed by the simple distillation?
2. Suggest a reason why a high-boiling solvent is added to the reaction mixture before the simple distillation is performed.
3. Why is the simple distillation necessary?
4. What is the purpose of adding methanol during the recrystallization of triptycene?
5. Explain, on the basis of stabilization (delocalization) energies, why the product of reaction between benzyne and anthracene is triptycene, rather than the compound shown below.

6. Predict the structure of the product resulting from addition of one mole of bromine to anthracene.
7. Write a mechanism for the reaction of propylene oxide with hydrochloric acid to

[8] *All* of the benzenediazonium-2-carboxylate hydrochloride that you have isolated *must* be used in this experiment. Therefore, adjust the amounts of reagents and solvents required so that all the diazonium salt will be used.

give propylene chlorohydrin. Suggest a reason for the preference for the formation of this compound rather than the compound shown below.

$$CH_3—CH—CH_2Cl$$
$$\underset{OH}{|}$$

REFERENCES

1. L. Friedman and F. Logullo, *Journal of the American Chemical Society*, **85**, 1549 (1963).
2. H. Heaney, *Chemical Reviews*, **62**, 81 (1962).

SPECTRA OF STARTING MATERIALS

FIGURE 12.4   IR spectrum of anthracene.

FIGURE 12.5   IR spectrum of propylene oxide.

FIGURE 12.6 NMR spectrum of propylene oxide.

Organometallic chemistry

Organometallic reagents represent perhaps one of the single most useful synthetic tools for laboratory reactions. This chapter is devoted to discussion and experimental application of the Grignard and the organolithium reagents to organic syntheses.

13.1 The Grignard Reagent: Its Preparation and Reactions

The preparation of the Grignard reagent, RMgX, is represented in equation (1),

$$RX + Mg \xrightarrow[\text{(solvent)}]{\text{dry ether}} RMgX \tag{1}$$

where RX represents an organic halide. The exact nature of the Grignard reagent in solution is not known. It is believed to be a mixture of numerous species, but some of the principle species are undoubtedly R_2Mg and MgX_2 in equilibrium with 2 RMgX. These species are highly solvated by ether and are complexed with one another. It is customary to represent the Grignard reagent by the formula RMgX when writing chemical equations, but it should be kept in mind that the species in solution are of a much more complex nature.

The ether solvent is an essential part of the Grignard reagent, for it is known to form a complex with the magnesium which is present in the reagent. Several cases are known where Grignard reagents have been prepared in the absence of ether, but the yields are not good. Satisfactory yields are usually obtained when ether is present. The most common ether solvent is ethyl ether, $(C_2H_5)_2O$, due to its low cost and ease of removal (its boiling point is 36°). Other ethers, for example n-butyl ether, bp 142°, have been used successfully when higher-boiling solvents are required. Some cyclic ethers, such as tetrahydrofuran, have been used with excellent success in the preparation of certain Grignard reagents (see below).

The organic halide may, in general, be of any organic moiety (alkyl or aryl) and the halide may be bromide, chloride, or iodide. Alkyl iodides, bromides, or chlorides and aryl iodides or bromides can be converted to the corresponding Grignard reagents using ethyl ether as the solvent. Aryl chlorides, on the other hand, do not react with magnesium in ethyl ether, but they react readily when tetrahydrofuran is used as the solvent.

The initiation of the reaction between the halide and magnesium is sometimes difficult, with the iodides being more reactive than the bromides, which are in turn more reactive than chlorides. Aryl halides are less reactive than their alkyl counterparts, and it has been found that the aryl bromides and the alkyl chlorides react about equally well. In cases where one is dealing with a particularly unreactive halide, the initiation of the reaction may sometimes be accomplished by the addition of a small amount of iodine to the reaction mixture. This is believed to convert a small amount of the halide to the iodide, which is more reactive.

The preparation of the Grignard reagent *must* be carried out under anhydrous conditions and if possible in the absence of oxygen. It is exceedingly important to maintain completely dry conditions throughout, for the presence of water inhibits the initiation of the reaction as well as destroys the reagent once it is formed.

The reaction that occurs when the Grignard reagent comes in contact with water is shown in equation (2). The reagent is a strong base, as one of the carbon atoms bears

$$2\ RMgX + 2\ H_2O \rightarrow 2\ RH + Mg(OH)_2 + MgX_2 \tag{2}$$

substantial negative charge ($R^-Mg^{++}X^-$). The Grignard reagent, acting as a base, removes a proton from water, which acts as an acid. The overall effect is the destruction of the reagent, with the formation of a hydrocarbon (RH) and magnesium salts. The hydrocarbon is inert, so that the net effect is the production of hydroxide ion in solution. In fact, it is possible to standardize a solution of a Grignard reagent by adding a known volume of it to a known volume of standardized mineral acid and back-titrating with standardized base.

In addition to the reaction with water [equation (2)], there are other side reactions that may occur during formation of the Grignard reagent, as shown in equations (3–5).

REACTION WITH OXYGEN:

$$2\ RMgX + O_2 \rightarrow 2\ ROMgX \tag{3}$$

REACTION WITH CARBON DIOXIDE:

$$RMgX + CO_2 \rightarrow RCO_2MgX \tag{4}$$

COUPLING:

$$RMgX + RX \rightarrow R{-}R + MgX_2 \tag{5}$$

It is possible to minimize these reactions by taking certain precautions when carrying out the experimental work. The reaction with oxygen and carbon dioxide may be avoided by carrying out the reaction under an inert atmosphere (such as nitrogen or helium gas). In research work, this is routinely done; however, when ethyl ether is used as a solvent, it excludes a certain amount of air from the reaction vessel due to its very high vapor pressure. The coupling reaction [equation (5)] is an example of a Wurtz reaction. It is not possible to eliminate this coupling reaction completely, but it may be minimized by using dilute solutions so as to avoid localized high concentrations of halide. This can be done by using very efficient stirring and by slowly adding the halide to the magnesium in ether. Normally the rate of addition of halide (dissolved in ether) and the rate of reflux (when ethyl ether is used) should be adjusted so that they are about equal. Alkyl iodides are much more prone to coupling reactions than are the bromides and chlorides, so that the latter are preferable for preparing Grignard reagents even though they are less reactive.

The experiments which appear in this chapter involve the preparation of phenylmagnesium bromide in ethyl ether, followed by some reactions of this Grignard reagent with various compounds. Many different reactions involving the Grignard reagent appear in textbooks on organic chemistry and will not be covered; the experiments included here are representative of many similar types of reactions.

The equation for the preparation of phenylmagnesium bromide indicates that one mole of magnesium must be added for each mole of bromobenzene which is used; the main reaction is given in equation (6) and the most important side reaction in equation

$$C_6H_5Br + Mg \xrightarrow{(C_2H_5)_2O} C_6H_5MgBr \qquad (6)$$

$$C_6H_5MgBr + C_6H_5Br \rightarrow C_6H_5–C_6H_5 + MgBr_2 \qquad (7)$$

(7). Once the reagent is prepared, it is used directly in subsequent reactions; it is not possible to separate the biphenyl ($C_6H_5–C_6H_5$) from the reagent before using it.

Reaction of phenylmagnesium bromide with methyl benzoate. When phenylmagnesium bromide is allowed to react with an ester such as methyl or ethyl benzoate, the chief product of the reaction is triphenylmethanol. The formation of this compound can be envisioned as arising from the two-step reaction shown by equations (8) and (9). The intermediate benzophenone, whose formation is shown in equation (8) as arising from the reaction of one mole of methyl benzoate with one mole of phenylmagnesium bromide, then reacts with one more mole of phenylmagnesium bromide to give the salt of triphenylmethanol, as shown in equation (9). Subsequent hydrolysis of

$$(8)$$

$$(9)$$

the salt gives triphenylmethanol, so that the reaction for the overall process becomes that given by equation (10).

$$C_6H_5C–OCH_3 + 2\ C_6H_5MgBr \xrightarrow[\text{(2) } H_2O]{\text{(1) ether}} (C_6H_5)_3COH + CH_3OH + 2\ HOMgBr \qquad (10)$$

The synthesis of triphenylmethanol is accomplished experimentally by first preparing the Grignard reagent, and then adding the methyl benzoate to it with stirring and cooling (the reaction is highly exothermic). In working up the reaction mixture, water is added to hydrolyze the magnesium salts to the final organic products; a little mineral acid is also commonly added to dissolve the magnesium hydroxide which is formed. If this is not done, the magnesium hydroxide will produce an unworkable emulsion. The organic products (triphenylmethanol, benzophenone, biphenyl, unchanged methyl benzoate, and benzene—the benzene coming from action of water upon unchanged phenylmagnesium bromide) all remain in the organic layer (consisting primarily of ethyl ether). The principal organic products present are triphenylmethanol and biphenyl, with the former being present in the major amount. It is possible to separate these two compounds from one another, owing to the solubility of the hydrocarbon product (biphenyl) in hydrocarbon solvents (such as petroleum ether, hexane, etc.). Recrystallization of the crude product mixture from these solvents leaves pure triphenylmethanol as the solid. This is because the polar, hydroxylic triphenylmethanol is much less soluble in hydrocarbon solvents than is biphenyl.

Reaction of phenylmagnesium bromide with carbon dioxide. The reaction that occurs when phenylmagnesium bromide is allowed to react with carbon dioxide and the resulting magnesium salt hydrolyzed is shown in equation (11); the major product is a carboxylic acid (benzoic acid). The principal side reactions which are encountered in

$$C_6H_5:^{\ominus}\overset{\frown}{MgBr}^{\oplus} + O{=}\overset{\frown}{C}{\overset{\frown}{=}}O \longrightarrow C_6H_5{-}\overset{\overset{O}{\|}}{C}{-}O^{\ominus}MgBr^{\oplus} \xrightarrow{H_2O} C_6H_5\overset{\overset{O}{\|}}{C}{-}OH \qquad (11)$$
$$+ \ HOMgBr$$

this process result in the formation of a ketone or a tertiary alcohol, as shown by the steps in equation (12). These side reactions can, however, be minimized in several ways.

$$C_6H_5CO_2MgBr + C_6H_5MgBr \longrightarrow (C_6H_5)_2CO + MgBr_2 + MgO$$
$$\downarrow {\scriptstyle C_6H_5MgBr}$$
$$(C_6H_5)_3COMgBr \xrightarrow{H_2O} (C_6H_5)_3COH \qquad (12)$$
$$+ \ HOMgBr$$

First of all, the bromomagnesium salt of the carboxylic acid, because it is moderately insoluble in ether, precipitates from solution and is thus effectively prevented from undergoing further reaction. Secondly, Dry Ice is used as the source of carbon dioxide, so that the temperature is maintained at $-78°$ throughout. Finally, a large excess of carbon dioxide is always used thus enhancing the probability that phenylmagnesium bromide will react with carbon dioxide rather than with the magnesium salt (to produce the ketone) or with the ketone (to yield the magnesium salt of the alcohol), both of which are present in lower concentrations than carbon dioxide and are also less reactive toward the Grignard reagent.

Experimentally, this reaction is carried out by preparing the Grignard reagent, which is then poured slowly and with swirling onto an excess of finely powdered Dry Ice. The excess Dry Ice is then allowed to evaporate, and on treatment of the resulting mixture (which contains ether as the solvent) with dilute mineral acid, the carboxylic acid is liberated and dissolves in the ether. All of the possible organic compounds, which include benzoic acid, benzophenone, triphenylmethanol, benzene, and biphenyl, are extracted into the ether layer. However the desired product, benzoic acid, is separated quite readily from the by-products by extracting it into a dilute solution of sodium hydroxide. The acid is converted to its soluble sodium salt, whereas the other compounds remain in the organic layer. Careful acidification of the aqueous layer regenerates the benzoic acid, which can be purified further by recrystallization.

It should be pointed out that the two experiments described here represent two possible ways in which organometallic compounds can be allowed to react with substrates. In the preparation of triphenylmethanol, the substrate (methyl benzoate) is added to the Grignard reagent; this mode of addition is termed *normal addition*. On the other hand, in the preparation of benzoic acid, the Grignard reagent is added to the substrate (carbon dioxide); this procedure is designated *inverse addition*. The method of addition is dictated by the chemical reactions involved. For example, in the preparation of benzoic acid, if one bubbled carbon dioxide gas into the Grignard reagent (normal addition), the reaction illustrated in equation (12) would be the predominant one. There would be an excess of the Grignard reagent present, which would react further with the bromomagnesium salt of benzoic acid once it formed. To obtain a good yield of benzoic acid, it is necessary to minimize this reaction, so the Grignard reagent should be added to a large excess of carbon dioxide. On the other hand, in the preparation of triphenylmethanol, one could equally well add the Grignard reagent to the methyl benzoate. However, to do this would involve transferring the Grignard reagent to a dropping funnel and adding it to a clean flask containing the substrate.

EXPERIMENTAL PROCEDURE

Phenylmagnesium bromide. Equip a 250-ml round-bottomed flask with a Claisen connecting tube to which are attached an efficient reflux condenser and a dropping funnel. (If a drying oven is available, dry the equipment before assembling.) Place 0.1 mole of magnesium turnings in the flask, and dry the apparatus by flaming it with a burner; do not have water in the condenser during the drying. After heating, immediately attach a calcium chloride drying tube to the condenser; use a drying tube which will also fit the flask. *Allow the flask to cool to room temperature before continuing.* (At this point, start water through the condenser.)

Note: *From this point on, it is imperative that all flames be extinguished. Care should be taken to insure that there are no open flames (yours or your neighbor's) present within several feet.*

As a precaution for controlling the reaction, prepare an ice bath and have it ready for immediate use if needed. Add 0.115 mole of bromobenzene to 15 ml of *anhydrous* ether [1] contained in the dropping funnel and swirl until the mixture is homogeneous.

Add about 25 ml of anhydrous ether to the flask containing the magnesium turnings, and then add a portion (*ca.* 5 ml) of the bromobenzene–ether mixture. Shake the flask gently. A change in the appearance of the reaction mixture, as evidenced by the presence of a slightly cloudy solution and by the formation of bubbles at the surface of the magnesium turnings, indicates that the reaction has commenced.[2] Once the reaction has

---

[1] Commercial anhydrous ether is normally dry enough that it can be used as is. Anhydrous ether may be prepared by drying technical grade ether over phosphorus pentoxide, but the hazards and cost of this procedure far exceed the cost of commercial anhydrous ether.

[2] If there is no apparent change after the portion of the solution of bromobenzene has been added, remove the condenser and mash a piece of magnesium firmly against the bottom of the flask with a *dry stirring rod.* (Be careful not to produce a hole in the flask!) This has the effect of providing a clean, metallic surface on magnesium, which often accelerates initiation of the reaction. If this does not work,

started, the ether will be observed to reflux and the flask will become slightly warm. At this time add 10–15 ml of ether to the reaction mixture through the condenser. The rest of the bromobenzene–ether mixture should be added over a period of *ca.* 15–20 min by dropwise addition. If the reaction becomes too vigorous, cool the flask in an ice bath. If the spontaneous refluxing of the reaction becomes slow or if it is slow, mount the flask on a steam bath and heat gently at reflux; be sure that the condenser and drying tube are attached. Continue heating at reflux until most of the magnesium has reacted; during this time, the solution normally acquires a tan to brown as well as a cloudy appearance. The completion of the reaction is indicated by the nearly complete disappearance of the magnesium turnings; usually residual bits of metal remain.

Use the Grignard reagent as soon as possible after its preparation. Two typical reaction procedures making use of it are given below. It is possible to divide the solution and do both preparations; consult the instructor about this possibility. *If you are to do both experiments, be sure to decrease the amount of the other starting materials used in each part by a factor of two.*

Triphenylmethanol. Dissolve 0.045 mole of methyl (or ethyl) benzoate in about 20 ml of anhydrous ether, and add this mixture to the dropping funnel used in the preparation of the Grignard reagent. Cool the flask containing the phenylmagnesium bromide solution with an ice-water bath (note that the *same* apparatus which was used for the preparation of the Grignard reagent is being used for this synthesis and that the phenylmagnesium bromide has not been removed from the flask). Add the ether solution slowly to the Grignard reagent, cooling the reaction mixture as needed with an ice bath in order to control the exothermic reaction which occurs on mixing. Gentle refluxing of the ether may occur. Quite often a white solid forms as the reaction proceeds; this is a further sign that the reaction is progressing normally. After the addition is complete, remove the dropping funnel and gently swirl the flask until its contents warm to room temperature and the exothermic reaction has subsided. To complete the reaction, either of the following may be done: (1) Heat the mixture at reflux for 30 min using the steam bath, or (2) equip the flask with the drying tube and allow the mixture to stand overnight (no reflux required).★

To a 250-ml Erlenmeyer flask add about 50 ml of 6 *M* sulfuric acid and about 25–30 g of ice; pour the entire reaction mixture into this flask with swirling. Swirl until the heterogeneous mixture is completely free of undissolved solids. It may be necessary to add more technical ether in order to dissolve all of the organic material. Transfer the entire mixture to a separatory funnel, shake vigorously but carefully (be sure to vent the funnel often), and remove the aqueous layer.★ Wash the organic layer with dilute sulfuric acid and then with saturated salt solution. Repeat the washing with salt solution until the resulting aqueous solution is no longer acidic. Dry the organic layer by filtering it through anhydrous sodium sulfate into an Erlenmeyer flask of suitable size. Remove the ether either in the hood or by attaching a funnel to an aspirator line and holding the funnel upside down over the Erlenmeyer flask while the ether is boiled off

the following remedies might be tried, in the order listed: (a) Warm the flask on a steam bath, with swirling. If boiling continues after the heat is removed, reaction has probably started. (b) Crush another piece of magnesium metal with a dry stirring rod. (c) Add a small crystal of iodine. Mix thoroughly, and if reaction does not appear to start, warm over a steam bath again. If iodine is added, then the ether solution containing the final product must be washed with a dilute solution of sodium bisulfite or sodium thiosulfate to remove the iodine. (d) Start the reaction over again; use greater care to insure complete drying of the equipment and reagents.

(use a steam bath for heating). The crude residue which is left as a solid should weigh about 11.5 g and may melt over a wide range. Determine its melting point.★

The crude solid may be purified by fractional crystallization from a 2:1 mixture of petroleum ether–ethanol.[3] Use a minimum amount of the solvent, and carry out this operation either in the hood or with a funnel inverted over the flask and attached to an aspirator. Once all of the material is in solution, evaporate the solvent slowly until small crystals of triphenylmethanol start to form. Remove the heat and allow crystallization to continue at room temperature; complete the crystallization by cooling to 0° until no further crystals appear to form. Isolate the crystalline triphenylmethanol by filtration; it should be colorless.[4] Determine the melting point and yield of the product. The yield should be about 5.5–6.0 g. The mother liquor may be concentrated to yield a second crop, although this will give only a very small amount of additional product. If the melting point is lower than 160° or if the melting range is wide, recrystallize the product once more from a minimum amount of 2:1 petroleum ether–ethanol.

Benzoic acid.   Place 25–30 g of solid carbon dioxide (Dry Ice), which has been coarsely crushed and which has been protected from moisture as much as possible, in a 500-ml Erlenmeyer flask. Add slowly, with gentle swirling, the phenylmagnesium bromide solution. The mixture becomes thick, and an additional 50–75 ml of technical ether should be added after the addition is complete. The excess carbon dioxide is allowed to escape. This is perhaps most easily done by allowing the flask to stand overnight and to come to room temperature on its own accord; it would be preferable to leave the flask in the hood, properly labeled with your name, rather than to leave it in a locker.[5] ★

After the excess Dry Ice is gone, treat the mixture with about 50 ml of 6 M sulfuric acid which has been mixed with 25–30 g of ice. The addition of the acid to the crude reaction mixture should be done with care in order to avoid foaming; if the ether has evaporated appreciably, more technical ether may be added so that the total volume of ether is about 100 ml. Transfer the entire mixture to a separatory funnel after mixing it thoroughly by swirling. Rinse the flask with a small portion of technical ether and add the rinse to the separatory funnel. Shake the funnel cautiously (with venting) and separate the layers. Extract the aqueous layer two additional times with 15-ml portions of ether and combine all of the ether extracts.

Extract the ethereal solution with three 20-ml portions of a 5% solution of sodium hydroxide. Treat the alkaline extract with decolorizing carbon and filter. Add 6 M hydrochloric acid until precipitation is complete and filter the solid.★ The solid (ben-

---

[3] Refer to footnote 15, p. 39 in Chapter 2.

[4] An alternative method of recovering the triphenylmethanol from the crude mixture containing it and biphenyl is as follows. Add about 150 ml of petroleum ether and about 50 ml of technical grade ether to the crude product. Heat the mixture on a steam bath (preferably in the hood) until the crystals dissolve; add more ether if needed. Then boil off the solvent *until* the first crystals appear, and then stop heating. Allow the flask to cool and collect the crystals which form. Proceed to determine the melting point of the crystals, and recrystallize if needed.

[5] The process of removing the excess CO_2 can be expedited in one of several ways: (1) Shake or swirl the flask while warming it *very slightly* in a warm water bath; (2) stir the viscous mixture that forms during the reaction; or (3) add small amounts of warm water to the reaction mixture. All of these methods may cause a sudden loss of CO_2 gas, and great care should be taken in hastening this process. At no time should the flask be stoppered. Any warming which is done also causes the ether to evaporate, and care should be taken to replace the ether before going to the next step.

zoic acid) may be recrystallized from water if needed. After drying the product, determine the melting point and yield; the yield should be about 6 g.

EXERCISES

1. Comment on the use of steam distillation as a possible alternative procedure for the purification of the crude triphenylmethanol. Consider the possible unchanged starting materials and the products which are formed, and indicate which of these would steam distil and which would not. Explain in detail how this method of purification would yield pure triphenylmethanol (if it would at all).

2. Ethyl alcohol quite often is present in technical grade ether. If this grade were used, what effect, if any, would the ethyl alcohol have on the formation of the Grignard reagent? Explain.

3. Give a plausible structure for $C_6H_5MgBr \cdot 2(C_2H_5)_2O$. How do you think the ether molecules are bound to C_6H_5MgBr?

4. Arrange the following compounds in increasing order of reactivity towards a Grignard reagent: Methyl benzoate, benzoic acid, benzaldehyde, acetophenone, benzoyl chloride. Explain the basis for your decision, making use of mechanisms where needed. What are the product(s) of each of the above carbonyl-containing compounds with *excess* phenylmagnesium bromide?

5. How might primary, secondary, and tertiary alcohols be prepared from an aldehyde or a ketone and an organometallic reagent? Suggest suitable carbonyl-containing compounds, as well as suitable organometallic reagents. Write chemical reactions for these preparations, and indicate stoichiometry where important.

SPECTRA OF STARTING MATERIALS AND PRODUCTS

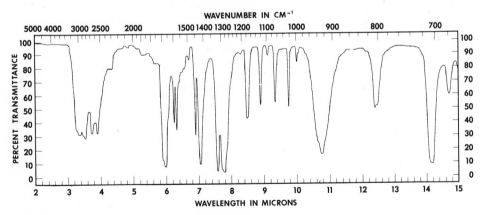

FIGURE 13.1   IR spectrum of benzoic acid.

12.8δ

FIGURE 13.2 NMR spectrum of benzoic acid.

FIGURE 13.3 IR spectrum of methyl benzoate.

FIGURE 13.4 NMR spectrum of methyl benzoate.

FIGURE 13.5 IR spectrum of bromobenzene.

FIGURE 13.6 NMR spectrum of triphenylmethanol.

13.2 *The Organolithium Reagent*

Organolithium reagents play an important role in organic synthesis, and because of their increasingly extensive use, it is desirable to discuss briefly the essential differences in the preparation and reactions of these from those of the Grignard reagent. These reagents both give similar types of reactions, but owing to difficulties of conducting experiments involving highly reactive lithium metal in the undergraduate organic laboratory, no experimental procedures for the preparation and reactions of an organolithium reagent are included here.

The preparation of the lithium reagent, RLi, is indicated in equation (13), where RX represents an alkyl or aryl halide. The solvent does not appear to play as important a

$$RX + 2\,Li \xrightarrow[\substack{\text{or} \\ \text{hydrocarbon solvent}}]{\text{dry ether solvent}} RLi + LiX \tag{13}$$

role in the preparation of organolithium reagents as it does in the preparation of Grignard reagents. This has been shown by the fact that it is possible to prepare organolithium reagents in both hydrocarbon solvents (such as pentane and hexane) and ether solvents. However, there is a difference in reactivity of the reagents toward the solvent. Ethereal solutions of the Grignard reagent may be stored under anhydrous conditions for long periods of time, but organolithium reagents may be kept only when a hydrocarbon solvent is used. The latter cleave ethers, even though one can prepare the reagent in ether solution. If this is done, the reagent must be used immediately after its preparation and may not be stored.

It should be noted that two moles of lithium are required for each mole of organic halide, in contrast to the one to one mole ratio of magnesium to halide. Pure lithium metal can be used effectively in preparing the lithium reagent, although it has been found that the use of lithium containing 0.8% by weight of sodium metal often increases the ease with which the reaction starts.

The reactivity of the alkyl and aryl halides toward lithium metal is the same as that for the Grignard reagent. The major exception is the necessity of using tetrahydrofuran as the solvent for the preparation of the Grignard reagent from aryl chlorides. Most alkyl halides form lithium reagents by reaction with lithium metal in either ethyl ether or in a hydrocarbon solvent.

The side reactions arising from the preparation of the lithium reagent are the same as those which are found for the Grignard reagents [see equations (3), (4), and (5)]. The same conditions of extreme dryness must be utilized, and the side reactions may be minimized by use of an inert atmosphere and dilute solutions.

The lithium reagents usually react in a manner similar to Grignard reagents. In some instances, lithium reagents are preferable due to the fact that they are generally more reactive than their halomagnesium counterparts. Furthermore, once the lithium reagent has reacted with a substrate, the resulting salt which forms is generally more soluble in the solvent than the corresponding salt resulting from the Grignard reagent.

An advantage of the greater reactivity of the lithium reagent is realized in *metalation reactions*, wherein the lithium reagent, RLi, abstracts a proton from a hydrocarbon, R′H, as shown in equation (14). This reaction has been used to determine the acidity of

$$RLi + R'H \rightarrow RH + R'Li \tag{14}$$

hydrocarbons and can also be used to introduce substituents into various aromatic compounds. *n*-Butyllithium [equation (15)] has been used extensively for this purpose, and some typical examples of metalation reactions are shown in equations (16)–(18).

$$n\text{-}C_4H_9Br + 2\,Li \xrightarrow[-10°\ 85\%\ \text{yield}]{\text{ether}} n\text{-}C_4H_9Li + LiBr \tag{15}$$

(16)

(17)

$$\text{(pyridine-CH}_3\text{)} + n\text{-C}_4\text{H}_9\text{Li} \xrightarrow{-\text{C}_4\text{H}_{10}} \text{(pyridine-CH}_2\text{Li)} \xrightarrow[-\text{LiBr}]{\text{Br--Br}} \text{(pyridine-CH}_2\text{Br)} \qquad (18)$$

14

Oxidation reactions of alcohols and carbonyl compounds

14.1 Preparation of Aldehydes and Ketones by Oxidation of Alcohols

The considerable utility of aldehydes and ketones in organic synthesis makes their efficient preparation of great importance. These substances can be synthesized from alkynes by acid-catalyzed hydration [equation (1)] or by hydroboration followed by

$$R-C\equiv C-R(H) \xrightarrow[\substack{H_2SO_4 \\ H_2O}]{HgSO_4} R-\overset{\displaystyle O}{\overset{\|}{C}}-CH_2-R(H) \tag{1}$$

oxidation [equation (2)], and from carboxylic acids or their derivatives by reaction with organometallics or a variety of reducing agents [equations (3)–(6)]. One of the most common synthetic schemes, however, is the oxidation of primary and secondary

$$R-C\equiv C-R(H) \xrightarrow[\substack{2)\ H_2O_2/HO^{\ominus}}]{1)\ B_2H_6} R-CH_2-\overset{\displaystyle O}{\overset{\|}{C}}-R(H) \tag{2}$$

$$R-\overset{\displaystyle O}{\overset{\|}{C}}-OH \xrightarrow[\substack{2)\ H_3O^{\oplus}}]{1)\ R'Li} R-\overset{\displaystyle O}{\overset{\|}{C}}-R' \tag{3}$$

$$R-\overset{\displaystyle O}{\overset{\|}{C}}-Cl + R'_2Cd \longrightarrow R-\overset{\displaystyle O}{\overset{\|}{C}}-R' \tag{4}$$

$$R-C\equiv N \xrightarrow[\substack{2)\ H_3O^{\oplus}}]{1)\ R'MgX} R-\overset{\displaystyle O}{\overset{\|}{C}}-R' \tag{5}$$

$$R-C\equiv N \xrightarrow[\substack{2)\ H_3O^{\oplus}}]{1)\ LiAlH_4/-80°} R-\overset{\displaystyle O}{\overset{\|}{C}}-H \tag{6}$$

214

alcohols with chromic acid, H_2CrO_4, or with potassium permanganate [equation (7)]. A description of the use of the former oxidizing agent for conversion of alcohols to aldehydes and ketones follows; procedures involving the use of both reagents are included in the experimental section.

$$\underset{\overset{|}{R-CH-R'(H)}}{OH} \xrightarrow[\text{or KMnO}_4]{H_2CrO_4} \underset{\overset{\parallel}{R-C-R'(H)}}{O} \tag{7}$$

Chromic acid is not stable for long periods of time and is therefore produced when required by reaction of sodium or potassium dichromate with an excess of an acid such as sulfuric or acetic [equation (8)] or by dissolution of chromic anhydride in water [equation (9)]. In the latter preparation, either sulfuric or acetic acid is also added since

$$Na_2Cr_2O_7 + 2\ H_2SO_4 \longrightarrow [H_2Cr_2O_7] \xrightarrow{H_2O} 2\ H_2CrO_4 + 2\ NaHSO_4 \tag{8}$$

$$CrO_3 + H_2O \longrightarrow H_2CrO_4 \tag{9}$$

the rate of oxidation of alcohols by chromic acid is much greater in acidic solutions. For preparation or oxidation of substances which would decompose under strongly acidic conditions, either chromic anhydride dissolved in pyridine or basic potassium permanganate can be used as the oxidizing agent.

The stoichiometry of the chromic acid oxidation of alcohols is such that two equivalents of acid oxidize three equivalents of an alcohol to the corresponding carbonyl compound [equation (10)]. A consideration of the mechanism of the oxidation will illustrate some of the individual steps which result in the overall stoichiometry given in equation (10).

$$\underset{\overset{|}{3\ RCHR'(H)}}{OH} + 2\ H_2CrO_4 + 3\ H_2SO_4 \longrightarrow \underset{\overset{\parallel}{3\ RCR'(H)}}{O} + Cr_2(SO_4)_3 + 8\ H_2O \tag{10}$$

In the presence of chromic acid, alcohols form esters, just as they do when allowed to react with carboxylic acids (*cf.* esterification, Chapter 16.2). Subsequent decomposition of the chromate ester, **1**, yields the carbonyl compound [equation (11)].

$$\underset{R}{\overset{R'(H)}{>}}CH-OH + HO\overset{\overset{O}{\parallel}}{\underset{\parallel}{O}}CrOH \xrightarrow{-H_2O} \underset{R}{\overset{R'(H)}{>}}CH-O\overset{\overset{O}{\parallel}}{\underset{\parallel}{O}}CrOH \longrightarrow \underset{R}{\overset{R'(H)}{>}}C=O + H_2CrO_3 \tag{11}$$

1

During the course of the oxidation of the alcohol, chromium undergoes reduction from the +6 to the unstable +4 valence state. A disproportionation rapidly occurs between chromium (VI) and chromium (IV) to produce chromium (V), a new oxidizing agent which we shall write as HCr^VO_3 [equation (12)]. The two moles of chromium (V) will oxidize two more moles of alcohol to the carbonyl compound, and chromium (III), a stable oxidation state of this metal, will be generated [equation (13)]. The sum of equations (11)–(13) corresponds to equation (10).

$$H_2Cr^{IV}O_3 + H_2Cr^{VI}O_4 \longrightarrow 2\ HCr^{V}O_3 + H_2O \tag{12}$$

$$2\ HCrO_3 + 2\ R{-}\overset{\overset{\displaystyle R'(H)}{|}}{C}HOH + 3\ H_2SO_4 \longrightarrow 2\ \overset{\displaystyle R'(H)}{\underset{\displaystyle R}{C}}{=}O + Cr_2(SO_4)_3 + 6\ H_2O \tag{13}$$

There are some important side reactions which complicate the oxidation of a primary alcohol to an aldehyde. Perhaps the most important of these is the ready oxidation of aldehydes to carboxylic acids by chromic acid [equation (14)]. This further, undesired, oxidation can be minimized by adding the chromic acid *to* the primary alcohol, so that an excess of the oxidizing agent is *not* present in the reaction mixture, and by distilling the aldehyde from the reaction mixture as it is formed. As a consequence, the desired aldehyde must be quite volatile, *i.e.*, boiling point less than about 150°, if it is to be prepared in high yield by chromic acid oxidation.

$$3\ R{-}\overset{\overset{\displaystyle O}{\|}}{C}{-}H + 2\ H_2CrO_4 + 3\ H_2SO_4 \longrightarrow 3\ R{-}\overset{\overset{\displaystyle O}{\|}}{C}{-}OH + Cr_2(SO_4)_3 + 5\ H_2O \tag{14}$$

Even in cases in which the aldehyde is volatile, a poor yield of product is sometimes obtained. This is due to the facile formation from the aldehyde of the hemiacetal, **2**, which is subsequently oxidized to the ester of a carboxylic acid [equation (15)]. Occasionally, this "side" reaction is turned into a useful synthesis of the ester as is the case in the preparation of *n*-butyl butyrate by the chromic acid oxidation of 1-butanol.

$$RCH_2OH \xrightarrow{H_2CrO_4} R{-}\overset{\overset{\displaystyle O}{\|}}{C}{-}H \xrightarrow[RCH_2OH]{H^\oplus} R{-}\overset{\overset{\displaystyle OH}{|}}{\underset{\underset{\displaystyle \mathbf{2}}{\displaystyle OCH_2R}}{C}}{-}H \xrightarrow{H_2CrO_4} R{-}\overset{\overset{\displaystyle O}{\|}}{C}{-}OCH_2R \tag{15}$$

Ketones are much more stable toward oxidizing agents in mildly acidic media than are aldehydes, so that the side reactions mentioned in the oxidation of primary alcohols do not occur to a significant extent in the converson of secondary alcohols to ketones. However, under alkaline or *strongly* acidic conditions, enolizable ketones will undergo oxidation with cleavage to give two carbonyl fragments. For example, cyclohexanone, which can be obtained in good yield by the chromic acid oxidation of cyclohexanol, can be converted to adipic acid by treatment with potassium permanganate under mild alkaline conditions [equation (16)]. The reaction undoubtedly requires initial conversion of the ketone to the enolate ion, **3**, which is then oxidized by the permanganate [equation (17)].

$$\tag{16}$$

$$\tag{17a}$$

$$4 \xrightarrow{\text{KMnO}_4} \quad \xrightarrow{\text{H}_3\text{O}^{\oplus}} \quad (17b)$$

Cyclohexanone is a symmetrical ketone and can give only the single enolate ion, **3**. If a ketone is not symmetrical, it can produce two different enolate ions, each of which will be oxidized by permanganate to different products [equation (18)]. The formation of complex mixtures of products when unsymmetrical ketones are oxidized is a complication which detracts from the synthetic utility of this reaction.

$$\text{RCH}_2\!-\!\overset{\text{O}}{\underset{\|}{\text{C}}}\!-\!\text{CH}_2\text{R}' \underset{}{\overset{\text{HO}^{\ominus}}{\rightleftharpoons}} \text{RCH}\!=\!\overset{\text{O}^{\ominus}}{\underset{}{\text{C}}}\!-\!\text{CH}_2\text{R}' + \text{R}\!-\!\text{CH}_2\!-\!\overset{\text{O}^{\ominus}}{\underset{}{\text{C}}}\!=\!\text{CHR}' \quad (18)$$

EXPERIMENTAL PROCEDURE

A. PREPARATION OF 2-METHYLPROPANAL

Prepare a solution of chromic acid by dissolving 0.20 mole of sodium dichromate dihydrate in 290 ml of water and then *slowly* adding with swirling 77 g of concentrated sulfuric acid. While this solution is cooling to room temperature, fit a 500-ml round-bottomed flask with a Claisen adapter equipped with a dropping funnel and a Hempel column bearing a still head with a thermometer. The still head. is connected to a water-cooled condenser which leads to a 100-ml graduated cylinder. Place 0.60 mole of isobutyl alcohol and 45 ml of water in the flask along with a boiling stone and heat the mixture with a Bunsen burner until gentle boiling begins. Remove the burner and immediately begin to add the red-orange solution of chromic acid from the dropping funnel to the hot mixture of alcohol and water, at a rate such that all of the acid will have been added in *ca.* 15–25 min. This rate of addition will cause the reaction mixture to boil vigorously, and a mixture of alcohol, aldehyde and water will steam distil, giving a head temperature of 80–85°. This temperature should be maintained by appropriate adjustment of the rate of addition of acid. After all the chromic acid has been added, replace the Bunsen burner and distil the dark green [chromium (III)] reaction mixture for an additional 15 min.★ Note the amount of water contained in the two-phase distillate, transfer the mixture to a separatory funnel, and add 1 g of sodium carbonate. Shake the funnel thoroughly, with venting to relieve any pressure, and then saturate the aqueous layer by adding about 0.2 g of sodium chloride for each milliliter of water in order to salt out any dissolved product. Shake the funnel again to effect solution of the salt, remove the organic layer and dry it over anhydrous sodium sulfate.★ Filter the crude organic product into a 100-ml distilling flask and isolate the 2-methyl-propanal, bp 63–66°, by fractional distillation. If the fractionation is continued after the aldehyde distils, unchanged alcohol, bp 107–108°, and some isobutyl isobutyrate, bp 148–149°, can be isolated.

Prepare the 2,4-dinitrophenylhydrazone of 2-methylpropanal, mp 186–188°, by the procedure given in Chapter 15.1. Apply the chromic acid in acetone test described in Section 14.2 to the 2-methylpropanal.

The preparation of the sodium bisulfite addition product of the aldehyde can be accomplished by adding 6 ml of saturated sodium bisulfite solution to 1 ml of the aldehyde contained in a 125-ml Erlenmeyer flask. Swirl the flask and then allow the mixture to stand for 10 min. Reaction will be indicated by warming of the solution. Add 25 ml of 95% ethanol, swirl well and then cool the solution in an ice-salt bath. Collect the sodium bisulfite addition product by suction filtration, washing the filter cake once with ethanol and once with ether. You should *not* attempt to determine the melting point of the solid. The 2-methylpropanal can be regenerated by adding 5–10 ml of either 10% sodium carbonate solution or dilute hydrochloric acid to the addition compound and gently warming the mixture.

Although sodium bisulfite addition compounds are not, in general, good derivatives for characterizing carbonyl compounds, they often are used to purify aldehydes, methyl ketones, and cyclic ketones such as cyclohexanone and cyclopentanone.

B. Preparation of Cyclohexanone

Prepare a solution of chromic acid by dissolving 0.081 mole of sodium dichromate dihydrate in 125 ml of water contained in a 250-ml Erlenmeyer flask and *carefully* adding, with swirling, 35 g of concentrated sulfuric acid. Cool the resulting orange-red solution of chromic acid to room temperature and then, in one portion, *add it to* a mixture of 0.20 mole of cyclohexanol and 75 ml of water in a 500-ml Erlenmeyer flask. Throughly mix the solutions by swirling and determine the temperature of the reaction mixture. The mixture should quickly become warm; when it reaches a temperature of 55°, cool the flask in an ice-water bath or a pan of cold water so that a temperature of 55–60° is maintained. When the temperature of the solution no longer exceeds 60° upon removal of the flask from the cooling bath, allow the flask to stand for one hour with occasional swirling.★

Transfer the reaction mixture to a 500-ml round-bottomed flask, add 100 ml of water and a boiling stone, and attach the flask to an apparatus set for fractional distillation. Distil the mixture until approximately 100 ml of distillate, which consists of an aqueous and an organic layer, has been obtained.★ Place the distillate in a separatory funnel, saturate the aqueous layer by adding sodium chloride (about 0.2 g of salt per ml of water will be required) and swirling to effect solution. Separate the layers and extract the aqueous layer with 15 ml of either dichloromethane or technical ether. Combine this extract with the organic layer and dry the solution over anhydrous magnesium sulfate.★ Filter the solution into a 50- or 100-ml flask and attach the flask to an apparatus set for simple distillation. Remove the flammable ether by heating with a water bath and then distil the remaining cyclohexanone with the aid of a Bunsen burner. If dichloromethane was used, the burner can be used throughout the distillation. Cyclohexanone is a colorless liquid, bp 152–155°.

Prepare the 2,4-dinitrophenylhydrazone of cyclohexanone, mp 162–163°, by the procedure outlined in Chapter 15.1. Also apply the chromic acid in acetone test described in Section 14.2 to cyclohexanone.

C. Oxidation of Cyclohexanone to Adipic Acid

Combine 0.10 mole of cyclohexanone and a solution of 0.20 mole of potassium permanganate in 250 ml of water in a 500-ml Erlenmeyer flask. To this mixture add 2 ml of 10% sodium hydroxide solution and note the temperature of the reaction mixture. The mixture will warm to 45° and should be held at that temperature by intermittent cooling with an ice- or cold-water bath. When the temperature no longer rises above 45° upon removal of the flask from the cooling bath, allow the flask to stand for an additional 5–10 min (a drop in the temperature should occur), and then complete the reaction by heating the mixture at gentle reflux for a few minutes with a Bunsen burner. Test for the presence of permanganate by placing a drop of the reaction mixture on a piece of filter paper; any unchanged permanganate will appear as a purple ring around the brown manganese dioxide. If permanganate remains, add *small* portions of *solid* sodium bisulfite to the reaction mixture until the spot test is negative.★ Filter the mixture by suction through a pad of filter-aid,[1] thoroughly wash the brown filter cake with water ★ and then concentrate the filtrate to a volume of about 65 ml by heating over a flame.★ If the concentrate is colored, add decolorizing carbon, reheat the solution to boiling for a few minutes and then filter. Carefully add concentrated hydrochloric acid to the filtrate until the solution tests acid to litmus paper and then add an additional 15 ml of acid.★ Allow the solution to cool to room temperature and isolate the precipitated adipic acid by suction filtration. Adipic acid is a colorless solid, mp 152–153°, and can be recrystallized from ethanol–water if necessary.

EXERCISES

1. What is the purpose of washing the steam distillate with sodium carbonate in the preparation of 2-methylpropanal?
2. Write the structure of the sodium bisulfite addition product of 2-methylpropanal.
3. Why is the chromic acid solution added *to* the alcohol, rather than the reverse, when the preparation of an aldehyde is being attempted?
4. Why are the two-phase distillates obtained in the oxidation of isobutyl alcohol and of cyclohexanol first saturated with sodium chloride before separation of the organic layer?
5. Point out what modifications might be employed in the oxidation of isobutyl alcohol with chromic acid in order to *maximize* the yield of isobutyl isobutyrate.
6. Write out the balanced equation for the oxidation of cyclohexanol to cyclohexanone by sodium dichromate and aqueous sulfuric acid.
7. The costs of sodium dichromate dihydrate and potassium dichromate are $1.28/lb and $1.20/lb, respectively. Determine which reagent would be more economical to use for oxidation of one mole of cyclohexanol to cyclohexanone.
8. Why is the reaction mixture made alkaline in the oxidation of cyclohexanone to adipic acid?
9. Write the balanced equation for the oxidation of cyclohexanone to adipic acid.
10. Give the products to be expected on oxidation of 2-methylcyclohexanone with alkaline permanganate.

REFERENCES

Oxidation in Organic Chemistry, K. B. Wiberg, editor, Academic Press, New York, 1965, Vol. 5-A, Chapters 1 and 2.

SPECTRA OF STARTING MATERIALS AND PRODUCTS

[1] The pad can be prepared by first making a slurry of 0.5–1 g of filter-aid in a few milliliters of water. The slurry is poured onto a filter paper contained in a Büchner funnel attached to a suction flask; suction is then slowly applied to remove the water from the slurry and leave a thin, even pad of filter-aid.

FIGURE 14.1 IR spectrum of isobutyl alcohol.

FIGURE 14.2 NMR spectrum of isobutyl alcohol.

FIGURE 14.3 IR spectrum of 2-methylpropanal.

FIGURE 14.4 NMR spectrum of 2-methylpropanal.

FIGURE 14.5 IR spectrum of adipic acid.

14.2 Classification of Alcohols and Carbonyl Compounds by Means of Chromic Acid

The greater ease of oxidation of aldehydes than of ketones (Section 14.1) is the basis of several classical qualitative tests. Fehling's and Benedict's reagents contain complex salts (tartrate and citrate, respectively) of cupric ion as oxidizing agents, and Tollens' reagent consists of a silver diammine salt. The first two have been used as tests for aliphatic aldehydes (although they are more often used to detect reducing sugars), and the Tollens' test has been widely employed to distinguish between aldehydes (both aliphatic and aromatic) and ketones.

Tertiary alcohols are much more resistant toward oxidation than primary and secondary alcohols. Potassium permanganate and certain other oxidizing agents have been used to distinguish tertiary alcohols from those of the other two types.

There are some disadvantages to all of these tests. A newer reagent which gives quick and definitive results is chromic acid, prepared by dissolving chromic anhydride in sulfuric acid (Section 14.1), and used in acetone solution. This reagent oxidizes primary and secondary alcohols and all aldehydes with a distinctive color change, and it gives no visible reaction with tertiary alcohols and ketones under the conditions of the test.

$$RCH_2OH \xrightarrow{H_2CrO_4} RCHO \xrightarrow{H_2CrO_4} RCO_2H$$
$$R_2CHOH \xrightarrow{H_2CrO_4} R_2CO \xrightarrow{H_2CrO_4}\!\!\!\!/\!/ \text{ no visible reaction}$$
$$R_3COH \xrightarrow{H_2CrO_4}\!\!\!\!/\!/ \text{ no visible reaction}$$

Primary and secondary alcohols and aliphatic aldehydes give a positive test within five seconds. Aromatic aldehydes require 30 seconds to one minute. Tertiary alcohols and ketones produce no visible change in several minutes. Phenols and aromatic amines give dark precipitates, as do aromatic aldehydes having hydroxyl or amino groups on the aromatic ring.

Thus, the chromic acid reagent gives a clear-cut distinction between primary and secondary alcohols and aldehydes, on the one hand, and tertiary alcohols and ketones, on the other. Aldehydes may be distinguished from primary and secondary alcohols by means of Tollens', Benedict's, or Fehling's tests, and lower-molecular-weight primary and secondary alcohols may be differentiated on the basis of their rates of reaction with concentrated hydrochloric acid containing zinc chloride—the Lucas reagent (Chapter 20, Section A.2 under "Alcohols").

EXPERIMENTAL PROCEDURE

If the chromic acid reagent is not supplied to you, prepare it as follows: Add 25 g of chromic anhydride (CrO_3) to 25 ml of concentrated sulfuric acid and stir until a smooth paste is obtained. Dilute the paste (*cautiously*) with 75 ml of distilled water and stir until a clear orange solution is obtained.

Place 1 ml of acetone (reagent grade, or solvent grade which has been distilled from potassium permanganate) in a test tube and dissolve one drop of a liquid or *ca.* 10 mg of a solid alcohol or carbonyl compound in it. Add one drop of the acidic chromic anhydride reagent to the acetone solution and shake the tube to mix the contents. A positive oxidation reaction is indicated by disappearance of the orange color of the reagent and the formation of a green or blue-green precipitate or emulsion.

Test a number of compounds of each type chosen from the following classes of compounds: primary alcohols, secondary alcohols, tertiary alcohols, aldehydes, ketones.

EXERCISES

1. Write the formula of the expected oxidation product from each alcohol or aldehyde tested.
2. What is the green precipitate formed in a positive test?
3. What structural feature do tertiary alcohols and ketones have in common, and what is the relationship of this feature to their resistance toward oxidation?
4. Why can the Lucas reagent *not* be used to distinguish between 1-octanol and 2-octanol? (Refer to Chapter 20, Section A.2 under "Alcohols".)

REFERENCES

1. F. G. Bordwell and K. M. Wellman, *Journal of Chemical Education*, **39**, 308 (1962).
2. J. D. Morrison, *ibid.*, **42**, 554 (1965).

14.3 Base-Catalyzed Oxidation-Reduction of Aldehydes; The Cannizzaro Reaction

Aldehydes which have no hydrogen atoms on the carbon atom adjacent to the carbonyl group (the α-carbon atom) undergo mutual oxidation and reduction in the presence of strong alkali. Those which *have* hydrogens on the α-carbon atom undergo other reactions preferentially, as described in Chapter 15.2. Since aldehydes are intermediate in oxidation state between alcohols and carboxylic acids, it is not too surprising to find that this reaction, called the Cannizzaro reaction, takes place. The mechanism, which has been supported by considerable evidence, follows quite logically in view of the well-known ease of addition of nucleophiles to the carbonyl group. The first step explains the function of the strong basic catalyst:

$$
R-C\overset{O}{\underset{H}{\big\langle}} + HO^{\ominus} \longrightarrow R-\underset{H}{\overset{O^{\ominus}}{\underset{|}{C}}}-OH \tag{19}
$$

$$
R-\underset{H}{\overset{O^{\ominus}}{\underset{|}{C}}}-OH + R-C\overset{O}{\underset{H}{\big\langle}} \longrightarrow R-C\overset{O}{\underset{OH}{\big\langle}} + R-\underset{H}{\overset{O^{\ominus}}{\underset{|}{C}}}-H \tag{20}
$$

$$
R-C\overset{O}{\underset{OH}{\big\langle}} + HO^{\ominus} \longrightarrow R-C\overset{O}{\underset{O^{\ominus}}{\big\langle}} + HOH \tag{21}
$$

$$
R-\underset{H}{\overset{O^{\ominus}}{\underset{|}{C}}}-H + HOH \longrightarrow R-\underset{H}{\overset{OH}{\underset{|}{C}}}-H + HO^{\ominus} \tag{22}
$$

Summation of equations (19) through (22), gives the equation for the overall reaction:

$$
2\,RCHO + HO^{\ominus} \longrightarrow RCO_2^{\ominus} + RCH_2OH \tag{23}
$$

Aromatic aldehydes are the most common type which undergo the Cannizzaro reaction, but formaldehyde and α, α, α-trisubstituted acetaldehydes also react in this way. Reaction between two *different* aldehydes of these types may occur, producing the acids and alcohols corresponding to each of the two aldehydes. Such reactions are referred to as *crossed-Cannizzaro reactions*.

EXPERIMENTAL PROCEDURE

Dissolve 2 g of solid potassium hydroxide in 2 ml of distilled water by swirling in a 100-ml beaker; cool the mixture to room temperature in a water bath. Put 2 ml of benzaldehyde in an 18 × 150-mm test tube and add the concentrated potassium hydroxide solution to it. Cork the tube securely and shake the mixture vigorously until an emulsion is formed. Allow the stoppered tube to stand in your desk until the next laboratory period.

Add about 1 ml of water to the mixture, stopper the tube and shake it. If all of the crystals obtained do not dissolve, add a little water, break up the solid mass with a glass rod, stopper, and shake again. Repeat this procedure until all the solid is in solution. Pour the solution into a separatory funnel and extract it three times with 10-ml portions of ether (shake the mixture *gently* to avoid forming an emulsion).

The ether solution may be dried over anhydrous magnesium sulfate and examined by gas chromatography [2] to determine the proportion of benzyl alcohol and unchanged benzaldehyde present. If the reaction is carried out on a larger scale, the ether solution may be shaken with aqueous sodium bisulfite solution to remove the benzaldehyde, the ether solution washed, dried, and distilled to yield benzyl alcohol as a high-boiling fraction, bp 205°. Alternatively, a crystalline derivative of benzyl alcohol may be prepared. The phenyl urethan (mp 78°) or 3,5-dinitrobenzoate (mp 113°) are suitable (see Chapter 20, Sections B.1 and B.2 under "Alcohols").

Following its extraction with ether, pour the alkaline aqueous solution into a mixture of 5 ml of concentrated hydrochloric acid, 4 ml of water and *ca.* 5 g of crushed ice, with vigorous stirring. Cool the resultant mixture by placing its container in an ice bath and then collect the crystals of benzoic acid.★ Dry the crystals and determine their melting point. Save some of this product for comparison with a product of an experiment in Chapter 15, p. 246.

EXERCISES

1. When the Cannizzaro reaction is carried out in D_2O solution, no deuterium becomes attached to carbon in the alcohol produced. How does this support the mechanism given above?

2. By what means would aqueous sodium bisulfite remove unchanged benzaldehyde from the reaction mixture?

3. Write an equation for the reaction of benzaldehyde and formaldehyde with concentrated potassium hydroxide solution. Show all products.

4. How would propionaldehyde react with potassium hydroxide solution under the conditions of this experiment? Answer the same question for the reaction of trimethylacetaldehyde.

---

[2] A 5-ft column with 5% silicone gum rubber as the stationary phase is satisfactory. In one experiment, about 20% of the original benzaldehyde was found present after 24 hr at room temperature.

SPECTRA OF STARTING MATERIAL AND PRODUCT

The ir and nmr spectra of benzoic acid are given in Figures 13.1 and 13.2.

FIGURE 14.6 IR spectrum of benzaldehyde.

FIGURE 14.7 NMR spectrum of benzaldehyde.

FIGURE 14.8 IR spectrum of benzyl alcohol.

Some typical reactions of carbonyl compounds

15.1 Reactions Involving Nucleophilic Addition to the Carbonyl Group

The addition of a nucleophilic reagent (Nu—E) to a carbonyl compound may be written in the general form

$$
\begin{array}{c}
\text{R} \\
\diagdown \\
\quad\text{C}=\text{O} + \text{Nu}^{\ominus} \longrightarrow \\
\diagup \\
\text{R}
\end{array}
\qquad
\begin{array}{cc}
\text{R} & \text{O}^{\ominus} \\
\diagdown & \diagup \\
& \text{C} \\
\diagup & \diagdown \\
\text{R} & \text{Nu}
\end{array}
\xrightarrow{\text{E}^{\oplus}}
\begin{array}{cc}
\text{R} & \text{O—E} \\
\diagdown & \diagup \\
& \text{C} \\
\diagup & \diagdown \\
\text{R} & \text{Nu}
\end{array}
\qquad (1)
$$

where Nu is a nucleophile and E is an electrophile. One of the most important examples of this reaction is the addition of Grignard reagents in which Nu is an alkyl group and E is MgX. In Chapter 13.1 experiments are described in which phenylmagnesium bromide adds to carbon dioxide and methyl benzoate; the reactions of Grignard reagents with aldehydes, ketones, and various other carbonyl compounds also provide extremely valuable synthetic procedures.

Addition of derivatives of ammonia. Nitrogen compounds that may be considered to be derivatives of ammonia constitute other important types of nucleophilic reagents that add to carbonyl compounds [equation (2)]. In the case of *primary amines* (R =

$$
(2)
$$

3   2

alkyl or aryl) the addition product, **2**, readily loses a molecule of water to produce an *imine*, **3**. Although aliphatic amines give imines that are unstable and polymerize, aromatic amines give stable imines, also called Schiff bases, that can be isolated. These imines are readily hydrolyzed to regenerate the carbonyl compound; that is, the reaction of equation (2) is reversible.

Imines may undergo addition of a mole of hydrogen in the presence of a nickel catalyst to produce secondary amines [equation (3)] in a process which is analogous to hydrogenation of alkenes to alkanes and of carbonyl compounds to alcohols. It is

$$\underset{\textbf{3}}{\overset{C_6H_5}{\underset{H}{\diagdown}}C=N\overset{}{\underset{R}{\diagup}}} \; + \; H_2 \; \xrightarrow{\text{Ni}} \; \underset{\textbf{4}}{C_6H_5-\overset{\overset{\displaystyle H}{|}}{C}H-\overset{\overset{\displaystyle H}{|}}{N}-R} \qquad (3)$$

not necessary to isolate the imine, and, in fact, reaction mixtures from an aldehyde or ketone and an amine (either aromatic or aliphatic) or even ammonia may be subjected directly to catalytic hydrogenation to produce primary or secondary amines in good yields. Presumably imines are transitory intermediates in these reactions.

One of the limitations to this procedure for the synthesis of amines is that any other functional group in the imine that is sensitive to catalytic hydrogenation, such as C=C, N=O (as in NO_2), or C—X (X = Cl, Br), may be reduced. This limitation has been removed by the discovery of the alkali metal hydrides, which are more selective reducing agents. For example, N-cinnamyl-*m*-nitroaniline (**9**) may be synthesized starting with cinnamaldehyde (**5**) and *m*-nitroaniline (**6**) using sodium borohydride as the reducing agent. Thus the carbon-nitrogen double bond of the intermediate imine (**7**) is reduced, but neither the carbon-carbon double bond nor the nitro group is affected [equations (4)–(6)].

$$\underset{\textbf{5}}{C_6H_5CH=CH-\overset{\overset{\displaystyle O}{\|}}{C}\diagdown_H} + \underset{\textbf{6}}{H_2N-\hspace{-0.3em}\underset{NO_2}{\bigcirc}} \rightleftharpoons \underset{\textbf{7}}{C_6H_5CH=CH-C\diagdown_H^{\diagup N}} + H_2O \qquad (4)$$

$$\textbf{7} \text{ (4 moles)} + Na^{\oplus}BH_4^{\ominus} \longrightarrow \left(C_6H_5CH=CH-CH_2-N\underset{\underset{NO_2}{\bigcirc}}{\diagup}\right)_4 B^{\ominus}Na^{\oplus} \qquad (5)$$

$$\textbf{8}$$

$$\textbf{8} + 3\,H_2O \longrightarrow 4\,\underset{\textbf{9}}{C_6H_5CH=CH-CH_2-\underset{\underset{NO_2}{\bigcirc}}{N}H} + NaH_2BO_3 \qquad (6)$$

The reaction of sodium borohydride with the imine [equation (5)] is analogous to the addition of a Grignard reagent to a carbonyl compound. A hydride ion ($H:^-$) is transferred from the borohydride anion (BH_4^-) to the electrophilic carbon of the carbon-nitrogen double bond, and the electron-deficient boron adds to nitrogen. All four hydrogens of the borohydride anion are transferred to imine carbons in this way, producing and organoboron anion, **8**, which is subsequently decomposed with water to yield the secondary amine[**9**, equation (6)].

Although the formation of the imine is a reversible reaction [equation (4)], it can be brought effectively to completion by removing the water from the reaction mixture as it is produced.[1] A convenient way to do this is to heat the carbonyl compound and primary amine in benzene solution, allowing the benzene to distil continuously. The benzene forms a minimum-boiling azeotrope with the water and removes it as it is formed.

Sodium borohydride is conveniently used in methanol solution. Although it reacts slowly with the solvent, its rate of reaction with the imine is much faster. In small scale reactions, it is more convenient to use a sufficient excess of the borohydride to allow for its partial decomposition by reaction with the solvent than to use a less reactive solvent in which the borohydride is less soluble.

Other useful derivatives of ammonia that react with carbonyl compounds according to equation (2) are hydroxylamine ($R = OH$), semicarbazide ($R = NHCONH_2$), and various arylhydrazines ($R = NHAr$). The reaction of semicarbazide with carbonyl compounds to produce semicarbazones was presented in Chapter 10 to illustrate the principle of kinetic and equilibrium control. However, the main importance of the reaction of carbonyl compounds with semicarbazide, hydroxylamine, and the arylhydrazines lies in the fact that the products (semicarbazones, oximes, and arylhydrazones) are almost invariably crystalline solids, whereas the carbonyl compounds from which they are derived are often liquids. Arylhydrazines which are commonly employed in making crystalline derivatives of carbonyl compounds include phenylhydrazine, *p*-nitrophenylhydrazine, and 2,4-dinitrophenylhydrazine. The formation of a 2,4-dinitrophenylhydrazone, **10**, is represented by equation (7).

$$R_2C{=}O + H_2NNH{-}\underset{NO_2}{\overset{NO_2}{\bigcirc}}{-}NO_2 \rightleftharpoons R_2C\overset{OH}{\underset{NH}{<}} \rightleftharpoons R_2C{=}N{<}_{NH} + H_2O \qquad (7)$$

10

As the oximes, semicarbazones, and arylhydrazones are generally very insoluble in the reaction medium, they precipitate from solution, and the reversibility of the reaction therefore does not prevent their isolation in high yields.[2]

Directions for the preparation of semicarbazones, oximes, and 2,4-dinitrophenyl-

---

[1] Compare with the similar problem encountered in esterification reactions, which are also equilibrium processes, Chapter 16.2.

[2] These reactions exhibit an interesting and important dependence of rate on the pH of the medium. Consult Chapter 10 for a discussion of this subject.

hydrazones are given in the Experimental Procedures section. The directions for semi-carbazones are more generally applicable than those given in Chapter 10.

   Addition of Wittig reagents. In 1953 G. Wittig discovered that certain organophosphorus compounds, called ylides **13**, add to carbonyl compounds to form unstable intermediates called betaines, **14**, which decompose to produce an alkene and a phosphine oxide, **15**. The highly reactive ylide is generated in the presence of the carbonyl compound by the action of a strong base, usually phenyllithium, on a phosphonium halide, **12**. The latter compound is produced most commonly by reaction of a primary or secondary organic halide with triphenylphosphine [**11**, equation (8); R may be alkyl, aryl, CN, $CO_2C_2H_5$, etc.].

$$(C_6H_5)_3P: \; + \; \underset{\underset{R_2}{|}}{\overset{\overset{R_1}{|}}{C}}H\!-\!Br \; \longrightarrow \; (C_6H_5)_3\overset{\oplus}{P}\!-\!\underset{\underset{R_2}{|}}{\overset{\overset{R_1}{|}}{C}}HBr^{\ominus} \tag{8}$$

$$\underset{\textbf{11}}{} \qquad\qquad\qquad\qquad \underset{\textbf{12}}{}$$

$$\textbf{12} \xrightarrow{C_6H_5Li} \left[(C_6H_5)_3P\!=\!C\overset{R_1}{\underset{R_2}{\diagdown}} \; \longleftrightarrow \; C_6H_5)_3\overset{\oplus}{P}\!-\!\overset{\ominus}{\underset{..}{C}}\overset{R_1}{\underset{R_2}{\diagdown}} \right] + C_6H_6 + LiBr \tag{9}$$

$$\underset{\textbf{13}}{}$$

$$\textbf{13} + O\!=\!C\overset{R_3}{\underset{R_4}{\diagdown}} \; \longrightarrow \; (C_6H_5)_3\overset{\oplus}{P}\!-\!\underset{\underset{\underset{R_4}{|}}{\overset{\ominus}{O}-\overset{|}{C}-R_3}}{\overset{\overset{R_1}{|}}{C}}\!-\!R_2 \; \longrightarrow \; (C_6H_5)_3\overset{P}{\underset{O}{\|}} + \underset{R_4\diagup \;\;\diagdown R_3}{\overset{R_1\diagdown \;\;\diagup R_2}{C \atop C}} \tag{10}$$

$$\qquad\qquad\qquad\qquad \underset{\textbf{14}}{} \qquad\qquad\qquad \underset{\textbf{15}}{}$$

   As may be judged from the polarized resonance structure of the ylide, the electron-rich carbon is nucleophilic, and it adds to a carbonyl group in the expected way to give the betaine. The decomposition of the betaine is thought to involve a four-center transition state; as the second C—C bond and the P—O bond are forming, the P—C and O—C bonds are breaking.

   The overall effect of reactions (8)–(10) is conversion of a carbon-oxygen double bond to a carbon-carbon double bond, *i.e.*, C=O → C=C. The Wittig synthesis thus represents a very general method of preparation of alkenes, one which has two important advantages over older procedures: (1) The carbonyl group is replaced *specifically* by a carbon-carbon double bond, without the formation of isomeric alkenes and (2) the reactions are carried out in basic rather than in acidic media under very mild conditions.

   Various modifications of the original Wittig synthesis have been made. One of these that is convenient for an introductory experiment (because it does not require an inert atmosphere and the use of the somewhat dangerous organolithium reagents) involves the use of triethyl phosphite (**16**) in place of triphenylphosphine. Activated organic halides (such as benzyl and allyl chlorides and α-bromo esters) will react satisfactorily

$$(C_2H_5O)_3P + C_6H_5CH_2-Cl \longrightarrow (C_2H_5O)_3\overset{\oplus}{P}-CH_2C_6H_5 \quad Cl^{\ominus} \qquad (11)$$
$$\mathbf{16} \qquad\qquad\qquad\qquad \mathbf{17}$$

$$\mathbf{17} \xrightarrow[\text{heat}]{} (C_2H_5O)_2\overset{\overset{O}{\|}}{P}-CH_2C_6H_5 + C_2H_5Cl \qquad (12)$$
$$\mathbf{18}$$

$$\mathbf{18} + CH_3O^{\ominus}Na^{\oplus} \longrightarrow (C_2H_5O)_2\overset{\overset{O}{\|}}{P}-\overset{\ominus}{C}HC_6H_5 \ Na^{\oplus} + CH_3OH \qquad (13)$$
$$\mathbf{19}$$

$$\mathbf{19} + O{=}CHC_6H_5 \longrightarrow (C_2H_5O)_2\overset{\overset{O}{\|}}{P}-\underset{\underset{\ominus O-\overset{}{C}HC_6H_5}{|}}{C}HC_6H_5 \ Na^{\oplus} \qquad (14)$$

$$(C_2H_5O)_2\overset{\overset{O}{\|}}{P}O^{\ominus}Na^{\oplus} + \overset{\overset{CHC_6H_5}{\|}}{C}HC_6H_5$$
$$\mathbf{21} \qquad\qquad \mathbf{20}$$

with esters of phosphorous acid [equation (11)]. The phosphonium salt, **17**, produced is unstable toward heat; for example, the product from benzyl chloride evolves ethyl chloride and yields the phosphonate ester, **18**. Treatment of this ester with sodium methoxide produces a nucleophile analogous to an ylide and, in the presence of benz-aldehyde, the reactions of equations (13) and (14) take place to yield an alkene, **20** (stilbene) and sodium diethylphosphate (**21**). The latter is water-soluble and hence easily separated from the alkene.

There are two geometric isomers of stilbene: The *trans* isomer is a solid at room temperature, mp 126–127°, whereas the *cis* isomer is a liquid, mp 6°. The phosphonate ester modification of the Wittig reaction is reported to give the pure *trans* isomer, even though the original Wittig procedure (using triphenylphosphine as starting material) is reported to give a mixture of 30% *cis*-, 70% *trans*-stilbene. The infrared spectra of the two isomers are distinctly different and may be used in both qualitative and quantitative analysis of mixtures.

Optional or additional experiments.   Other aldehydes may be used in place of ben-zaldehyde in the phosphonate modification of the Wittig reaction. With *trans*-cin-namaldehyde (**22**), the product is *trans,trans*-1,4-diphenyl-1,3-butadiene (**23**). If some

$$C_6H_5CH{=}CHCHO$$
$$\mathbf{22}$$

23

students prepare this diene, it will be interesting to have the *ultraviolet* absorption spectrum of this product compared with that of the stilbene product to see the effect of an additional conjugated double bond. If it is not practical to record the spectra of the products in your laboratory, the literature may be consulted (see references 3 and 4).

When 2-furaldehyde **(24)** is substituted for benzaldehyde, the product is *trans*-2-styrylfuran **(25)**.

25

EXPERIMENTAL PROCEDURE

A. ADDITION OF DERIVATIVES OF AMMONIA

1. Addition of primary amines to produce imines. Synthesis of secondary amines by sodium borohydride reduction of imines.   a. Cinnamaldehyde and m-nitroaniline. In a 100-ml round-bottomed flask, place 0.022 mole of cinnamaldehyde, 0.020 mole of *m*-nitroaniline, 25 ml of benzene and several boiling stones. Set up apparatus for simple distillation, using a graduated cylinder as a receiver, and heat the reaction mixture on a steam cone or in a boiling water bath. (If a burner is used to heat a water bath, take proper precautions against the flammability of the benzene.) Distil until almost all of the benzene has been removed; 22–24 ml of distillate should be obtained in about 25–30 min.

During the distillation, prepare a solution of 0.02 mole of sodium borohydride in 15 ml of methanol.[3] When the distillation has been completed, remove the flask from the steam cone and either pour out about 0.5 ml of the residual liquid or, preferably, insert a micropipet and take a 0.5-ml sample of the liquid. Add 3–4 ml of methanol to this sample in a test tube, swirl to dissolve it, and then place the test tube in ice. (Go to the directions of the next paragraph while the solution is cooling.) Collect any crystals which separate, dry them, and determine their melting points. The reported melting point of the imine, N-cinnamylidene-*m*-nitroaniline, is 92–93°. The imine may be purified by recrystallization from methanol.

To the remainder of the residue from the distillation (crude imine) add 20 ml of methanol. Attach a Claisen connecting tube equipped with a water-cooled reflux condenser and an addition funnel.[4] Pour the methanolic solution of sodium borohydride into the addition funnel (stopcock closed), and then add this solution to the imine solution dropwise, or in several portions, at a rate such that the addition is completed within five minutes. Swirl the reaction mixture while the addition is being made.

[3] The solution should be prepared no more than 5 or 10 min before it is used, because the sodium borohydride reacts with the methanol. Do not stopper the flask containing the solution, as pressure will build up from hydrogen gas produced by the reaction:

$$NaBH_4 + 4 CH_3OH \rightarrow 4 H_2 + NaOCH_3 + B(OCH_3)_3$$

[4] If a Claisen adapter is not available, the condenser may be attached directly to the flask and an addition funnel inserted in the top of the condenser through a *slotted* cork or rubber stopper. (The slot is to prevent a closed system.)

After all of the borohydride solution has been added, heat the reaction mixture at reflux for 15 min.

Cool the reaction mixture to room temperature and pour it into 50 ml of water. Stir the mixture and allow it to stand with occasional stirring for 10–15 min. Collect the orange crystals by suction filtration and wash them with water.★ Weigh the dry product and calculate the yield. Determine the melting point of the product, N-cinnamyl-*m*-nitroaniline. It may be recrystallized for 95% ethanol; the melting point of the pure secondary amine is 106–107°.

b. Cinnamaldehyde and aniline. Aniline may be substituted for *m*-nitroaniline and the above procedure followed, with the following modification: After the reduction step has been completed, pour the reaction mixture into 50 ml of 10% *hydrochloric acid* instead of water. The free secondary amine, N-cinnamylaniline, is a liquid at room temperature; for ease of handling, the product is converted by hydrochloric acid to the salt, which is only slightly soluble in cold water.

The melting point of the intermediate imine, N-cinnamylideneaniline (which may be isolated as described above for the corresponding imine from *m*-nitroaniline) is reported to be 109° (yellow leaflets from ethanol). The reported melting point of N-cinnamyl-anilinium chloride is 185°.

The picric acid salt (picrate) of N-cinnamylaniline (mp 137°) may be prepared as follows: Add 0.3 g of N-cinnamylanilinium chloride to 3 ml of 5% sodium hydroxide solution and 2 ml of chloroform in a test tube. Shake the mixture, allow it to settle, and then remove the aqueous layer with a pipet. Add 2 ml of water, shake, and again remove the aqueous layer. Add a small amount of anhydrous magnesium sulfate to the chloroform solution, shake the mixture intermittently during 5–10 min, and then filter the solution from the magnesium sulfate.★ Evaporate the chloroform solution and dissolve the residue (the free secondary amine) in 2 ml of ethanol. Add 3 ml of a cold, saturated solution of picric acid, prepared by shaking excess picric acid with ethanol at 0° and filtering the mixture. Warm the ethanolic solution in a water bath at 100° for five minutes and then cool it in an ice bath. Collect the yellow crystals and determine their melting point (reported, 137°).

c. Benzaldehyde and m-nitroaniline. The imine and the amine may be prepared from these reactants by the procedure used with cinnamaldehyde and *m*-nitroaniline. The melting point of the imine, N-benzylidene-*m*-nitroaniline (crystallized from methanol) is 70°. The melting point of N-benzyl-*m*-nitroaniline is 106°.

d. Benzaldehyde and aniline. The imine and the amine–hydrochloric acid salt may be prepared from these reactants by the procedure used with cinnamaldehyde and aniline. The imine, N-benzylideneaniline, is more soluble in ethanol than the corresponding *m*-nitroaniline derivative, and its melting point is lower (54°), so it may be more difficult to crystallize. The reported melting point of N-benzylanilinium chloride is 214–216°.

2. The preparation of crystalline derivatives of carbonyl compounds. a. Semicarbazones. Dissolve 0.5 g of semicarbazide hydrochloride [5] and 0.8 g of sodium acetate in 5 ml of water in a test tube and then add about 0.5 ml of the carbonyl compound.

---

[5] Neither semicarbazide nor hydroxylamine is stable as the free base, so they are usually stored in the form of the hydrochloric acid salt, or "hydrochloride."

Stopper and shake the tube vigorously, remove the stopper, and place the test tube in a beaker of boiling water. Discontinue heating the water and allow the test tube to cool to room temperature in the beaker of water. Remove the test tube to an ice bath and scratch the side of the tube with a glass rod at the interface between the liquid and air. The semicarbazone may be recrystallized from water or aqueous ethanol.

If the carbonyl compound is insoluble in water, dissolve it in 5 ml of ethanol. Add water until the solution becomes turbid, then add a little ethanol until the turbidity disappears. Add the semicarbazide hydrochloride and sodium acetate and continue as above from this point.

b. Oximes.   Dissolve 0.5 g of hydroxylamine hydrochloride in 5 ml of water and 3 ml of 10% sodium hydroxide solution and then add 0.5 g of the aldehyde or ketone. If the carbonyl compound is insoluble in water, add just enough ethanol to give a clear solution. Warm the mixture on a steam bath (or boiling water bath) for ten minutes and then cool it in an ice bath. If crystals do not form immediately, scratch the side of the container at and below the liquid level with a glass rod. The oxime may be recrystallized from water or aqueous ethanol.

In some cases the use of 3 ml of pyridine and 3 ml of absolute ethanol in place of the 3 ml of 10% sodium hydroxide solution and 5 ml of water will be found to be more effective. A longer heating period is often necessary. After the heating is finished, pour the mixture into an evaporating dish and remove the solvent with a current of air in a hood. Triturate (grind) the solid residue with 3–4 ml of cold water and filter the mixture. Recrystallize the oxime from water or aqueous ethanol.

c. 2,4-Dinitrophenylhydrazones.   If the reagent is not supplied, it is prepared by dissolving 1 g of 2,4-dinitrophenylhydrazine in 5 ml of concentrated sulfuric acid. This solution is then added, with stirring, to 7 ml of water and 25 ml of 95% ethanol. After stirring vigorously, the solution is filtered from any undissolved solid.

Dissolve one or two drops of a liquid (or about 100 mg of a solid) in 2 ml of 95% ethanol and add this solution to 2 ml of the 2,4-dinitrophenylhydrazine reagent. Shake the mixture vigorously; if a precipitate does not form immediately, let the solution stand for 15 min.

If more crystals are desired for a melting point, dissolve 200–500 mg of the carbonyl compound in 20 ml of 95% ethanol and add this solution to 15 ml of the reagent. The product may be recrystallized from aqueous ethanol.

B. Addition of Wittig reagents

Caution: Organic phosphorus compounds are toxic, and benzyl chloride is irritating and lachrymatory. Avoid contact of these substances with the skin and inhalation of their vapor.

1. *trans*-Stilbene.   Weigh 0.045 mole of triethyl phosphite into a 25- or 50-ml round-bottomed flask and add 0.045 mole of benzyl chloride and several small boiling stones. Connect a water-cooled condenser to the flask and heat the mixture at gentle reflux for 1 hr. While this reaction mixture is cooling to room temperature, place 0.045 mole of sodium methoxide in a 125-ml Erlenmeyer flask and immediately add 20 ml of N,N-dimethylformamide. Pour the *cool* phosphonate reaction mixture into the Erlenmeyer flask and rinse the round-bottomed flask with 20 ml of N,N-dimethylformamide, adding the rinse solution to the rest of the reaction mixture. Cool the Erlenmeyer flask and its contents in an ice-water bath, stirring with a thermometer until the temperature of the solution is below 20°. While continuing to stir the cooled solution, slowly add 0.025 mole of benzaldehyde, cooling as necessary so that the temperature of the mixture does not rise above 35°. Remove the flask from the cooling bath and allow it to stand at room temperature for about 10 min. Add 15 ml of water with stirring, collect the crystals on a Büchner funnel, and wash them with cold 1:1 methanol-water. Determine the weight and the melting point of the crystalline product. A mixture of ethanol and ethyl acetate may be used to recrystallize impure stilbene. Submit a sample of the product in carbon disulfide or carbon tetrachloride solution for infrared analysis and compare the spectrum with those of Figures 15.13 and 15.14.

2. 1,4-Diphenyl-1,3-butadiene.   This diene may be prepared by following the procedure above up to the point of addition of benzaldehyde. Use 0.04 mole of cinnamaldehyde in place of benzaldehyde, and continue to follow the procedure up to the point of collecting the crystals. Wash the crystals first with water and then with methanol until the filtrate is colorless. Weigh the product and determine its melting point. The melting point of *trans, trans*-1,4-diphenyl-1,3-butadiene is reported to be 150–151°.

3. *trans*-**2-Styrylfuran.**   Follow the procedure for *trans*-stilbene exactly except substitute an equimolar amount of 2-furaldehyde for benzaldehyde. The melting point of *trans*-2-styrylfuran is reported to be 54–55°. This product may be recrystallized from methanol.

EXERCISES

A. ADDITION OF DERIVATIVES OF AMMONIA

1. What causes the turbidity in the distillate collected during the heating of the carbonyl compound and primary amine in benzene solution?
2. Why was the use of *dry* benzene not specified?
3. Although 95% ethanol is a satisfactory solvent for recrystallization of N-cinnamyl-*m*-nitroaniline, it is not suitable for recrystallization of the corresponding imine. Why?
4. Note the molar proportion of NaBH₄ to imine used in your experiment. How does this compare with the stoichiometric requirement? Account for the proportion used.
5. Sodium borohydride may be "stabilized" toward reaction with solvent methanol by addition of sodium hydroxide. Explain.
6. In Figures 15.3 and 15.4, assign as many of the ir absorption peaks to structural components of the molecules as you can. Which peaks give evidence of the

   conversion of $-\overset{|}{C}=N-$ to $-\overset{|}{\underset{H}{C}}-\overset{|}{\underset{H}{N}}-$ and of retention of the carbon-carbon

   double bond in the secondary amine?
7. Can you identify the ir absorptions of the NO₂ group by comparing Figures 15.3 and 15.4 with Figure 15.6? Explain the basis of your identification.
8. By comparing the nmr spectra of Figures 15.5 and 15.11, identify the vinyl hydrogen (—CH═CH—) signals in Figure 15.5 and the methylene (—CH₂—) signals in both Figures 15.5 and 15.11.

9. By calculating the integrations in Figures 15.5, 15.8, and 15.11, and, particularly, by considering the signals of Figure 15.8, see if you can discover the "hiding place" of the elusive N—H signal.

10. There are two NH_2 groups in the semicarbazide molecule, yet only one of them is nucleophilic. Note which one this is, and explain.

11. The melting points of 2,4-dinitrophenylhydrazones are generally higher than those of the corresponding phenylhydrazones. Suggest an explanation for this behavior.

12. How does scratching the side of a container with a glass rod help to induce crystallization?

B. Addition of Wittig reagents

1. Compare the mechanism of the addition of Grignard and Wittig reagents to the carbonyl group, pointing out similarities and differences.

2. Write equations for the preparation of the following alkenes by the original Wittig synthesis from triphenylphosphine and the necessary organic halides and carbonyl compounds:
   (a) $C_6H_5CH{=}C(CH_3)_2$ (b) $C_6H_5C(CH_3){=}CHCH_3$ (c) $CH_3CH_2C(CH_3){=}CH_2$

3. Write equations for Wittig syntheses of the compounds of Exercise 2 using alternative organic halides and carbonyl compounds.

4. Write equations for the preparation of the following alkenes by the phosphonate ester modification (starting with triethyl phosphite):
   (a) $C_6H_5C(CH_3){=}C(CH_3)C_6H_5$ (b) $CH_2{=}CH{-}CH{=}CH{-}C_6H_5$
   (c) $(CH_3)_2C{=}CH{-}CO_2C_2H_5$

5. Would you expect an ylide, **13**, in which R = CN to be more or less stable than one in which R = alkyl?

6. Why should the sodium methoxide not be exposed to the atmosphere for more than a few minutes?

7. Why should the aldehydes used as starting materials in Wittig syntheses be free of carboxylic acids?

8. What peaks in the ir spectra of the stilbenes (Figures 15.13 and 15.14) would be most useful for quantitative analysis of a mixture of *cis* and *trans* isomers?

REFERENCES

1. E. J. Seus and C. V. Wilson, *Journal of Organic Chemistry*, **26**, 5243 (1961).
2. G. Wittig and U. Schollkopf, *Chemische Berichte*, **87**, 1318 (1954).
3. J. L. Bills and C. R. Noller, *Journal of the American Chemical Society*, **70**, 958 (1948).
4. J. H. Pinckard, B. Willie, and L. Zechmeister, *ibid.*, **70**, 1939 (1948).
5. A. Maercker, in *Organic Reactions*, A. C. Cope, editor, John Wiley & Sons, Inc., New York, N. Y., 1965, Vol. 14, Chap. 3. A good general review.

SPECTRA OF STARTING MATERIALS AND PRODUCTS

The ir and nmr spectra of 2-furaldehyde are given in Figures 10.3 and 10.4. The ir and nmr spectra of benzaldehyde are given in Figures 14.6 and 14.7. The ir and nmr spectra of aniline are given in Figures 16.3 and 16.4.

FIGURE 15.1 IR spectrum of cinnamaldehyde.

FIGURE 15.2 NMR spectrum of cinnamaldehyde.

FIGURE 15.3 IR spectrum of N-cinnamylidene–*m*-nitroaniline (in CCl₄ solution).

FIGURE 15.4   IR spectrum of N-cinnamyl–*m*-nitroaniline (in CCl₄ solution).

FIGURE 15.5   NMR spectrum of N-cinnamyl–*m*-nitroaniline.

FIGURE 15.6   IR spectrum of N-cinnamylideneaniline (in CCl₄ solution).

FIGURE 15.7 IR spectrum of N-cinnamylaniline (in CS₂ solution).

FIGURE 15.8 NMR spectrum of N-cinnamylaniline.

FIGURE 15.9 IR spectrum of N-benzylidene–*m*-nitroaniline (in CCl₄ solution).

FIGURE 15.10   IR spectrum of N-benzyl–*m*-nitroaniline (in CCl₄ solution).

FIGURE 15.11   NMR spectrum of N-benzyl–*m*-nitroaniline.

FIGURE 15.12   IR spectrum of benzyl chloride.

FIGURE 15.13   IR spectrum of *cis*-stilbene.

FIGURE 15.14   IR spectrum of *trans*-stilbene.

FIGURE 15.15   NMR spectrum of *cis*-stilbene.

FIGURE 15.16 NMR spectrum of *trans*-stilbene.

FIGURE 15.17 IR spectrum of *trans, trans*-1,4-diphenyl–1,3-butadiene.

FIGURE 15.18 NMR spectrum of *trans, trans*-1,4-diphenyl–1,3-butadiene.

15.2   *Reactions of Carbonyl Compounds Involving the α- and β-Carbon Atoms*

The haloform reaction. The reactions studied in Chapter 14 and in the preceding sections of this Chapter all take place at the carbonyl group, *i.e.*, oxidation to a carboxyl group, reduction to an alcohol group, or addition of various reactants across the carbon-oxygen π-bond. The nature of the carbonyl group also affects its neighboring α- and β-carbon atoms. The partial positive charge on the carbonyl carbon tends to make the hydrogens on the α-carbons (the "α-hydrogens") easily removed as protons, and the resulting carbanion is stabilized by resonance with the carbonyl π-electron system to produce what is termed an enolate anion [**26**, equation (15)].

$$
R-\underset{\underset{H}{|}}{\overset{\overset{R}{|}}{C}}-C\overset{O}{\underset{R}{\diagup}} + B: \rightleftharpoons \left[\underset{R}{\overset{R}{\diagdown}}\underset{\ominus}{C}-C\overset{O}{\underset{R}{\diagup}} \longleftrightarrow \underset{R}{\overset{R}{\diagdown}}C=C\overset{O^{\ominus}}{\underset{R}{\diagup}} \right] + BH^{\oplus} \tag{15}
$$

26

When an aldehyde or ketone which has α-hydrogens is treated with a halogen in basic medium, the enolate anion reacts rapidly with the halogen [equation (16)]. In the

$$
26 + X_2 \longrightarrow R-\underset{\underset{X}{|}}{\overset{\overset{R}{|}}{C}}-C\overset{O}{\underset{R}{\diagup}} + X^{\ominus} \tag{16}
$$

case of acetaldehyde or a methyl ketone, all three of the α-hydrogens of the methyl group are replaced by halogen to give **27** [equation (17), R = H or alkyl].

$$
H-\underset{\underset{H}{|}}{\overset{\overset{H}{|}}{C}}-\overset{O}{\overset{\|}{C}}-R + 3\,NaOH + 3\,X_2 \longrightarrow X-\underset{\underset{X}{|}}{\overset{\overset{X}{|}}{C}}-\overset{O}{\overset{\|}{C}}-R + 3\,NaX + 3\,H_2O \tag{17}
$$

27

When the first α-hydrogen is replaced by halogen, the remaining hydrogens attached to this α-carbon atom become more acidic because of the inductive effect of the halogen, so that further substitution by halogen at this site occurs more rapidly than at the other α-carbon atom. The inductive effect of the three halogens makes the carbon atom of the carbonyl group particularly susceptible to nucleophilic addition of hydroxyl. The intermediate adduct **28** readily undergoes cleavage as shown in equation (18) and the fragments **29** and **30** are immediately converted to the observed products, a haloform **31**, and a carboxylic acid salt **32**, as shown in equations (19) and (20). (**31** and **32** may also be formed simply by transfer of a proton from **30** to

$$
X-\underset{\underset{X}{|}}{\overset{\overset{X}{|}}{C}}-\overset{O}{\overset{\|}{C}}-R + HO^{\ominus} \longrightarrow X-\underset{\underset{X}{|}}{\overset{\overset{X}{|}}{C}}\overset{O^{\ominus}}{\underset{\overset{|}{OH}}{C}}-R \longrightarrow X_3C^{\ominus} + HO-\overset{O}{\overset{\|}{C}}-R \tag{18}
$$

28   **29**   **30**

$$X_3C^{\ominus} + H_2O \longrightarrow HCX_3 + HO^{\ominus} \tag{19}$$
$$\mathbf{31}$$

$$R\overset{\overset{\displaystyle O}{\|}}{-}C-OH + HO^{\ominus} \longrightarrow R\overset{\overset{\displaystyle O}{\|}}{-}C-O^{\ominus} + H_2O \tag{20}$$
$$\mathbf{32}$$

29). The overall equation for the reaction of a methyl ketone, or acetaldehyde (R = H), is given by equation (21).

$$CH_3\overset{\overset{\displaystyle O}{\|}}{C}R + 4\ NaOH + 3\ X_2 \longrightarrow HCX_3 + NaO\overset{\overset{\displaystyle O}{\|}}{C}R + 3\ NaX + 3\ H_2O \tag{21}$$

Although all carbonyl compounds having hydrogens on the α-carbons undergo halogenation at these positions, only those with *methyl* groups adjacent to the carbonyl group undergo carbon-carbon cleavage, presumably because *three* halogens attached to one carbon atom are required to weaken the bond to the point of rupture. Thus, although nearly all carbonyl compounds react with halogens in basic media, only those with α-methyl groups produce haloforms and carboxylic acid salts. This fact is exploited in two important ways: (1) Since iodoform (HCI_3) is a highly insoluble crystalline yellow solid with a characteristic odor, it is readily detected, and its formation is used as a *qualitative test* for the structural moieties shown below (R = H, alkyl, or

$$CH_3\overset{\overset{\displaystyle O}{\|}}{-}C-R \qquad CH_3\overset{\overset{\displaystyle OH}{|}}{-}CH-R$$

aryl).[6] (2) The conversion of a methyl ketone to a carboxylic acid with one less carbon atom is often useful in synthesis. In this case chlorine is the halogen of choice because it is the least expensive and most readily available.

$$R\overset{\overset{\displaystyle O}{\|}}{-}C-CH_3 \xrightarrow[\text{(2) }H_3O^{\oplus}]{\text{(1) }Cl_2,\ NaOH} R\overset{\overset{\displaystyle O}{\|}}{-}C-OH \tag{22}$$

Aldol additions and condensations. Another important and general reaction of enolate anions (besides reaction with halogens) is addition to the carbonyl group of the aldehyde or ketone from which the enolate anion was derived. In this way a dimeric anion **33** is produced [equation (23)], which stabilizes itself by abstracting a proton from the solvent (water or alcohol), as shown in equation (24). The β-hydroxy

$$\tag{23}$$
$$\mathbf{26} \qquad\qquad\qquad\qquad\qquad \mathbf{33}$$

---

[6] Alcohols of this structural type respond to the test because of the oxidizing power of the reagent, which converts them partially to the corresponding carbonyl compounds.

$$33 + H_2O \longrightarrow -\overset{|}{\underset{|}{C}}-\overset{OH}{\underset{|}{C}}-\overset{|}{\underset{|}{C}}-\overset{O}{\underset{H}{C}} \tag{24}$$

$$34$$

carbonyl compound of type 34 is called an "aldol," since it is both an aldehyde and an alcohol. The term "aldol *addition*" is also applied generally to the base-catalyzed self-addition of ketones as well as of aldehydes. The overall change for the reaction of acetaldehyde is given by equation (25). Most β-hydroxy aldehydes and ketones undergo dehydration readily to α, β-unsaturated aldehydes and ketones [equation (26)]. In this

$$2 \ CH_3CHO \xrightarrow{\ HO^{\ominus}\ } CH_3\overset{OH}{\underset{|}{C}}H-CH_2CHO \tag{25}$$

$$\overset{OH}{\underset{|}{C}}H_3CH-CH_2CHO \longrightarrow CH_3CH{=}CHCHO + H_2O \tag{26}$$

case the overall reaction may be referred to as an "aldol *condensation*," since a molecule of water is eliminated from the adduct.

In general, ketones do not undergo self-addition as readily as aldehydes; in fact, special conditions must usually be employed to obtain good yields in such reactions. (Refer to a textbook for a discussion of the reaction of acetone, for example). "Mixed (or crossed) aldol condensations" of two different aldehydes or of an aldehyde with a ketone are possible. Such mixed condensations are practical for synthesis in the case of two aldehydes *if* one of the aldehydes has no α-hydrogens (and hence only serves as a carbonyl acceptor of the enolate anion derived from the other aldehyde) and in the case of a ketone and aldehyde under conditions such that the ketone undergoes no appreciable self-condensation. A good example of the latter case is the reaction of benzaldehyde with acetophenone in the presence of dilute sodium hydroxide solution [equations (27)–(30)]. The product, 38, is called benzalacetophenone. Under the

$$C_6H_5\overset{O}{\overset{\|}{C}}-CH_3 + HO^{\ominus} \xrightarrow{\ -H_2O\ } \left[C_6H_5\overset{O}{\overset{\|}{C}}-\overset{\ominus}{C}H_2 \longleftrightarrow C_6H_5\overset{O^{\ominus}}{\overset{|}{C}}{=}CH_2 \right] \tag{27}$$

$$35$$

$$35 + C_6H_5\overset{O}{\overset{\|}{C}}H \longrightarrow C_6H_5\overset{O^{\ominus}}{\overset{|}{C}}H-CH_2\overset{O}{\overset{\|}{C}}C_6H_5 \tag{28}$$

$$36$$

$$36 + H_2O \longrightarrow C_6H_5\overset{OH}{\overset{|}{C}}H-CH_2\overset{O}{\overset{\|}{C}}C_6H_5 + HO^{\ominus} \tag{29}$$

$$37$$

$$37 \longrightarrow C_6H_5CH{=}CH\overset{O}{\overset{\|}{C}}C_6H_5 + H_2O \tag{30}$$

$$38$$

conditions of the experiment, the dehydration of the "aldol," 37, is spontaneous [equation (30)].

Reactions of benzalacetophenone. As expected, benzalacetophenone gives reactions typical of both the ketone function and the alkenic double bond. For example, a semicarbazone and a 2,4-dinitrophenylhydrazone may be obtained by the procedures of Section 15.1.

Bromine adds readily to the carbon-carbon double bond. Since two asymmetric carbon atoms are produced in this reaction, there should be four possible stereoisomers of the dibromide, two enantiomeric pairs (see Chapter 19.1):

$$
\begin{array}{c}
\quad\ \text{H}\ \ \text{H} \\
\quad\ \ |\ \ \ | \\
\text{C}_6\text{H}_5\text{---C}\text{—C---COC}_6\text{H}_5 \ \text{and enantiomer} \\
\quad\ \ |\ \ \ | \\
\quad\ \text{Br}\ \text{Br}
\end{array}
$$

erythro

$$
\begin{array}{c}
\quad\ \text{H}\ \ \text{Br} \\
\quad\ \ |\ \ \ | \\
\text{C}_6\text{H}_5\text{---C}\text{—C---COC}_6\text{H}_5 \ \text{and enantiomer} \\
\quad\ \ |\ \ \ | \\
\quad\ \text{Br}\ \text{H}
\end{array}
$$

threo

The higher-melting pair (mp 159–160°) is thought to have the *erythro* configuration, and the lower-melting pair (mp 123–124°) the *threo* configuration. It should be noted that benzalacetophenone is capable of geometric (*cis*-*trans*) isomerism; the mixed aldol condensation gives predominantly the more stable *trans* isomer. If the addition of bromine to benzalacetophenone were stereospecific, only one of the pairs of enantiomers should result from each geometric isomer (see Chapter 6.2); however, the higher melting dibromide is the major product of addition of bromine to both *cis*- and *trans*-benzalacetophenone.

As mentioned in Chapter 6.2, nucleophiles will add to compounds having carbon-carbon double bonds conjugated with carbonyl groups, although they do not add to simple alkenes. The addition of aniline to benzalacetophenone represents a good example of the addition of a nucleophile to an α,β-unsaturated ketone [equation (31)].

$$
\begin{array}{ccc}
& \quad\ \overset{\text{O}}{\underset{\|}{}} & \qquad\qquad\qquad \overset{\text{O}}{\underset{\|}{}} \\
\text{C}_6\text{H}_5\text{CH=CHCC}_6\text{H}_5 + \text{C}_6\text{H}_5\text{NH}_2 & \longrightarrow & \text{C}_6\text{H}_5\underset{|}{\text{CH}}\text{—CH}_2\text{CC}_6\text{H}_5 \qquad (31) \\
& & \quad \text{NHC}_6\text{H}_5 \\
\textbf{38} & & \qquad\qquad \textbf{40}
\end{array}
$$

The driving force for this addition is the formation of the resonance-stabilized intermediate, **39** (or the anion formed from it by loss of one of the protons from the nitrogen atom). The final product, β-anilino-β-phenylpropiophenone (**40**), may be formed directly from **39** or via the 1,4-addition product **41**, the less stable enol form of **40**.

EXPERIMENTAL PROCEDURE

A. THE HALOFORM REACTION

1. Sodium hypoiodite—the "iodoform test."   Apply the following procedure to small samples of isopropyl alcohol, acetophenone, cyclohexanone, and pinacolone.

If the substance is water-soluble, dissolve 2–3 drops of a liquid or an estimated 50 mg of a solid in 2 ml of water in a small test tube, add 2 ml of 10% sodium hydroxide, and then slowly add 3 ml of iodine solution.[7] In a positive test the brown color disappears and yellow iodoform separates. If the substance tested is insoluble in water, dissolve it in 2 ml of dioxane, proceed as above and at the end dilute with 10 ml of water.

Iodoform can be recognized by its odor and yellow color and, more definitely, by its melting point, 119°. The substance can be isolated by suction filtration of the test mixture or by adding 2 ml of chloroform, shaking the stoppered test tube to extract the iodoform into the small lower layer, withdrawing the clear part of this layer with a capillary dropping tube, and evaporating it in a small tube on the steam bath. The crude solid is recrystallized from methanol-water.

2. Sodium hypochlorite—preparation of benzoic acid.   To 40 ml of 5% sodium hypochlorite solution ("Clorox," "Purex," etc.) in a 150-ml beaker add 1 ml of acetophenone. Swirl the mixture vigorously and notice whether any heat of reaction may be detected. Place the beaker on the steam cone and heat gently, swirling from time to time, for about 10 min to volatilize the chloroform. Add about 10 drops of acetone to the warm solution to destroy any excess hypochlorite.

Add a small amount of decolorizing carbon to the warm solution, heat it to boiling while swirling gently but continuously, and filter the hot mixture. To the hot filtrate add 5 ml of concentrated hydrochloric acid with stirring, allow the mixture to cool to room temperature, and finally cool it in an ice bath. Collect the crystals, wash them with a little cold water, dry them, and determine their melting point.

Determine the melting point of a mixture of this product with the one from the Cannizzaro reaction (Chapter 14.3).

B. THE ALDOL CONDENSATION

1. Preparation of *trans*-benzalacetophenone.   In a 125-ml Erlenmeyer flask mix 25 ml of 10% sodium hydroxide solution, 15 ml of 95% ethanol, and 0.05 mole of acetophenone. Cool the mixture in an ice-bath and, while swirling, add an equimolar amount of benzaldehyde. Allow the mixture to warm to room temperature and occasionally shake it (stoppered) vigorously during one or two hours. When a yellow oil separates, it will be helpful to induce crystallization by adding a few seed crystals of benzalacetophenone and/or scratching the side of the flask with a stirring rod. *Caution:* Do not allow benzalacetophenone to come in contact with the skin as it is a skin irritant.

Cool the reaction mixture in an ice bath until crystallization appears to be complete and collect the product by vacuum filtration. Wash the crystals with cold water and

---

[7] The iodine reagent is prepared by dissolving 25 g of iodine in a solution of 50 g of potassium iodide in 200 ml of water.

then with a little ice-cold 95% ethanol. Reserve a few crystals to be used for seeding and recrystallize the crude product from 95% ethanol, using 4–5 ml of solvent per gram of crystals. Determine the melting point of the recrystallized material. The melting point of pure *trans*-benzalacetophenone is reported to be 58–59°.

2. Reactions of benzalacetophenone.   Benzalacetophenone dibromide.   Dissolve 1 g of dry *trans*-benzalacetophenone in 4 ml of carbon tetrachloride. Cool the solution in an ice bath and add dropwise, while swirling, 4 ml of a solution of bromine in carbon tetrachloride, prepared by dissolving 1 g of bromine in 6 ml of carbon tetrachloride. Allow the mixture to stand for five or ten minutes and collect the crystals, wash them with a little cold 95% ethanol, dry them, and determine their melting point.

Addition of aniline to benzalacetophenone.   Dissolve 1 g of benzalacetophenone in 20 ml of 95% ethanol and add an equimolar amount of aniline. Stir or shake until the solution is homogeneous and allow it to stand (stoppered) overnight. If crystals have not separated, scratch the side of the tube or flask with a glass rod, and cool the mixture in an ice bath. Collect the crystals, dry them, and determine their melting point (the reported melting point of β-anilino-β-phenylpropiophenone is 175°).

EXERCISES

A. THE HALOFORM REACTION

1. Ethanol gives a clearly positive "iodoform test," yet when sodium hypoiodite is used to oxidize ethanol to acetaldehyde, the yield is found to be less than 10%. Explain how one is able to obtain a good qualitative test in spite of this fact. (Hint: Calculate the molecular weight of iodoform.)

2. Suggest an explanation for the failure of ethyl acetoacetate, $CH_3COCH_2CO_2C_2H_5$, to give iodoform on treatment with sodium hypoiodite. (Hint: Consider the mechanism of the base-catalyzed Claisen condensation, which is a reversible reaction.)

3. Dibenzoylmethane, $C_6H_5COCH_2COC_6H_5$, gives iodoform when treated with sodium hypoiodite, even though the compound is not a methyl ketone. Explain this behavior, using equations.

B. THE ALDOL CONDENSATION

1. Why is the mixed aldol condensation, which produces benzalacetophenone, the main reaction that occurs in this experiment rather than the two possible side reactions, self-condensation of acetophenone and Cannizzaro reaction of benzaldehyde?

2. Write equations to show the major products from reaction of the following carbonyl compounds and bases:
   (a) propionaldehyde, formaldehyde, dilute sodium hydroxide
   (b) benzaldehyde, concentrated potassium hydroxide
   (c) benzaldehyde, ethyl acetate, sodium ethoxide in ethanol
   (d) acetone, benzaldehyde (2 moles), dilute sodium hydroxide

3. Suggest a reason for the fact that the *trans* isomer of benzalacetophenone predominates in the reaction mixture from the aldol condensation.

4. How may *cis*-benzalacetophenone be isomerized to the *trans* isomer?

SPECTRA OF STARTING MATERIALS AND PRODUCTS

The spectra of benzaldehyde are given in Figures 14.6 and 14.7.

FIGURE 15.19   IR spectrum of acetophenone.

FIGURE 15.20   NMR spectrum of acetophenone.

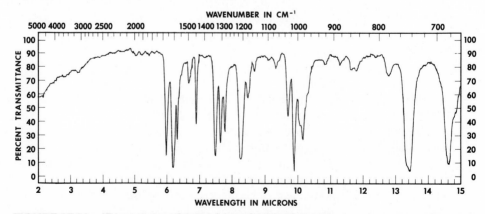

FIGURE 15.21   IR spectrum of *trans*-benzalacetophenone.

FIGURE 15.22   NMR spectrum of *trans*-benzalacetophenone.

15.3   *Identification of an Alcohol or Carbonyl Compound by Means of Three Reagents*

The reactions of alcohols, aldehydes and ketones with the reagents, chromic acid in acetone, 2,4-dinitrophenylhydrazine, and sodium hypoiodite, have been discussed in Sections 14.2, 15.1, and 15.2. By use of only these three reagents, with proper interpretation of the results of their reactions, it is possible to identify an unknown alcohol or carbonyl compound, if there is a known limitation on the number of possibilities for the unknown.

As an illustration, it is possible to decide on the identity of any one of the compounds in the following sets of five by using only the three reagents. In some instances only two tests may be required. Your instructor will provide you with an unknown compound labeled as belonging to one of the sets. Identify the compound by using a *minimum* number of tests.

Set A	Set B	Set C
1-pentanol	acetone	1-octanol
2-butanol	2-methylpropanol	cinnamaldehyde
2-butanone	2-methylpropanal	acetophenone
benzaldehyde	2-propanol	4-methyl-2-pentanol
2-methyl-2-butanol	2-methyl-2-propanol	cyclohexanone

EXERCISE

Write out balanced equations for all positive reactions which you observe and give a careful explanation of their meaning in terms of the structure of your unknown compound.

16

Carboxylic acids and their derivatives

Organic compounds which possess the general structural formula, RCOX, in which R is an alkyl or aryl substituent and X represents a functional group having an atom other than carbon or hydrogen bonded to the carbonyl carbon, are classified as carboxylic acids or their derivatives. Thus, X is OH in a carboxylic acid, OR′ in an ester, OCOR′ in an anhydride and NR_2 in an amide. Nitriles, RCN, are also sometimes classified as derivatives of carboxylic acids because their hydrolysis produces these acids.

It is of interest that many naturally occurring products are carboxylic acids or their derivatives. For example, fats are esters of straight-chain carboxylic acids containing from twelve to eighteen carbon atoms and glycerol (1,2,3-trihydroxypropane), and proteins are basically polyamides. In addition, volatile esters are partially responsible for the pleasant odors of many fruits, whereas carboxylic acids are often the cause of unpleasant odors such as those of rancid butter and of goats. Soaps are the sodium or potassium salts of the same carboxylic acids contained in fats.

The chemistry of carboxylic acids and their derivatives in many ways parallels that of aldehydes and ketones (see Chapters 14 and 15) in that reactions at both the carbonyl carbon and the α-carbon are known. An example of attack at the α-carbon is the Hell-Volhard-Zelinsky reaction in which a carboxylic acid is converted to an α-haloacid [equation (1), X = Cl or Br]. The Claisen ester condensation exemplifies reaction at the α-carbon of one molecule of an ester and at the carbonyl carbon of a second [equation (2)]. These two conversions are analogous to the base-catalyzed halogenation of aldehydes and ketones and to the base-catalyzed aldol condensation, respectively. Of course, a third site of reaction, not present in aldehydes and ketones, is the functional group X, present in carboxylic acids and their derivatives.

$$R{-}\underset{\underset{R'}{|}}{\overset{\overset{H}{|}}{C}}{-}COOH + X_2 \xrightarrow[\text{or PX}_3\text{ (cat.)}]{\text{P (cat.)}} R{-}\underset{\underset{R'}{|}}{\overset{\overset{X}{|}}{C}}{-}COOH + HX \tag{1}$$

$$2\ RCH_2\overset{\overset{O}{\|}}{C}OR' \xrightarrow{R'O^{\ominus}} RCH_2\overset{\overset{O}{\|}}{C}{-}\underset{\underset{R}{|}}{\overset{\overset{O}{\|}}{C}H}COR' + R'OH \tag{2}$$

The types of reactions to be considered in this chapter will be limited to those occurring at the carbonyl carbon or the substituent X. Although the specific experiments to be described all involve benzoic acid or its derivatives, C_6H_5COX, the student should bear in mind that other carboxylic acids and their derivatives are prepared or react in a similar manner. As a rule, however, aliphatic derivatives of acids are more reactive than their aryl counterparts, undergoing hydrolysis much more readily, for example.

16.1 Carboxylic Acids

Carboxylic acids can be synthesized in a variety of ways, which include the hydrolysis of nitriles or of any derivative of the acid, and oxidation of an aldehyde or ketone. The carbonation of a Grignard reagent, described in Chapter 13.1, is another useful synthetic route to these acids.

A special method available for the preparation of aryl carboxylic acids involves the oxidation of the side chain R of an alkyl-substituted benzene with alkaline permanganate or with chromic acid solution [equation (3)]. The carbon atom bonded to

$$ArCH_2R \xrightarrow[\text{or } H_2CrO_4]{\text{KMnO}_4/\text{base}} ArCOOH \qquad (3)$$

the aromatic ring must bear *at least one* hydrogen for this method of preparation to be efficient. If more than one alkyl side chain is present, each will be oxidized to a carboxyl group. Identification of the resulting acid will allow determination of the positions of attachment of the side chains on the aromatic ring, if this is not known.

In the preparation of a carboxylic acid described in this experiment, ethylbenzene, which is readily available from the Friedel-Crafts alkylation of benzene, is oxidized with alkaline permanganate to benzoic acid [equation (4)]. The mechanism of this conversion is not well understood but probably involves the removal of the benzylic hydrogen in a free-radical process initiated by permanganate ion; subsequent reactions lead first to the alcohol and then to the ketone, which is oxidized further to the carboxylic acid. The mechanism of this final oxidation step parallels that given for the oxidation of cyclohexanone to adipic acid (Chapter 14.1).

As a group, carboxylic acids are considered weak acids as compared to mineral acids such as hydrochloric, and have K_a's of the order of 10^{-5}. Nevertheless, carboxylic acids are sufficiently acidic to give a test with litmus paper and to readily undergo reaction with bases to form salts.

Use is often made of the acidity of the carboxylic acids in their removal from a reaction mixture by extraction of the mixture with base. The organic acid can be regenerated from the basic solution by acidification with mineral acid. A good way to differentiate carboxylic acids from phenols (ArOH, another type of weak organic acid) depends on the ability of carboxylic acids to react with the weak base bicarbonate to produce the carboxylate salt and carbon dioxide, which bubbles out of the solution. Most phenols, being weaker acids (K_a's $\approx 10^{-10}$), will not react with this base. (Recall the selective extraction experiment in Chapter 3.1.)

The relatively high acidity of carboxylic acids enables ready determination of the *equivalent weight* (sometimes called "neutralization equivalent") of the acid by titration with standard base. The equivalent weight of an acid is that weight, in grams, of acid which reacts with one equivalent of base. As an example, suppose that 0.1000 g of an unknown acid required 16.90 ml of 0.1000 N sodium hydroxide solution to be titrated to a phenolphthalein endpoint. This means that 0.1000 g of the acid corresponds to (16.90 ml) (0.1000 equivalent/1000 ml) or 0.00169 equivalent of the acid, or that one equivalent of the acid weighs 0.1000/0.00169 or 59.2 grams. Thus, the following expression applies:

$$\text{Equivalent weight} = \frac{\text{grams of acid}}{(\text{volume of base consumed in liters})(N)}$$

where N is the normality of the standard base.

Because each carboxylic acid function in a molecule will be titrated with base, the equivalent weight corresponds to the molecular weight of the acid divided by n, where n is the number of acid functions present in the molecule. Thus, for the example above, the molecular weight may be 59.2, if a single acid function is present, 118.4 if two are present, and 177.6 if three are present. If the molecular weight of an unknown compound is known, then the number of acid groups in the molecule can be calculated by dividing the molecular weight by the equivalent weight. Hence, if the molecular formula of the unknown compound is $C_4H_6O_4$ and its equivalent weight is 59.2, the unknown must have two titratable acid functions.

16.2   Carboxylic Acid Esters

The esters of carboxylic acids are most commonly obtained by allowing the acid and an alcohol to react in the presence of a mineral acid [equation (5)]. The role of the catalyst undoubtedly is to promote attack of the nucleophilic oxygen of the alcohol on the carbonyl carbon of the acid by protonation of the carbonyl oxygen to give **1**. The process is one of equilibria between a variety of compounds, and, in order that the esterification be a reaction of high yield, the equilibria must be shifted toward the products, the desired ester and water. This can be accomplished either by removing one or more of the products from the reaction mixture as formed or by employing a large excess of one of the starting reagents.

$$\tag{5}$$

The effect of the latter approach can be understood by consideration of the mass law relating starting materials and products [equation (6)]. Increasing the amount of either the alcohol or the carboxylic acid will result in an increase in the amount of products formed since the equilibrium constant, K, for the reaction must remain constant, at a given temperature, no matter what quantity of either reagent is used.

$$RCOOH + R'OH \overset{K}{\rightleftarrows} RCOOR' + H_2O$$

$$K = \frac{[RCOOR'][H_2O]}{[RCOOH][R'OH]} \qquad (6)$$

Occasionally it is economically unfeasible to use a large excess of either reagent, and other methods must be employed. These may involve removal of water by azeotropic distillation with the esterifying alcohol, effective removal of the water from the reaction mixture by use of a large excess of acid catalyst (which converts water to its conjugate acid, H_3O^+) or by use of a solvent such as ethylene chloride in which the ester is soluble, but water is not. The procedure adopted in the esterification of benzoic acid with methanol [equation (6), $R = C_6H_5$, $R' = CH_3$], a reaction described in the experimental section of this chapter, is to use a large excess of the alcohol which is quite inexpensive.

If the carboxylic acid has bulky substituents on the carbons α and/or β to the carboxyl group, the esterification either may be very slow or may not occur at all because of steric hindrance to the attack of the alcohol on the intermediate **1** [equation (5)]. An alternative method for the preparation of esters of such hindered acids involves initial conversion of the acid to its acid chloride (see below) and subsequent reaction of this compound with an alcohol to produce the ester [equation (7)].

$$\underset{\text{R—C—OH}}{\overset{O}{\|}} \longrightarrow \underset{\text{R—C—Cl}}{\overset{O}{\|}} \overset{R'OH}{\longrightarrow} \underset{\text{R—C—OR'}}{\overset{O}{\|}} + HCl \qquad (7)$$

Hydrolysis of esters to carboxylic acids, the reverse of esterification, can be accomplished by heating the ester with aqueous acid or base, or, in the case of esters which are insoluble in water, by heating with methanolic or ethanolic potassium hydroxide [equation (8)]. Catalysis by base rather than acid is generally preferred because the rate of hydrolysis of an ester is greater in basic medium. Moreover, if at least a molar equivalent of base is used, the acid will be converted to its salt, which makes the reaction irreversible.

$$\underset{\text{R—C—OR'}}{\overset{O}{\|}} + H_2O \xrightarrow{H_3O^\oplus \text{ or } HO^\ominus} \underset{\text{R—C—OH}}{\overset{O}{\|}} + R'OH \qquad (8)$$

It is possible to carry out the hydrolysis of an ester with alkali in a quantitative manner so that a value, termed the *saponification equivalent*, can be derived. This value is analogous to the equivalent weight of an acid in that it is the molecular weight of the ester divided by the number of ester functions in the molecule. Therefore, the saponification equivalent is the number of grams of ester required to react with one gram-equivalent of alkali. This equivalent is determined by hydrolyzing a weighed amount of the ester with an excess of standardized alkali and then titrating the excess alkali to a phenolphthalein end-point with standardized hydrochloric acid. The saponification equivalent is then given as follows:

$$\text{Saponification Equivalent} = \frac{\text{grams of ester}}{\text{equivalents of alkali consumed}}$$

$$= \frac{\text{grams of ester}}{(\text{volume of alkali in liters})\,(N) - (\text{volume of acid in liters})\,(N')}$$

where N is the normality of the standard base and N' is the normality of the standard acid.

16.3 Carboxylic Acid Halides (Acyl Halides) and Anhydrides

Acid halides, substances in which the hydroxyl function of the carboxylic acid has been replaced by a halogen, are exceedingly useful and reactive compounds. Some of the more important reactions of acid halides are those with various nucleophiles, and a few of these are shown in Chart 16.1. Thus, acid halides are converted to carboxylic acids upon hydrolysis, to esters upon reaction with alcohols, to amides when treated with ammonia or amines, and to anhydrides when allowed to react with carboxylic acids or their salts.

CHART 16.1 Some reactions of acid halides.

· A generalized mechanism, which is common to almost all reactions of derivatives of carboxylic acids with nucleophiles, can be applied to the conversions shown in Chart 16.1. In this mechanism, the nucleophile, NuH (or Nu $^-$), adds to the carbonyl carbon of the derivative, RCOX, to produce the tetrahedral intermediate, **2**, which then decomposes with elimination of HX (or X $^-$) to give the product [equation (9)]. Although the overall reaction is one of *substitution* of Nu for X, the mechanism of the conversion involves an addition step and an elimination step, and is therefore markedly different from the familiar substitutions of the S_N1 and S_N2 types. The success of such a

substitution reaction might be expected to depend to a great extent upon the relative basicities of the reagent, NuH (or Nu⁻), and the leaving group, XH (or X⁻), and such a dependence is observed experimentally. Thus, it is impossible to convert an ester (X = OR) to an acid chloride (X = Cl) by reaction of the ester with chloride ion because chloride ion is a much weaker base than is alkoxide, RO⁻.

The rate at which the substitution reaction occurs with acid halides is enhanced compared to other acid derivatives because the halogen is electron-withdrawing and therefore increases the electrophilicity of the carbonyl carbon. In addition, halide ions, being very weak bases, are good leaving groups.

Acid chlorides can be prepared by treatment of the carboxylic acid with thionyl chloride [equation (10)], phosphorus trichloride [equation (11)] or phosphorus penta-

$$RCOOH + SOCl_2 \rightarrow RCOCl + HCl + SO_2 \tag{10}$$

$$3\ RCOOH + PCl_3 \rightarrow 3\ RCOCl + H_3PO_3 \tag{11}$$

$$RCOOH + PCl_5 \rightarrow RCOCl + POCl_3 + HCl \tag{12}$$

chloride [equation (12)]. Acid bromides may be produced by using the corresponding bromine-containing reagents but seldom are because, in sharp contrast to the relative reactivities of *alkyl* bromides and chlorides in S_N1 and S_N2 reactions, acid bromides are *less* reactive than the corresponding chlorides.

The phosphorus pentahalides are more reactive halogenating agents than the others but are more difficult to handle in the laboratory. Consequently they are usually employed only in instances in which the other reagents fail, as is the case in the conversion of some aryl carboxylic acids and sterically hindered acids to their acid halides. It should be noted that the most complete utilization of the halogens present in the halogenating agent is accomplished in the reaction of the carboxylic acid with phosphorus trichloride, but use of this reagent is sometimes precluded because of possible experimental difficulties.

Thionyl chloride is used for the preparation of benzoyl chloride from benzoic acid [equation (10), R = C_6H_5] as described in the Experimental Procedure. It is used in preference to phosphorus trichloride because the latter reagent, although somewhat less expensive, leads to reaction mixtures containing phosphorous acid, H_3PO_3, from which it is difficult to separate the benzoyl chloride. In contrast, the by-products from the reaction of thionyl chloride with benzoic acid are gases (HCl and SO_2) and therefore are readily removed from the reaction mixture. Furthermore, thionyl chloride has a low boiling point (79°) so that it may be separated from the product by simple distillation.

Methods are also given for the conversion of benzoyl chloride to esters, amides and an anhydride. Although the reactions are typical of acid chlorides in general, it is important to note, once again, that the reaction conditions are somewhat more vigorous than those required for aliphatic acid chlorides. Thus, hydrolysis of benzoyl chloride requires gentle heating, whereas acetyl chloride, CH_3COCl, reacts almost violently with cold water.

The preparation of benzoic anhydride from benzoyl chloride deserves some special comment. The reaction of benzoic acid with benzoyl chloride, using pyridine, C_5H_5N, as a catalyst, produces a high yield of the anhydride [equation (13)]. The role of the

catalyst is twofold: (1) It activates the acid chloride by formation of a pyridinium salt, **3**, which is more susceptible to nucleophilic attack than is the acid chloride itself, and

(2) it converts benzoic acid to the more nucleophilic benzoate ion. It is possible to obtain a still higher yield of benzoic anhydride if the benzoic acid is generated *in situ* by reaction of the acid chloride-pyridine complex, **3**, with 0.5 equivalent of water. The benzoic acid resulting from hydrolysis of the complex reacts immediately with a second mole of the complex to produce the desired anhydride.

Because of their greater ease of handling in the laboratory, anhydrides are often used in place of the more reactive acid halides for the acylation of alcohols and of amines to produce esters and amides, respectively [equations (14) and (15)]. Use of acetic anhydride to acetylate an amine is described in Chapter 17.2 and 17.3.

$$\underset{\text{O}}{\overset{\text{O}}{\text{R—C—O—C—R}}} + \text{R'OH} \longrightarrow \text{R—C—OR'} + \text{R—C—OH} \tag{14}$$

$$\text{R—C—O—C—R} + \text{R'NH}_2 \longrightarrow \text{R—C—NHR'} + \text{R—C—OH} \tag{15}$$

16.4   Carboxylic Acid Amides and Nitriles

The amides of carboxylic acids can be prepared by reaction of an acid halide or anhydride with either anhydrous or aqueous ammonia, if an unsubstituted amide is desired, or with an amine, if an N-alkylated amide is required. Thermal decomposition of the ammonium salt of a carboxylic acid also generates the corresponding amide [equation (16)].

$$\text{RCOOH} + \text{NH}_3 \longrightarrow \text{RCOO}^{\ominus}\text{NH}_4^{\oplus} \xrightarrow{\text{heat}} \text{R—C—NH}_2 + \text{H}_2\text{O} \tag{16}$$

Amides can also be prepared by treatment of esters with amines [equation (17)]. This conversion generally requires heat and, in the case of volatile amines or ammonia, must be carried out under pressure.

$$\text{R—C—OR'} \xrightarrow{\text{HNR}_2} \text{R—C—NR}_2 + \text{R'OH} \tag{17}$$

Although amides undergo a variety of reactions, two of the more important are their reduction to amines with lithium aluminum hydride [equation (18)] and their conversion to nitriles with dehydrating agents such as phosphorus oxychloride, POCl₃, phosphorus pentoxide, P₂O₅, or thionyl chloride [equation (19)]. The fact that amides

$$\text{R—C—NH}_2 \xrightarrow[\text{2) H}_3\text{O}^{\oplus}]{\text{1) LiAlH}_4} \text{R—CH}_2\text{—NH}_2 \tag{18}$$

$$\text{R—C—NH}_2 \xrightarrow[\text{or SOCl}_2]{\overset{\text{POCl}_3}{\text{or P}_2\text{O}_5}} \text{R—C}\equiv\text{N} \tag{19}$$

are reduced to amines is somewhat unexpected since all other derivatives of carboxylic acids and, in fact, the acids themselves are reduced to primary alcohols on treatment with lithium aluminum hydride [equation (20)]. In short, the functional group X is retained on reduction of amides but is lost in the case of the other derivatives of acids.

$$R-\overset{\overset{\displaystyle O}{\|}}{C}-X \xrightarrow[\text{2) } H_3O^{\oplus}]{\text{1) } LiAlH_4} R-CH_2-OH \qquad (20)$$
$$(X \neq NR_2)$$

The mechanism of dehydration of an amide to a nitrile is interesting. As an example, the conversion of benzamide to benzonitrile with thionyl chloride probably involves attack of the amide oxygen on the sulfur atom of thionyl chloride to produce an intermediate salt, **4**, which then decomposes to the nitrile by loss of sulfur dioxide and two moles of hydrogen chloride [equation (21), $R = C_6H_5$].

$$(21)$$

4

Nitriles can also be prepared by the S_N2 reaction between cyanide ion and a primary or secondary alkyl halide. These substances can be hydrolyzed to amides, which are isolable in some cases, and then to the corresponding carboxylic acids by treatment with aqueous acid or base [equation (22)]. They can also be converted into esters by heating with an alcohol in the presence of one molar equivalent of water and a catalytic amount of mineral acid [equation (23)].

$$RC\equiv N \xrightarrow[H_2O]{H^{\oplus} \text{ or } HO^{\ominus}} \left[R\overset{\overset{\displaystyle O}{\|}}{C}-NH_2 \right] \longrightarrow R\overset{\overset{\displaystyle O}{\|}}{C}-OH \qquad (22)$$

$$RC\equiv N \xrightarrow[R'OH]{HCl} R\overset{\overset{\displaystyle O}{\|}}{C}-OR' \qquad (23)$$

EXPERIMENTAL PROCEDURE

A. OXIDATION OF AN AROMATIC SIDE CHAIN; PREPARATION OF BENZOIC ACID FROM ETHYLBENZENE

Place 0.071 mole of potassium permanganate, 120 ml of water, 1.5 ml of 10% aqueous sodium hydroxide and 0.016 mole of ethylbenzene in a 500-ml round-bottomed flask fitted with a reflux condenser, and gently heat this mixture at reflux for two hours.★ Severe bumping will occur if strong heating is used. Test the hot solution for unchanged permanganate by placing a drop of the reaction mixture on a piece of filter paper; if a purple ring appears around the brown spot of manganese dioxide, permanganate remains. Destroy any excess permanganate by adding small amounts of solid sodium bisulfite to the mixture until the spot test is negative. Do not add a large excess of bisulfite. Add 2 g of filter-aid to the hot mixture to promote its rapid filtration; manganese dioxide is so finely dispersed that it would tend to clog the pores of the filter paper otherwise. Filter the mixture by suction and rinse the reaction flask and filter cake with two 10-ml portions of hot water.★ Concentrate the filtrate to about 30 ml by performing a simple distillation; if the concentrate is turbid, clarify it by a gravity filtration.* Acidify the aqueous residue with concentrated hydrochloric acid, adding acid until no more benzoic acid separates (about 5–10 ml of acid will be required).* Cool this mixture in an ice-water bath and isolate the benzoic acid by suction filtration. The crude acid can be purified by recrystallization from hot water or by sublimation. If the latter procedure is to be used, the crude acid must be *dry*. Determine the melting point and yield of benzoic acid obtained.

B. REACTIONS OF BENZOIC ACID

1. Acidity. Saturate 1 ml of hot water with benzoic acid and test the solution with litmus paper, noting the result. Add a small amount of solid sodium bicarbonate or a few drops of 10% aqueous sodium bicarbonate to the saturated solution and note the result. Acidify the resulting solution with concentrated hydrochloric acid and note the result. Compare the solubility of benzoic acid both in cold water and in cold 10% aqueous sodium hydroxide.

2. Esterification; preparation of methyl benzoate. Place 0.082 mole of benzoic acid and 0.62 mole of methanol in a 100-ml round-bottomed flask and carefully pour 3 ml of concentrated sulfuric acid down the walls of the flask. Swirl the flask to mix the components thoroughly, attach a reflux condenser and gently heat the mixture at reflux for one hour.* Cool the solution and transfer it to a separatory funnel containing 50 ml of water. Rinse the flask with 40 ml of ether and add the rinsings to the funnel. Shake the funnel thoroughly to facilitate extraction of methyl benzoate into the ether layer, venting the funnel occasionally. Drain off the aqueous layer and wash the organic layer with a second 25-ml portion of water. Separate the layers, add 25 ml of 5% aqueous sodium bicarbonate to the funnel (*Caution! Foaming may occur.*) and shake the mixture, *frequently* venting the funnel. Drain off the aqueous washes and test to see that they are basic. If not, repeat the washing of the organic layer with aqueous bicarbonate until basic washes are obtained. After a final wash with saturated sodium chloride solution, dry the ether solution thoroughly over anhydrous magnesium sulfate.*

Filter the solution and remove the ether by evaporation on the steam bath (*No flames!*). Then decant the methyl benzoate into a 100-ml round-bottomed flask, and attach the flask to an apparatus set up for simple distillation. Distil the ester using an air-cooled rather than a water-cooled condenser (which might crack because of the high boiling point, 199°, of the ester), collecting the material boiling above 190° in a tared Erlenmeyer flask.

Recover unchanged benzoic acid by acidifying the basic washes with concentrated hydrochloric acid and collecting any precipitate that results. Dry and weigh the precipitate so that recovered starting material can be allowed for when calculating the percentage yield in this reaction. Ascertain that the precipitate is benzoic acid by determining its melting point.

3. Preparation of benzoyl chloride. (Carry out this reaction in a hood if possible. If not, use the gas trap called for.)[2] Equip a dry 100-ml round-bottomed flask with a water-cooled condenser connected to a gas trap like that shown in Figure 5.1, and place 0.20 mole of dry benzoic acid in the flask. Measure out 0.70 mole of thionyl chloride *in the hood*, pour it into the flask and quickly attach a reflux condenser.

---

[2] Acid halides generally are lachrymators, and exposure to their vapors should consequently be avoided. Care must also be taken to keep these compounds off the skin since they can produce burns.

Bring the mixture to a gentle reflux and continue heating until the vigorous evolution of HCl and SO_2 ceases (about 30 min).★ Quickly remove the flask from the condenser, attach it to an apparatus set for simple distillation and add a fresh boiling stone. Distil the excess thionyl chloride using a burner, collecting the distillate in a receiver *attached to a vacuum adapter* carrying a calcium chloride tube. This tube should be connected to a gas trap if the distillation is not performed in a hood. In this way, the distillation can be carried out in an open system without exposure of the distillate to atmospheric moisture. After no more thionyl chloride distils, drain the water from the condenser and continue the distillation. Material boiling between 85–192° should be collected in a second flask, and the product, bp 192–198°, in a third. Return the low boiling fraction to a bottle labeled "Recovered Thionyl Chloride," and pour the intermediate fraction down a sink in the hood, flushing with copious amounts of water. Store the benzoyl chloride (*Caution:* Lachrymatory) in an appropriately labeled *glass stoppered* bottle. Determine the yield of product.

C.   REACTIONS OF BENZOYL CHLORIDE

Caution: Benzoyl chloride is a strong lachrymator. Perform the reactions in a hood if possible.

1. Hydrolysis. Add 5 drops of benzoyl chloride to 5 ml of water contained in an 18 × 150-mm test tube and warm the mixture gently, with occasional shaking, until it is clear. Cool the solution and collect the precipitated benzoic acid. Compare the reactivity of this acid chloride to that of acetyl chloride by *cautiously* adding 2–3 drops of the latter to 5 ml of ice-cold water contained in a test tube, carefully shaking the tube, and observing the result.

2. Ammonolysis. Add 1 ml of benzoyl chloride dropwise to 5 ml of cold concentrated ammonium hydroxide in an 18 × 150-mm test tube, stopper the tube and shake for *ca.* 2–3 min, venting the test tube periodically to relieve any pressure. Pour the mixture into 10 ml of water contained in an Erlenmeyer flask, and collect the precipitated benzamide. Determine the yield and melting point of this product. If you are to prepare enough benzamide to use in the preparation of benzonitrile described below, the amounts of reagents used in the ammonolysis must be increased by a factor of 15.

Benzanilide, $C_6H_5CONHC_6H_5$, can be prepared by combining 0.027 mole of aniline and 0.017 mole of benzoyl chloride in a 25-ml Erlenmeyer flask and adding, in portions and with vigorous shaking, 12 ml of 10% aqueous sodium hydroxide. Isolate the product by suction filtration and recrystallize the crude benzanilide from 95% ethanol. The reported melting point of this compound is 163°.

3. Alcoholysis. Methyl benzoate can be prepared by adding 0.01 mole benzoyl chloride to 0.05 mole of methanol contained in an 18 × 150-mm test tube. Warm the tube gently on a steam cone for 2–3 min and then add 2 ml of saturated sodium chloride solution. Note the odor of the layer which separates.

To prepare phenyl benzoate, add 1.5 ml of 90% aqueous phenol to 10 ml of $3N$ aqueous sodium hydroxide contained in a 25-ml Erlenmeyer flask and to this solution add 1 ml of benzoyl chloride. Shake the mixture for 5–10 min, cool it in an ice-water bath, and isolate the product by suction filtration. Phenyl benzoate can be recrystallized from 95% ethanol-water and has a reported melting point of 72°.

4. Conversion to benzoic anhydride. Add 0.056 mole of benzoyl chloride to 15 ml of dry dioxane contained in a 50-ml Erlenmeyer flask, stopper the flask and cool the solution to about 5°. Add 0.060 mole of pyridine, which has been dried over potassium hydroxide, and swirl the mixture for a few minutes, keeping the temperature below 10°. Introduce 0.028 mole of water with vigorous swirling and then allow the solution to stand with occasional swirling in the ice-water bath for *ca.* 10 min. Pour the reaction mixture into a well-stirred mixture of 25 ml of concentrated hydrochloric acid, 40 g of cracked ice and 100 ml of water, and collect the solid benzoic anhydride by suction filtration. (Should the product "oil out" rather than crystallize, cool the solution in an ice-water bath and scratch to induce crystallization.) Shake the solid with 15 ml of cold 5% aqueous sodium bicarbonate for a minute to remove any residual acid and then filter the mixture. Wash the filter cake with cold water. The crude benzoic anhydride can be recrystallized by dissolving it in 10–15 ml of 95% ethanol, adjusting the temperature of the solution to 40°, and adding water until the mixture just becomes turbid. Allow the solution to cool to room temperature and scratch to induce crystallization. It may be necessary to add a seed crystal and cool the solution in an ice-water bath to obtain the product. Complete the recrystallization by cooling the solution thoroughly in an ice-water bath and isolate the product. Pure benzoic anhydride is colorless, mp 42–43°.

D. PREPARATION OF BENZONITRILE; DEHYDRATION OF AN AMIDE

Dissolve 0.083 mole of finely pulverized benzamide in 50 ml of benzene contained in a 100-ml round-bottomed flask. Equip the flask with a reflux condenser and connect the latter to a gas trap. Add 0.084 mole of thionyl chloride. Observe the resulting solution to see if some white crystals can be seen suspended in it. Bring the mixture to a gentle reflux and continue the reflux until the evolution of hydrogen chloride and sulfur dioxide ceases; about an hour will be needed.★ Attach the flask to an apparatus set for simple distillation, and distil the benzene using a microburner (*Caution: Benzene is flammable. All connections in the apparatus must be tight*).★ Transfer the residue to a 25-ml flask, rinsing the larger flask with a little benzene and adding the rinse to the 25-ml flask, and continue the distillation. After all the benzene has distilled, drain the water from the condenser and distil the benzonitrile (bp 188–192°). Record the yield of product.

EXERCISES

1. Write a balanced equation for conversion of ethylbenzene and permanganate to benzoic acid and manganese dioxide in the presence of hydroxide ion.
2. Write possible structures for a dicarboxylic acid which has the molecular formula, $C_4H_6O_4$.
3. Calculate the equivalent weight of an acid, 0.2370 g of which could be titrated to a phenolphthalein end-point with 27.50 ml of 0.0750 N sodium hydroxide solution.
4. Write a mechanism for the conversion of a carboxylic acid to an acid chloride with thionyl chloride.
5. Would it be feasible to attempt the preparation of benzaldehyde by permanganate oxidation of toluene? Why or why not?
6. Would you expect a carboxylic acid to be more or less soluble in water than in water containing mineral acids? Explain.

7. Write a balanced equation for reaction of benzoic acid with sodium bicarbonate.

8. Suggest a reason why phenols are less acidic than carboxylic acids.

9. Assuming that the equilibrium constant for esterification of benzoic acid with methanol is 3, calculate the yield of methyl benzoate expected using the molar amounts employed in this experiment.

10. Show how the expression for the saponification equivalent is derived.

11. Calculate the saponification equivalent that would be expected for dimethyl succinate, $CH_3O_2CCH_2CH_2CO_2CH_3$.

12. Give a mechanism for the hydrolysis of an ester in acidic and in basic media.

13. Why is it important that all apparatus used in the preparation of an acid chloride be as dry as possible?

14. Why is it possible to prepare unsubstituted amides by reaction of an acid chloride with ammonium hydroxide, which is nothing more than an *aqueous* solution of ammonia?

15. Suggest why acetyl chloride is significantly more reactive than benzoyl chloride.

16. Why must one carry out the reaction of benzoyl chloride with phenol in basic solution when no base is required for reaction of the acid halide with methanol?

17. Explain why anhydrides are less susceptible to nucleophilic attack than are acid chlorides. What evidence is there from experiments described in this chapter that this may be true? (Hint: Consider the method of purification of benzoic anhydride.)

18. Why might benzene be less suitable than dioxane as the solvent for the preparation of benzoic anhydride described in this chapter?

19. What might the crystals be that can sometimes be observed when benzamide and thionyl chloride are combined?

SPECTRA OF STARTING MATERIALS AND PRODUCTS

For spectra of benzoic acid and of methyl benzoate, see Figures 13.1, 13.2, 13.3, and 13.4.

FIGURE 16.1 IR spectrum of benzamide.

FIGURE 16.2 NMR spectrum of benzamide.

FIGURE 16.3 IR spectrum of aniline.

FIGURE 16.4 NMR spectrum of aniline.

FIGURE 16.5 IR spectrum of benzanilide.

FIGURE 16.6 NMR spectrum of benzanilide.

FIGURE 16.7 IR spectrum of benzoic anhydride.

FIGURE 16.8 IR spectrum of benzonitrile.

17

Multi-step organic syntheses

The synthesis of complex molecules from simpler starting materials is one of the most important aspects of organic chemistry. Some syntheses require as many as 20 or 30 sequential reactions or "steps." Multi-step syntheses of such lengths are usually economically feasible only for end-products which are biologically active, and which cannot be obtained readily from natural sources (*e.g.*, cortisone). This is because of the cumulative effect on overall yield of less-than-quantitative yields of individual steps, when there are many such steps. For example, even a five-step synthesis in which each step takes place in 80% yield gives an overall yield of only 33% $[(0.8)^5 \times 100 = 32.8]$. Although thirty-step syntheses are rare, five-step syntheses are common in both academic and industrial laboratories; because of the cumulative effect of multiple steps on overall yield, the organic chemist searches for reaction sequences capable of high yields, and he seeks to develop optimum experimental techniques to minimize losses in each step.

In this chapter we present three examples of multi-step syntheses. These will provide experience in using the product of one reaction in subsequent steps, an experience which will emphasize the importance of good experimental technique and will give some insight into the excitement of producing a complex organic substance from simpler starting materials.

In some of the procedures there are alternatives of either isolating intermediate products in pure form, or using crude reaction mixtures in the next step. This is the sort of choice the synthetic organic chemist faces constantly. It will be instructive for some students in a laboratory section to be assigned to one procedure and other students to the other; the two groups will then have an opportunity to compare the results of the alternative procedures.

17.1   Polystyrene

In this section a synthesis of polystyrene from benzene in four steps is described. An outline of the synthesis is presented in Chart 17.1.

The main industrial method of producing polystyrene, outlined in Chart 17.2, involves one less step than this laboratory synthesis, but it is not easily adapted for a

CHART 17.1 Laboratory synthesis of polystyrene.

CHART 17.2 Industrial synthesis of polystyrene.

student experiment for several reasons. It utilizes ethylene and benzene as starting materials and, since ethylbenzene (**5**), the initial product, is even more easily alkylated than benzene, a large ratio of benzene to ethylene must be used to minimize the formation of diethylbenzenes, **6**. Even under these conditions, some diethylbenzenes form and must be separated from the ethylbenzene. The catalytic dehydrogenation step also requires the difficult separation of the product, styrene (**3**), from unchanged ethylbenzene.

Acetophenone. Instead of the Friedel-Crafts *alkylation* of benzene employed in the industrial method, the laboratory synthesis utilizes *acylation* in the first step. The acylation of benzene is accomplished with acetic anhydride and aluminum chloride and is an example of a Friedel-Crafts ketone synthesis. The quantity of aluminum chloride used in the acylation, which in this case may be referred to as an acetylation, is much more than a catalytic amount such as was used in the alkylations of Chapter 7.2. This is because some aluminum chloride is "used up" in reaction with the acetate anion produced in the first step [equation (1)], which generates the active electrophile, the acylium (or "acylonium") cation, **7**.

Furthermore, the acetophenone, as it is formed, interacts with a mole of aluminum chloride to form a strong complex, **8** [equation (2)]. Thus, more than two moles of aluminum chloride per mole of acetic anhydride are used in order to insure an adequate supply of active catalyst. The acetophenone is set free from the complex at the end of

$$CH_3\overset{O}{\overset{\|}{C}}-O-\overset{O}{\overset{\|}{C}}CH_3 + AlCl_3 \longrightarrow CH_3\overset{\oplus}{C}=O + CH_3\overset{O}{\overset{\|}{C}}-O-\overset{\ominus}{AlCl_3} \qquad (1a)$$

7

$$CH_3\overset{O}{\overset{\|}{C}}-O-AlCl_2 + Cl^{\ominus} \qquad (1b)$$

$$7 + \text{benzene} \longrightarrow \text{intermediate} \xrightarrow{AlCl_3} \text{product } \mathbf{8} + H^{\oplus} \qquad (2)$$

8

$$\downarrow H_2O, HCl$$

$$AlCl_3(H_2O) + \text{acetophenone} \qquad (3)$$

1

the reaction by the addition of water [equation (3)]; precipitation of basic aluminum salts is prevented by the addition of concentrated hydrochloric acid.

Acylations are free of two undesirable features of alkylations: (1) The tendency toward polysubstitution, and (2) rearrangement of the alkylating agent. The introduction of an acyl group into an aromatic ring deactivates the ring toward further electrophilic substitution. The lack of rearrangement of acyl moieties makes acylation, followed by reduction, one of the most reliable synthetic procedures for the preparation of alkylbenzenes of authentic structures; *e.g.*, *n*-propylbenzene [**9**, equation (4)].

$$\text{benzene} + (CH_3CH_2CO)_2O \xrightarrow{AlCl_3} \text{ketone } CH_2CH_3 \xrightarrow[HClO_4]{H_2/Pd} \text{propylbenzene } CH_2CH_2CH_3 + H_2O \qquad (4)$$

9

1-Phenylethanol. The second step of the synthesis illustrates the use of one of the versatile metal hydrides as a reducing agent [equation (5)]. Both lithium aluminum

$$\text{acetophenone } CH_3 + NaBH_4 \xrightarrow{C_2H_5OH} \left(\text{PhCH(CH}_3)-O \right)_4 B^{\ominus} Na^{\oplus} \qquad (5a)$$

1   **10**

$$\text{PhCH(CH}_3)-OH + NaBO_2 \xleftarrow{\quad} \xleftarrow[H_2O, HCl]{} \qquad (5b)$$

2

hydride and sodium borohydride reduce aldehydes and ketones to alcohols; the virtue of the latter (milder) reagent is that it may be used in alcoholic and even aqueous solution, in contrast to lithium aluminum hydride. Although sodium borohydride reacts slowly with alcohols and water, it reacts so much more rapidly with carbonyl compounds that its reaction with these solvents is not troublesome.

The sodium borohydride transfers a hydride ion to the carbonyl carbon in the first step of the mechanism [equation (6)]. All of the hydrogens attached to boron are

$$
Na^{\oplus} \left[H-\underset{\underset{H}{|}}{\overset{\overset{H}{|}}{B}}-H \right]^{\ominus} \overset{R}{\underset{R}{>}}C=O \longrightarrow Na^{\oplus} \left[H-\underset{\underset{R}{|}}{\overset{\overset{R}{|}}{C}}-O-\underset{\underset{H}{|}}{\overset{\overset{H}{|}}{B}}-H \right]^{\ominus} \tag{6}
$$

transferred in this way to produce the intermediate borate salt, **10** [equation (5a)], which is decomposed upon the addition of water and acid to yield 1-phenylethanol [**2**, equation (5b)]. The 1-phenylethanol must then be separated from water, ethanol, inorganic acids, and salts. This is accomplished by evaporating the ethanol on a steam bath and then extracting the 1-phenylethanol with ether. The ether solution is dried by a dessicant and the ether is evaporated. The 1-phenylethanol may be isolated by distillation under reduced pressure; the next step (dehydration) may be carried out on the crude product, however.

Styrene.   The dehydration of 1-phenylethanol is accomplished by heating the alcohol with a mild dehydrating agent, potassium hydrogen sulfate, and collecting the water and styrene (**3**) as they distil at atmospheric pressure [equation (7)]. The crude styrene is

$$
\underset{\textbf{2}}{\overset{\overset{\displaystyle OH}{|}}{\underset{}{C6H5-CH-CH_3}}} \xrightarrow[\text{heat}]{KHSO_4} \underset{\textbf{3}}{C6H5-CH=CH_2} + H_2O \tag{7}
$$

then either extracted into ether, washed and distilled under reduced pressure, or else is diluted with toluene, and the solution is dried by azeotropic distillation. The choice depends on whether one wishes to carry out the fourth step, polymerization, on neat styrene or on a toluene solution of styrene.

Polystyrene.   The polymerization of styrene is represented most simply by equation (8). The mechanism of this free radical-catalyzed polymerization should be compared

$$
n \; C6H5-CH=CH_2 \xrightarrow[\text{heat}]{(C_6H_5COO)_2} C6H5-(\!-CH-CH_2-\!)_n \tag{8}
$$

with that of chlorination of hydrocarbons discussed in Chapter 5.1. The reaction is initiated by thermal decomposition of benzoyl peroxide, as depicted in equations (4) and (5) of Chapter 5, p. 83. The free radicals so produced (In·) *initiate* the polymerization as shown in equation (9).

$$\text{In} \cdot + \text{CH}_2{=}\underset{\underset{\text{C}_6\text{H}_5}{|}}{\text{CH}} \longrightarrow \text{In}{-}\text{CH}_2{-}\underset{\underset{\text{C}_6\text{H}_5}{|}}{\text{CH}} \cdot \qquad (9)$$

$$\text{In}{-}\text{CH}_2{-}\underset{\underset{\text{C}_6\text{H}_5}{|}}{\text{CH}} \cdot + \text{n}\,\text{CH}_2{=}\underset{\underset{\text{C}_6\text{H}_5}{|}}{\text{CH}} \longrightarrow \text{In}{-}\text{CH}_2{-}\underset{\underset{\text{C}_6\text{H}_5}{|}}{\text{CH}}{\left(\text{CH}_2{-}\underset{\underset{\text{C}_6\text{H}_5}{|}}{\text{CH}}\right)_{\overline{n}}} \cdot \qquad (10)$$

11

$$\textbf{11} \longrightarrow \text{In}{-}\text{CH}_2{-}\underset{\underset{\text{C}_6\text{H}_5}{|}}{\text{CH}}{\left(\text{CH}_2{-}\underset{\underset{\text{C}_6\text{H}_5}{|}}{\text{CH}}\right)_{\overline{n}-1}}\text{CH}_2{-}\underset{\underset{\text{C}_6\text{H}_5}{|}}{\text{CH}_2}$$

12

$$+\ \text{In}{-}\text{CH}_2{-}\underset{\underset{\text{C}_6\text{H}_5}{|}}{\text{CH}}{\left(\text{CH}_2{-}\underset{\underset{\text{C}_6\text{H}_5}{|}}{\text{CH}}\right)_{\overline{n}-1}}\text{CH}{=}\underset{\underset{\text{C}_6\text{H}_5}{|}}{\text{CH}} \qquad (11)$$

13

$$\textbf{11} + \text{R} \cdot \longrightarrow \text{In}{-}\text{CH}_2{-}\underset{\underset{\text{C}_6\text{H}_5}{|}}{\text{CH}}{\left(\text{CH}_2{-}\underset{\underset{\text{C}_6\text{H}_5}{|}}{\text{CH}}\right)_{\overline{n}}}\text{R} \qquad (12)$$

Equation (10) represents the *propagation* of the growing polymer chain. Equations (11) and (12) show possible *termination* processes. In equation (11) the free radical end of one growing polymer chain abstracts a hydrogen atom, with its electron, from the carbon atom next to the end of another polymer radical to produce one polymer molecule which is saturated at the end and one polymer molecule which is unsaturated at the end, a process termed *disproportionation*. In equation (12), an R· may be one of the initiating radicals, In·, or another growing polymer chain.

In the procedure below directions are given for producing the polymer in the form of an amorphous solid, a film, and a clear "glass."

EXPERIMENTAL PROCEDURE

Acetophenone. Connect a reflux condenser to a 500-ml round-bottomed flask and attach a Claisen adapter and separatory funnel at the top of the condenser, as shown in Figure 17.1. By means of a simple adapter and a rubber tube, connect the sidearm of the Claisen adapter to a gas trap; water is fed to this trap from the condenser outlet. The trap consists of a bent adapter partially filled with broken pieces of glass tubing; a screw-clamp on its outlet tube is adjusted so that the water level in the trap stays at about the half-full point. Place a few boiling stones in the reaction flask.

When the apparatus has been set up and adjusted correctly, weigh out 0.41 mole of anhydrous powdered aluminum chloride (quickly so as to avoid atmospheric moisture) and introduce it into the reaction flask, having temporarily disconnected the condenser. A good technique to prevent the solid from sticking in the neck of the flask, particularly if a greased ground-jointed flask is being used, is to roll a piece of paper into a tube, insert it in the neck of the flask, and then to pour the aluminum chloride through this paper tube. Remove the paper and immediately pour 70 ml of dry benzene [1] into the reaction flask so as to cover the aluminum chloride as quickly as is possible. Now reconnect the condenser and the rest of the apparatus. Mix 0.20 mole of acetic anhydride and 10 ml of dry benzene in the separatory funnel. Begin dropwise addition of the acetic

[1] Benzene may be dried satisfactorily by azeotropic distillation as described in Chapter 7.2.

Water Level

Water In

To Drain

FIGURE 17.1 Apparatus for addition with refluxing and trapping of a water-soluble gas.

anhydride solution to the benzene–aluminum chloride mixture; loosen the clamp around the neck of the reaction flask so that the mixture can be swirled continuously during the addition operation. Hydrogen chloride will be evolved (fumes will be observed in the trap) and the reaction mixture will become hot. If the benzene begins to boil, reduce the rate of addition and, if necessary, bring a cold water bath up around the reaction flask. The addition should be completed in about 15 to 20 min.

After all of the acetic anhydride solution has been added and the vigor of the reaction has decreased, move the apparatus carefully to a steam bath and heat the flask for *ca.* 20 min. Then remove the apparatus from the steam bath and place a pan of cold water around the flask. While the reaction mixture is cooling, place 120 g of crushed ice in a 600-ml beaker and add 100 ml of concentrated hydrochloric acid with stirring. Now pour the cool reaction mixture slowly into the ice-acid mixture. If any basic aluminum salts remain undissolved after the mixture has been stirred well, add enough hydrochloric acid to dissolve them.

Transfer the two liquid layers to a large separatory funnel and rinse the beaker with 25 ml of ether, adding the ether to the mixture in the separatory funnel. Shake the mixture gently and then remove the organic layer. Extract the aqueous layer with 25 ml of ether and combine the ether solution with the main organic layer.★ Wash the organic layer with 35 ml of 10% sodium hydroxide solution and then with 25 ml of water. Separate the organic solution and dry it with two 5-g portions of anhydrous magnesium sulfate, decanting from the drying agent the first time and filtering the second time.★ Remove the solvent by distillation, using a steam bath, and pour the residue into a 50-ml round-bottomed flask. Add boiling stones and then distil, taking care to keep flames away from the receiver while the residual ether and benzene distil. After these solvents have been collected and the distillation temperature has begun to rise, change the receiver and drain the water from the condenser. Collect the aceto-phenone boiling at 195–200°. The yield should be 65–85%. Weigh the product and calculate the percentage yield on this step.

1-Phenylethanol. In a 150-ml beaker, dissolve 0.032 mole of sodium borohydride (*Caution: Do not allow this strongly caustic reagent to come in contact with your skin*) in 25 ml of 95% ethanol and add 0.10 mole of acetophenone dropwise. Stir the mixture with a thermometer during the addition, and keep the temperature below 50° by de-creasing the rate of addition and by external cooling with an ice bath, if necessary. After the addition of the acetophenone is complete, allow the reaction mixture (which con-tains a white precipitate) to stand for 15 min at room temperature. Add *ca.* 10 ml of 10% hydrochloric acid solution dropwise and with stirring; hydrogen will be evolved and most of the white solid will dissolve. Place the beaker containing the reaction mix-ture on a steam bath and concentrate the solution by evaporation of the ethanol until two liquid layers separate. Add 20 ml of ether and transfer the mixture to a 125-ml separatory funnel. Shake the mixture gently, allow the layers to separate, and remove the ether layer.★ Extract the aqueous layer with a 10-ml portion of ether and add this to the other ether solution. Dry the combined ether solutions over anhydrous mag-nesium sulfate; briefly swirl the solution with the drying agent and then decant the solution into a dry flask. Add another portion of drying agent and repeat the swirling; filter the ether solution into a tared 100-ml round-bottomed flask.★

If pure 1-phenylethanol is to be isolated, add about 1 g of anhydrous potassium carbonate and arrange for vacuum distillation.[2] (The potassium carbonate present in the distillation pot helps to avoid acid-catalyzed dehydration during the distillation.) Remove the ether at atmospheric pressure (steam bath) and then reduce the pressure to below 20 mm, if possible, and distil using an oil bath; the boiling point of 1-phenyl-ethanol is 102.5–103.5° (19 mm), 97° (13 mm).

If pure 1-phenylethanol is not to be isolated by vacuum distillation, proceed in the following way. Connect the flask to an aspirator by means of an adapter or a one-hole rubber stopper fitted with a short glass tube. (The aspirator must be connected through a safety trap with a pressure release such as that shown in Figure 2.13.) Apply gentle suction to the flask by closing the stopcock partially or completely, but momentarily; swirl the flask containing the ether solution so that the ether boils, but does not bump. As the ether boils away, the pressure may be gradually lowered until the full aspirator vacuum is applied. Warm the flask with your hand or with a water bath at room tem-perature. Calculate the yield of crude 1-phenylethanol.

Styrene. At this stage, two or more of the students should combine their yields of 1-phenylethanol so that a total of at least 0.09 mole may be used in the dehydration step. This may be done by pouring the crude products together in one 100-ml round-bottomed flask with the aid of a little dry ether, which is then removed as described above.

[2] See Chapter 2.5 and references at the end of Chapter 2.

To *ca.* 0.09 mole of crude 1-phenylethanol, add 1.0 g of anhydrous potassium hydrogen sulfate and *ca.* 0.1 g of copper powder.[3] Attach a distilling head and a condenser to the 100-ml flask and distil the mixture by heating in an oil bath whose temperature is gradually (during about 10 min) raised to 200–220°. The temperature of the distilling vapor should not exceed 130°. Collect the distillate and calculate the· approximate yield of crude styrene from the volume of the organic layer. If the layers do not separate well, add a little sodium chloride to saturate the water layer.★

If you are to isolate pure styrene, add the mixture of styrene and water obtained as distillate above to 30 ml of ether in a 125-ml separatory funnel. Shake gently, allow the layers to separate, and drain off the aqueous layer. Wash the ether solution first with 15 ml of saturated sodium carbonate solution and then with 10 ml of water. Dry the ether solution by swirling with *ca.* 1 g of anhydrous calcium chloride.★ Decant the ether solution into a 50-ml round-bottomed flask, wash the drying agent with 5 ml of ether, and combine the wash with the main ether solution. Attach a still head and condenser, add some boiling stones, and distil using a steam bath. Remove the ether at atmospheric pressure, then arrange for vacuum distillation, using an aspirator.[2] The boiling point of styrene is 55° (30 mm), 48° (20 mm). Collect the styrene in a weighed flask and calculate the yield, based on 1-phenylethanol (one step), on acetophenone (two steps), and on acetic anhydride (three steps), considering the actual amount of material used in each step.

Polystyrene. (*a*) *Polymerization of pure styrene.*   Place about 5 g of pure, freshly-distilled styrene in an 18 × 150-mm soft glass test tube and add *ca.* 0.1 g of benzoyl peroxide. *Caution: Benzoyl peroxide is sensitive to shock and to static electricity. It should be dispensed from cardboard containers and transferred by means of a folded 3 × 5-inch card. Do not use a metal spatula.* Stopper the tube *loosely* and heat it on a steam bath for several hours; allow the tube to stand at room temperature for 24 hr, or until the next laboratory period. After the viscosity has increased sufficiently, one may suspend a small object (such as a coin) in the viscous liquid and allow the polymerization to proceed until a hard, transparent "glass" is obtained. The test tube may then be cracked away from the polystyrene.

(*b*) *Solution polymerization of crude styrene.*   Add the mixture of styrene and water collected in the step above to three times its volume of toluene or xylene (commercial mixture of isomers) and drain off the small water layer in a 125-ml separatory funnel. Distil the toluene or xylene solution until the distillate dropping from the end of the condenser is clear (all the water will have been removed by azeotropic distillation). Add 0.1 g of benzoyl peroxide [see *Caution* in (*a*)] for each 5 g of styrene (the amount estimated in the crude distillate from the dehydration reaction above) in the solution, connect a reflux condenser, and heat to reflux with a burner for 1 to 3 hr. (Since the boiling temperature of xylene is considerably higher than that of toluene, a comparable degree of polymerization will be obtained in a shorter time in xylene solution than in toluene solution.) Cool the solution to room temperature and then pour it into about 150 ml of methanol. Collect the white precipitate of polystyrene by decantation, or by suction filtration if decantation is not practical. Resuspend the polystyrene in fresh methanol and stir it vigorously, then collect it on a filter and allow it to dry.

Redissolve a portion of the polystyrene in toluene (the *minimum* amount required to dissolve it when heated on a steam bath) and pour the solution on a watch glass or the

---

[3] Copper powder has been reported to reduce the tendency of the styrene (produced in this step) to polymerize during the distillation.

bottom of a large inverted beaker and allow it to evaporate. A clear film of polystyrene should result.

EXERCISES

1. The water level in the gas trap should not be allowed to rise above the limit shown in Figure 17.1. What danger is involved if the level reaches the top of the trap?

2. Why is the water drained from the condenser after the solvent has been removed from the acetophenone and before the acetophenone distils?

3. Suggest a formula for the white precipitate formed in the reaction of acetophenone with sodium borohydride, and write an equation for its reaction with water and hydrochloric acid.

4. When 1-phenylethanol is purified by distillation, reduced pressure is always used. Why?

5. Styrene is distilled under reduced pressure. Why?

6. Calculate the overall percentage yield of cortisone, an important drug, produced in a 33-step synthesis, assuming an average yield of 90% on each step.

7. Suggest an explanation for the fact that acylation takes place without rearrangement, whereas rearrangement often accompanies alkylation; for example, compare reaction of CH_3CH_2COCl and $CH_3CH_2CH_2Cl$ with benzene and aluminum chloride.

8. What other reagents (besides $NaBH_4$ and $LiAlH_4$) will reduce acetophenone to 1-phenylethanol?

9. What reagents will reduce acetophenone to ethylbenzene?

10. Suppose that in the first step of this synthesis an appreciable amount (*ca.* 10%) of diacetylation had occurred, and the other steps had been carried out without purification of the intermediates. What difference in the physical properties of the resultant polystyrene might result?

SPECTRA OF STARTING MATERIALS AND PRODUCTS

The ir and nmr spectra of acetophenone are given in Figures 15.19 and 15.20, respectively.

FIGURE 17.2 IR spectrum of 1-phenylethanol.

FIGURE 17.3 NMR spectrum of 1-phenylethanol.

FIGURE 17.4 IR spectrum of styrene.

FIGURE 17.5 IR spectrum of polystyrene.

17.2 The Synthesis of Sulfathiazole

Some of the most effective drugs belong to a general class called "sulfa" drugs, the members of which have a common structural feature in that they are derivatives of the parent compound, *p*-aminobenzenesulfonamide (14), often called sulfanilamide. It should be noted that this compound contains two different $-NH_2$ moieties, one as the amino group and the other as the sulfonamido ($-SO_2NH_2$) group. Different substituents on the sulfonamido nitrogen give rise to the family of sulfa drugs.

One of the more familiar examples of a sulfa drug is sulfathiazole (15), and the current experiment presents its total synthesis. This preparation can be started, if desired, from benzene and simple aliphatic compounds; the steps of the total synthesis are outlined in Chart 17.3. The preparation of several interesting intermediate compounds, such as sulfanilamide itself, are included, and many of them can be isolated if desired or used in subsequent steps without extensive purification. A discussion of the reactions involved in each step is included.

CHART 17.3 Laboratory synthesis of sulfanilamide (14) and sulfathiazole (15).

Aromatic nitration: nitrobenzene (16).   Another typical electrophilic aromatic substitution reaction (Chapter 7) is found in nitration, where a nitro group ($-NO_2$) replaces a hydrogen atom on an aromatic ring. The exact conditions required for nitration depend on the aromatic compound. In extremely reactive compounds, such as phenols, dilute nitric acid can be used. In other less reactive compounds, such as benzene, concentrated nitric acid must be used in the presence of concentrated sulfuric acid. In very unreactive compounds, such as nitrobenzene, fuming sulfuric acid and fuming nitric acid must be used in order to effect a good yield of the desired product. Not only can the concentration of the nitrating agent be changed, but also the temperature may be varied so as to assist in controlling the reaction.

In the present work, the nitration reaction will be applied to benzene itself. The main reactions for the preparation of nitrobenzene from benzene are illustrated in equations (13) and (14). The first step is believed to be an equilibrium reaction, in which the concentration of nitronium ion, NO_2^+ (**23**), is increased by the presence of sulfuric acid. This reaction is followed by a slow, rate-determining step in which NO_2^+ attacks the benzene ring; this attack is followed by a fast reaction, which results in the loss of a proton to give nitrobenzene.

EQUILIBRIUM:

$$HO{-}NO_2 + 2\,H_2SO_4 \rightleftarrows NO_2^{\oplus} + H_3O^{\oplus} + 2\,HSO_4^{\ominus} \qquad (13)$$

23

SLOW STEP:

$$(14a)$$

FAST STEP:

$$(14b)$$

16

The principal side reaction is the formation of *m*-dinitrobenzene [equation (15)], but the formation of this compound is very slow under the reaction conditions. This is because of the fact that the nitro group is strongly deactivating, a trend characteristic of groups which orient a new group *meta* to the first group. It is possible to isolate pure nitrobenzene by separating the organic layer, washing to remove any acid, and fractionally distilling.

$$(15)$$

16

Reduction of aromatic nitro compounds: aniline (17).   A variety of methods are available for the reduction of an aromatic nitro compound to the corresponding amine. It should be noted that reduction of aliphatic nitro compounds is seldom found, owing to the difficulty in obtaining the starting compounds; on the other hand, aromatic nitro compounds are easily obtained. The general methods of reduction include catalytic hydrogenation and electrolytic and chemical reduction. In the laboratory, the most commonly used method is chemical reduction, in which various metals are used in acidic solution; the most important commercial method is probably catalytic hydrogenation.

The mechanisms of these reductions are not well understood. The fact that various stable intermediate compounds have been isolated suggests that it is a stepwise reaction. Some of the intermediates for the reduction are shown in equation (16). Differences in rates of reduction of the various intermediates may be important, but the nature of the reducing agent dictates the final product. For example, the reduction of nitrobenzene with tin metal and hydrochloric acid gives only aniline, whereas the use of zinc metal and ammonium chloride gives only N-phenylhydroxylamine **(25)**. Note that in equation (16) [H] means reduction by any of the possible reducing agents.

$$
\underset{\textbf{16}}{\overset{NO_2}{\bigcirc}} \xrightarrow{[H]} \underset{\textbf{24}}{\overset{N=O}{\bigcirc}} \xrightarrow{[H]} \underset{\textbf{25}}{\overset{NH-OH}{\bigcirc}} \xrightarrow{[H]} \underset{\textbf{17}}{\overset{NH_2}{\bigcirc}} \tag{16}
$$

In this experiment, iron powder with a small amount of hydrochloric acid will be used to reduce nitrobenzene to aniline. The main reaction is shown in equation (17), in unbalanced form. This is a typical oxidation-reduction equation which can perhaps

$$
\overset{NO_2}{\bigcirc} + Fe \xrightarrow[\text{(HCl as catalyst)}]{H_2O} \overset{NH_2}{\bigcirc} + Fe(OH)_3 \tag{17}
$$

most easily be balanced by considering the two half-reactions given in equations (17a) and (17b).

$$
6\,e^{\ominus} + 6\,H^{\oplus} + \overset{NO_2}{\bigcirc} \longrightarrow \overset{NH_2}{\bigcirc} + 2\,H_2O \tag{17a}
$$

$$
3\,H_2O + Fe = Fe(OH)_3 + 3\,H^{\oplus} + 3\,e^{\ominus} \tag{17b}
$$

Several possible side reactions take place but only to a minor extent under the reaction conditions used. These side reactions [equation (18)] occur due to the presence of some of the intermediate products [equation (16)], which then react with one another. Nitrosobenzene **(24)** can react with aniline to give azobenzene **(26)**, which can be reduced to hydrazobenzene **(27)** by iron. Hydrazobenzene can also undergo rearrangement in the presence of the acid catalyst to yield benzidine **(28)**; this rearrangement is often referred to as the **benzidine rearrangement**. N-Phenylhydroxylamine **(25)** can undergo acid-catalyzed rearrangement to p-aminophenol **(29)**. Thus, some of the anticipated side products are compounds **25, 26, 27, 28,** and **29**; their separation from aniline is discussed in the following paragraph.

Isolation of the aniline and its purification represents an interesting example of how solubility differences and steam distillation can be used. It is necessary to remove the aniline from its principal impurities, which are unchanged nitrobenzene, benzidine, and p-aminophenol. When the reaction is completed, some of the aniline will be present as its hydrochloric acid salt. The reaction mixture is made distinctly basic and then steam distilled. The phenol, which will have been converted to its water-soluble, non-volatile salt, will remain in the aqueous residue along with the high-molecular-weight benzidine. The aniline and nitrobenzene will steam distil together. To separate these two from one another, it is necessary to convert the amine to its soluble hydrochloric acid salt, and steam distil again. The nitrobenzene will now steam distil, whereas the salt will remain in the aqueous residue. The aniline may be recovered by making the acidic residue basic and extracting the amine into an organic solvent such as dichloromethane.

Alternatively, one could isolate the aniline from the unchanged nitrobenzene by treating the mixture of the two (obtained from steam distillation) with dilute acid and extracting the neutral nitrobenzene (the aniline will have been converted to the salt and will be in the aqueous layer) with an organic solvent, such as dichloromethane. Making the aqueous layer basic, followed by extraction with dichloromethane will render the aniline in the second dichloromethane layer.

Acetanilide (18).   Aniline (17) is acetylated in aqueous solution, using a mixture of acetic anhydride and sodium acetate [equation (19)]. The water-insoluble aniline is first converted to the water-soluble hydrochloric acid salt; to this salt in water is added acetic anhydride, followed by the addition of sodium acetate to free the aniline from its salt. The free amine is then rapidly acetylated by the acetic anhydride. This provides an easy method of acetylation, and the overall yields are high. Under the conditions used, the acetanilide is produced as a solid, which can be filtered and recrystallized from water.

A side reaction which might occur is diacetylation, shown in equation (20). It appears that this is minimized by the general method described above, whereas acetylation in pure acetic anhydride (using the free amine) frequently leads to the production of the diacetyl compound.

$$+ \; 2 \; CH_3CO_2H + NaCl \qquad (19)$$

$$+ \; CH_3CO_2H \qquad (20)$$

Chlorosulfonation reactions: *p*-**Acetamidobenzenesulfonyl chloride (19).** A simple, one-step reaction can be used to introduce the sulfonyl chloride group, $-SO_2Cl$, into an aromatic ring. This reaction, known as chlorosulfonation, involves the use of chlorosulfonic acid ($ClSO_3H$) and an aromatic compound. As applied to acetanilide **(18)**, the major product is *p*-acetamidobenzenesulfonyl chloride [**19**, equation (21)]. Two moles of the chlorosulfonic acid are needed. The reaction is known to proceed through the sulfonic acid, which is converted to the sulfonyl chloride on further reaction with chlorosulfonic acid [(equation (22)].

$$+ \; 2 \; ClSO_3H \longrightarrow \qquad + \; HCl + H_2SO_4 \qquad (21)$$

$$+ \; HCl \qquad + \; H_2SO_4 \qquad (22)$$

The sulfonation of acetanilide is a typical electrophilic aromatic substitution reaction, in which SO_3 is probably the electrophile. The SO_3 may arise from chlorosulfonic acid according to the equilibrium shown in equation (23). It should be noted that the

$$ClSO_3H \rightleftharpoons SO_3 + HCl \qquad (23)$$

acetamido group orients an incoming group predominantly *para*, and indeed virtually none of the *ortho* isomer is observed. Presumably this is due to the size of the acetamido group.

To isolate the product, the reaction mixture is poured into ice water, and the product is obtained as a precipitate. The water serves to hydrolyze the excess chlorosulfonic acid [equation (24)].

$$ClSO_3H + H_2O \rightarrow HCl + H_2SO_4 \tag{24}$$

The preparation of sulfanilamide may be carried out without drying or purifying the *p*-acetamidobenzenesulfonyl chloride, since the latter will be treated with aqueous ammonia. However, if the sulfonyl chloride is to be converted to sulfathiazole, it must be dried, since this reaction requires anhydrous conditions. In general, sulfonyl chlorides are much less reactive towards water than are acid chlorides, but they do hydrolyze slowly to give the corresponding sulfonic acid [equation (25)]. In order to purify

$$\tag{25}$$

p-acetamidobenzenesulfonyl chloride, the wet product is dissolved in chloroform; the small amount of water which is present forms as a second layer owing to its very low solubility in chloroform. The organic layer containing the desired product is then drawn off and the product is recrystallized from a suitable solvent. This represents a generally useful technique for purifying solids which contain small amounts of water.

Sulfanilamide [*p*-aminobenzenesulfonamide (14)].   The preparation of sulfanilamide starts with *p*-acetamidobenzenesulfonyl chloride **(19)**. The moist product can be used without purification, but then it must be used *immediately*. The amide is produced by treatment of the sulfonyl chloride with an excess of aqueous ammonia, followed by removal of the acetyl group. This latter process can be carried out in acidic or basic solution without affecting the sulfonamido group, which hydrolyzes much less rapidly than the acetamido group. Acid hydrolysis will be used here, and the resulting amine forms a soluble hydrochloric acid salt under these conditions. In order to obtain free sulfanilamide, the acidic solution must be neutralized with dilute base (sodium bicarbonate). These reactions are shown in equation (26). The sulfanilamide is purified further

$$\tag{26}$$

by recrystallization from water. A possible side reaction, though one which is not important under the conditions of the experiment, is the hydrolysis of the sulfonyl chloride during sulfonamide formation.

An important synthetic procedure has been introduced in the preparation of sulfanilamide—the use of a protective group (the acetyl group) which is subsequently removed at a later stage. It would appear that one reasonable method for preparing sulfanilamides would be to allow *p*-aminobenzenesulfonyl chloride (**30**) to react with ammonia or with some substituted amine, as is the case in the preparation of sulfathiazole. However, it is not practical to generate a sulfonyl chloride group in the presence of an amino group contained in the same molecule. The amino group of one molecule would react with the sulfonyl chloride group of another molecule and give a polymeric material containing sulfonamide linkages, as illustrated in equation (27). Thus, in order

$$\text{n H}_2\text{N} \underset{\mathbf{30}}{-\!\!\!\bigcirc\!\!\!-}\text{SO}_2\text{Cl} \xrightarrow{-\text{n HCl}} \left[-\text{NH}-\bigcirc-\text{SO}_2\text{NH}-\bigcirc-\text{SO}_2- \right]_n \qquad (27)$$

to prepare the sulfanilamides, it is necessary first to "protect" the free amino group so as to allow the other functional group to be introduced. In the present synthesis, this is most easily done by acetylating the amine. This general practice must be followed when a molecule contains functional groups which are reactive toward one another. Of course, the free amine can be regenerated by removing the acetyl group (called either the "protective group" or the "blocking group") after the sulfonamido group has been introduced. In using such a technique, care must be taken to insure that the blocking group can be removed without affecting the second functional group; in the present case, the acetyl group can easily be removed without hydrolyzing the sulfonamido group.

2-Aminothiazole (21).   The thiazole ring system, **31**, is shown below. The parent skeleton is found in various natural products (*e.g.*, Vitamin B$_1$) and in sulfathiazole. Many different substituted thiazole ring systems are known, and we shall now examine one such typical compound, 2-aminothiazole.

$$\begin{array}{c} \text{HC}\underset{4}{\text{—}}\underset{3}{\text{N}} \\ \text{HC}^5 \quad {}^2\text{CH} \\ \underset{1}{\diagdown} \underset{\text{S}}{\diagup} \end{array}$$

31

Since the thiazole ring system is not nearly as stable as the analogous carbon ring system, cyclopentadiene, it is not possible to substitute directly on the thiazole ring without destroying the molecule. Consequently, most thiazole systems are made by simple condensation reactions. The preparation of 2-aminothiazole is typical of the general synthetic route to this type of compound, namely, condensation of an α-haloketone or aldehyde with a thioamide.

The main reaction leading to the preparation of 2-aminothiazole is shown in equation (28). Chloroacetaldehyde is condensed with thiourea by heating an aqueous solution of the two at reflux.

$$\underset{\text{ClCH}_2\overset{\text{O}}{\overset{\|}{\text{C}}}\text{H}}{} + \underset{\text{H}_2\text{N}\overset{\text{S}}{\overset{\|}{\text{C}}}\text{NH}_2}{} \longrightarrow \underset{\mathbf{21}}{\left[\begin{array}{c}\text{N}\\ \diagdown \\ \text{S} \end{array}\!\!\!\text{NH}_2\right]} + \text{HCl} + \text{H}_2\text{O} \qquad (28)$$

Chloroacetaldehyde is extensively hydrated in aqueous solution and can be obtained commercially as a 30–40% solution in water. Alternatively, it can be prepared *in situ* from acid hydrolysis of either an α, β-dichloroethyl alkyl ether [equation (29)] or of a dialkyl chloroacetal [equation (30)].

$$\underset{\text{OR}}{\text{ClCH}_2\overset{|}{\text{CH}}-\text{Cl}} \xrightarrow{\text{H}_3\text{O}^{\oplus}} \text{ClCH}_2\overset{\text{O}}{\overset{\|}{\text{CH}}} + \text{HCl} + \text{ROH} \tag{29}$$

$$\underset{\text{OR}}{\overset{\text{OR}}{\text{ClCH}_2\text{CH}}} \xrightarrow{\text{H}_3\text{O}^{\oplus}} \text{ClCH}_2\overset{\text{O}}{\overset{\|}{\text{CH}}} + 2\,\text{ROH} \tag{30}$$

Chloroacetaldehyde is believed to condense with thiourea through the enol form of the latter (perhaps more properly called an "iminethiol"). Nucleophilic addition of the imino group to the carbonyl function, followed by the loss of a molecule of water, gives 2-aminothiazole.[4] A plausible mechanism for these steps is shown in equation (31).

$$\tag{31}$$

Important side reactions include the polymerization of chloroacetaldehyde, which produces materials which are water soluble and removed easily during filtration.

Sulfathiazole (15). The final reaction in the synthesis of sulfathiazole involves the condensation of *p*-acetamidobenzenesulfonyl chloride (19) with 2-aminothiazole (21), which occurs with the liberation of hydrogen chloride [equation (32)]. As the reaction proceeds, the hydrogen chloride which is given off would react with 2-aminothiazole and convert it to its unreactive hydrochloric acid salt. Thus, it is important to remove the hydrogen chloride as it is produced. Pyridine is most often used for this purpose because it is inexpensive, and it is easily removed from the reaction mixture. In this preparation, pyridine will be used both to remove hydrogen chloride and to serve as a solvent for the reaction. Precautions must be taken to insure that all equipment and reagents are dry in order to minimize hydrolysis of the sulfonyl chloride.

Another important principle of organic synthesis is involved in this preparation, this being the conservation of the more (or most) valuable reagent. 2-Aminothiazole is more expensive than is *p*-acetamidobenzenesulfonyl chloride,[5] so the latter is used in

---

[4] Hydrogen chloride is produced as the condensation reaction [equation (31)] proceeds and converts the 2-aminothiazole into its hydrochloric acid salt.

[5] This may not seem to be a valid statement to the student who started the preparation with benzene. However, acetanilide is available at very low cost, so that it is only one simple step to convert acetanilide to *p*-acetamidobenzenesulfonyl chloride. On the other hand, chloroacetaldehyde, which is needed for the preparation of 2-aminothiazole, is very expensive.

$$(32)$$

slight excess to maximize conversion of the more expensive reagent to product.

Important side reactions occur when the sulfonyl chloride reacts with 2-amino-thiazole in alternative ways, and some of these reactions are shown in equation (33). None of the side products so obtained have any hydrogen atoms on a sulfonamide nitrogen, and they will therefore be insoluble in the alkaline solution which is used in the work-up procedure (see the Hinsberg reaction, Chapter 20, Section A.1 under "Amines").

$$(33a)$$

$$(33b)$$

Purification of the sulfathiazole is accomplished by treating the reaction mixture with water, collecting the solid which forms, and treating the solid with excess base and filtering. Compound **22** is soluble in base, whereas the side products indicated by equation (33) are not. After removal of the side products, the basic solution is heated to effect hydrolysis of the amide. Under the reaction conditions, the sulfonamide group is not affected. The basic solution is acidified and then treated with sodium acetate until the pH of the solution is about 5. The acetate reacts with the mineral acid present to produce acetic acid, which in the presence of acetate ion yields a buffer solution. In strongly acidic solution, sulfathiazole is protonated and is water soluble, but after the solution is adjusted to pH 5, the sulfathiazole is liberated.

EXPERIMENTAL PROCEDURE

The experimental procedures start with benzene and carry through to sulfathiazole. The instructor will indicate at which point to start, and how far along the synthesis to proceed. Remember to adjust quantities of reagents according to the amount of starting material which is available. Beginning with benzene, for example, some of the reactions may give better yields than those indicated, whereas others may give poorer yields. In going to the next step, this must be taken into account; however, do not change the reaction times unless told to do so. It may be convenient to use different size flasks.

Nitrobenzene.[6]   Place 36 ml of concentrated sulfuric acid in a 250-ml Erlenmeyer flask and cool the flask in an ice bath. To the cool acid, add 28 ml of concentrated nitric acid, with continued cooling and shaking. After the exothermic reaction subsides and the mixture has cooled to room temperature, transfer it to a 250-ml separatory funnel. Place 0.25 mole of benzene in a 500-ml Erlenmeyer flask and then add about 7 ml of the sulfuric-nitric acid mixture. Insert a thermometer in the reaction mixture and leave it there during the duration of the nitration. Gently swirl the flask, and maintain the temperature of the reaction mixture between 50 and 60° by means of a water bath. As the exothermic reaction ceases, add the acid mixture in small portions until it has all been added, being careful to control the temperature between 50 and 60° throughout the additions.[7] After the addition is complete, gently swirl the mixture until the temperature drops below 40°. Cool the flask and transfer the mixture to a separatory funnel. Remove the organic layer and wash it with about 50 ml of water, 50 ml of 0.5 M sodium hydroxide, and another 50 ml of water. Discard the aqueous washings and transfer the nitrobenzene to a clean Erlenmeyer flask. Add about 5 g of anhydrous calcium chloride and warm the flask with a water bath until the cloudiness of the organic layer disappears.★ Filter the crude nitrobenzene into a distilling flask, and fractionally distil. Collect the product boiling 205–212°.[8] The presence of a small amount of water, as evidenced by cloudiness in the nitrobenzene fraction, is a result of incomplete drying. Such a cloudy fraction should be collected separately, redried, and redistilled. *When distilling, do not heat the flask to dryness.* The residue, which contains mostly *m*-dinitrobenzene, may decompose if heated above the boiling point of nitrobenzene. Only a small amount of dinitrobenzene forms, but it still may be enough to cause a small explosion. The yield of nitrobenzene should be 83–85%.

Aniline.   Place 30 g of iron powder[9] in a 500-ml round-bottomed flask and add to it 0.24 mole of nitrobenzene and 100 ml of water. Attach a reflux condenser and add about 0.5 ml of concentrated hydrochloric acid through the top of the condenser. Shake the reaction mixture vigorously, and if the reaction does not start soon (as evidenced by the production of heat), heat it gently with a small flame (be prepared to cool the flask in a pan of water should it appear that the reaction is going to get out of control). After the reaction has started, add another 0.5 ml of concentrated hydrochloric acid and shake the flask vigorously. When the reaction has subsided, bring the mixture to gentle reflux. After 15 min, add 1 ml of concentrated hydrochloric acid and then continue the heating under reflux for an additional 45 min.[10]

---

[6] Nitrobenzene is toxic. Care should be taken to avoid inhalation of its vapors.

[7] The yield of dinitrobenzene increases if the reaction temperature is allowed to rise above 60°.

[8] Unless special distillation columns are used, high-boiling materials, such as nitrobenzene, are normally collected over a wide range. This range is caused by air-cooling of the column, and the boiling points observed under these conditions are normally slightly lower than the "true" boiling point.

[9] Iron powder should be finely divided and should be free of an oxide coating. The reduction reaction is a heterogeneous surface reaction, so that a large area of metal surface is needed. Iron metal obtained by reduction of the oxide with hydrogen is termed "reduced with hydrogen" and is most suitable for this reduction reaction.

[10] If the reaction required heating with a burner for initiation, heat under reflux for at least 90 additional min.

When the reflux period is complete, add 5 ml of 6 *M* sodium hydroxide solution directly to the reaction mixture and equip the flask for steam distillation (Chapter 2.6 and Figures 2.9, 2.10, and 2.11). Steam distil the mixture until the distillate dropping from the condenser no longer contains any visible amount of organic product. (The distillate may still be slightly cloudy, but if there is no visible amount of oil in it, it may be assumed that the distillation is complete.) ★ Add 20 ml of concentrated hydrochloric acid to the distillate, and steam distil this mixture until the *residue* in the distilling flask is clear and free from oily material.★

Make the acidic distillation residue basic with a minimum volume of 12 *M* sodium hydroxide solution. (*Use care:* Heat is evolved.) Saturate the basic solution with sodium chloride (roughly 25 g of salt per 100 ml of solution), cool the mixture and transfer it to a separatory funnel. Extract the organic product with two 50-ml portions of carbon tetrachloride using the first portion to rinse the flask in which the neutralization was done. Separate the aqueous layer from the organic layer as thoroughly as possible; transfer the combined organic extracts to a small Erlenmeyer flask and dry with several sodium hydroxide pellets until the solution is clear.★ Transfer the solution by decantation into a distilling flask and distil it. Collect three fractions, one boiling between 70 and 90°, the second boiling between 90 and 180° and the third boiling between 180 and 185°. Discard the first fraction and, if the second fraction is of significant volume, redistil it to obtain more product. Pure aniline is colorless but may darken immediately following distillation owing to air oxidation. Aniline is often redistilled just before use to remove colored oxidation products. The yield of aniline should be 85–90%.

Acetanilide.   Dissolve 0.215 mole of aniline in 500 ml of water to which 18 ml of concentrated hydrochloric acid has been added; use a 1000-ml Erlenmeyer flask. Swirl the mixture to aid in dissolving the aniline. (If the solution is dark-colored, add about 1 g of decolorizing carbon to it, swirl and filter.) Prepare a solution containing 0.24 mole of sodium acetate trihydrate dissolved in 100 ml of water, and measure out 24 ml of acetic anhydride. Warm the solution containing the dissolved aniline to 50°, and add the acetic anhydride; swirl the flask to dissolve the anhydride. Then add the sodium acetate solution *immediately and in one portion.* Cool the reaction mixture in an ice bath with stirring while the product crystallizes. Collect the acetanilide by suction filtration, wash it with a small portion of cold water, and dry it.★ Determine the melting point and yield of product; the yield should be between 65 and 75%. If impure or slightly colored acetanilide is obtained, it may be recrystallized from a minimum volume of hot water, using decolorizing carbon if necessary to give a colorless product.

p-**Acetamidobenzenesulfonyl chloride.**   To a dry 250-ml Erlenmeyer flask, add 0.92 mole of chlorosulfonic acid, ClSO₃H. *Use care in handling chlorosulfonic acid. Be careful to avoid contact with the skin and with moisture. Chlorosulfonic acid reacts vigorously with water; care should be used in washing any equipment that has contained the acid.* If possible, carry out the following reaction in the hood or use a gas trap to keep hydrogen chloride or sulfur dioxide gases, which are by-products of the reaction, from being given off in the room. A simple gas trap can be constructed by attaching a filter flask containing some 12 *M* sodium hydroxide solution to the reaction flask, as shown in Figure 17.6. The base will react with the sulfur dioxide and hydrogen chloride.

Cool the flask to 10 –15° in a water bath containing a little ice, but do not cool below 10°. Add 0.155 mole of dry, finely powdered acetanilide to the chlorosulfonic acid in

FIGURE 17.6 Gas trap arrangement.

small portions and with good mixing. Do not allow the temperature to rise above 20°. After all of the acetanilide has been added and most of it has dissolved, allow the mixture to warm to room temperature. Then heat the mixture on a steam bath for *ca.* 10 min to complete the reaction. Cool the reaction mixture to room temperature (or slightly lower), using an ice bath. Place 600 g of cracked ice and 100 ml of water in a large beaker, and pour the reaction mixture slowly and with stirring onto the ice. *Use care in this step; do not add the mixture too quickly and avoid splattering of the chlorosulfonic acid.* After the addition is complete, rinse the flask with a little cold water and transfer this to the beaker. The precipitate which forms in the beaker is crude *p*-acetamidobenzenesulfonyl chloride, which may be white to pink in color. It soon may become a hard mass, and any lumps that form should be broken up with a stirring rod. Collect the crude material by suction filtration. Wash the solid with a small amount of cold water, and press it as dry as possible with a cork.

If sulfanilamide is to be prepared from the sulfonyl chloride, the crude product does not need to be purified but may instead be used directly. *If sulfathiazole is to be made, the sulfonyl chloride must be purified.*

For purification, dissolve the crude product in a minimum amount of boiling chloroform (the approximate volume required is 135 to 185 ml). Preheat a separatory funnel on the steam bath and transfer the boiling chloroform mixture to it. Remove the organic layer from the water layer as quickly as possible; be very careful in separating the layers. Rinse the funnel with an additional 40-ml portion of hot chloroform, and combine this rinse with the main organic layer. On cooling, colorless crystals of pure product are obtained.★ The yield should be 77–90%; the reported melting point of pure *p*-acetamidobenzenesulfonyl chloride is 149°. Air-dry the product and store it in a closed container.

Sulfanilamide. Transfer the crude *p*-acetamidobenzenesulfonyl chloride to a 500-ml Erlenmeyer flask and add 75 ml of concentrated ammonium hydroxide (28%). A very rapid exothermic reaction should occur. Break up any lumps of solid which may remain with a stirring rod; the reaction mixture should be thick but homogeneous insofar as is possible. Heat the mixture on a steam bath for about 30 min. Heating should be done in a well-ventilated room or preferably in a hood; some ammonia vapors will be released into the air. Cool the flask in an ice bath and add 6 *M* sulfuric acid to the cool reaction mixture until it is acidic to Congo red paper. Cool the reaction mixture again in an ice bath, and collect the product by suction filtration. Wash the crystals with cold water and dry.* The yield should be 89–96%. The product is pure enough for the hydrolysis reaction, but if desired it may be purified as follows. Dissolve the crude *p*-acetamidobenzenesulfonamide in a minimum amount of hot water. If needed, use decolorizing carbon. After filtering, cool the aqueous solution in an ice bath and collect and dry the crystals.* The melting point of the pure product is 220°.

Weigh the *p*-acetamidobenzenesulfonamide and transfer it to a 250-ml round-bottomed flask. Prepare a solution of dilute hydrochloric acid by mixing equal volumes of concentrated hydrochloric acid and water. Add to the amide an amount of dilute hydrochloric acid solution *twice* the weight of the amide. Attach a reflux condenser to the flask and heat at a gentle reflux for 30 min. Boil gently and swirl the reaction mixture when first heating so as to aid in the dissolution of the organic material. To the homogeneous reaction mixture, add an equal quantity of water, and transfer the new mixture to a 600-ml beaker. Neutralize the excess acid which is present by the addition of small quantities of solid sodium carbonate; continue addition of the base until the solution is just alkaline to litmus paper. (*Caution:* Add the sodium carbonate in small quantities, for foaming will occur!) A precipitate should form during neutralization. After making the solution just basic, cool the mixture in an ice bath to complete the precipitation of the product. Collect the crystals by suction filtration, wash with a small amount of cold water, and allow the product to air-dry.* Purify the crude product by recrystallization from hot water (12 to 15 ml of hot water per gram of compound will be required). Decolorize the hot solution, if necessary, filter and cool the filtrate in an ice bath. It may be necessary to preheat the filter funnel so that the product will not crystallize in the funnel. Sulfanilamide will form on cooling to give long, white needles; collect the product by suction filtration. Determine the melting point (reported 163°) and the yield, which should be about 0.7 g per gram of starting *p*-acetamidobenzene-sulfonamide.

Test the solubility of sulfanilamide in 5% hydrochloric acid solution and in 5% sodium hydroxide solution.

2-Aminothiazole.   Prepare a solution of 0.2 mole of thiourea and 90 ml of water in a 500-ml round-bottomed flask; the water may need to be warmed to effect complete solution. Add 0.2 mole of chloroacetaldehyde dimethyl acetal and 1.5 ml of 85% phosphoric acid,[11] and equip the flask with an efficient reflux condenser. Heat the mixture under reflux by means of a water bath for 2.5 hr. Fit the flask for simple distillation and slowly collect and then discard 60 ml of distillate. Cool the reaction mixture in an ice-water bath and make it slightly basic to litmus by slow addition of 12 *M* sodium hydroxide solution. Avoid a large excess of base. After cooling for an additional 15 min, collect the precipitated solid by suction filtration. Concentrate the mother liquor

[11] Alternative ways of obtaining chloroacetaldehyde are discussed in the text; if one of them is to be used, *substitute* the following reagents for the chloroacetaldehyde dimethyl acetal and phosphoric acid.

1. *Use of α,β-dichloroethyl ethyl ether:* Substitute 0.2 mole of α,β-dichloroethyl ethyl ether for the above reagents.

2. *Use of 40% aqueous chloroacetaldehyde:* Dissolve the thiourea in 44 ml of water and add 36 ml of a 40% aqueous solution of chloroacetaldehyde.

to about two-thirds of its initial volume, readjust the pH by adding more base until just basic to litmus, and recool. Collect the second crop of crystals and combine it with the first crop. Wash the product twice with 20-ml portions of ice-cold half-saturated sodium bisulfite solution, and press as dry as possible.*

For purification of this crude product, dissolve it in hot benzene (10 ml per gram of product), treat with decolorizing carbon, and filter by gravity while hot (*Caution:* Extinguish flames!); filter-aid will aid in the complete removal of the decolorizing carbon. Collect the filtrate in a distilling flask, and distil about one-half of the benzene from the mixture. If the pressing of the initial precipitate has been thorough, the small amount of water retained is removed at this point by azeotropic distillation with benzene. When the remaining benzene solution is cooled, 2-aminothiazole (mp 89–90°) is obtained. Dry the solid, and determine its melting point and the yield.

Sulfathiazole.   Dry about 40 ml of reagent grade pyridine over 2–3 g of potassium hydroxide pellets.[12] To a 250-ml round-bottomed flask equipped with a Claisen connecting tube, add 0.075 mole of 2-aminothiazole and about 30 ml of dry pyridine. Transfer the dry pyridine by use of a graduated pipet. Equip the flask with a calcium chloride drying tube and a thermometer. Add 0.082 mole of purified *p*-acetamido-benzenesulfonyl chloride to the reaction flask in small portions; swirl gently to insure mixing, and make the addition at such a rate that the temperature inside the flask does not exceed 40°. Do not cool the reaction mixture at any time during the addition. After the addition is complete, heat the mixture on a steam bath for *ca.* 30 min. Then cool the mixture and pour it into about 200 ml of warm water while stirring vigorously with a stirring rod. The oil which forms initially should solidify if a stirring rod is used to induce crystallization. Remove the solid by filtration,* wash it with a small amount of water, and dissolve it in 38 ml of 2 *M* sodium hydroxide solution; warm on a steam bath, if needed, to bring the solid into solution. If any solid remains, filter it off and discard it. Make the resulting filtrate acidic to litmus by the addition of glacial acetic acid, collect the solid which forms by suction filtration, and press as dry as possible.* Weigh the solid, add 10 times its weight of 2 *M* sodium hydroxide solution, and heat the mixture at reflux for one hour. After heating, cool the mixture and slowly add concentrated hydrochloric acid; at first a solid will form, which will then redissolve as excess acid is added. Add only enough acid so that the solid redissolves; do not add excess acid. After decolorizing and filtering the solution, neutralize it with 2 *M* sodium hydroxide, and then add solid sodium acetate in small portions until the solution is basic to litmus paper. Bring the heterogeneous mixture to boiling and then cool in an ice bath; collect the solid by suction filtration * and recrystallize it from a minimum amount of boiling water. Collect the sulfathiazole by suction filtration and air-dry it. Determine the melting point (reported 201–202°) and the yield.

EXERCISES

1. Outline a possible synthesis for the compound shown below, using benzene as the only source of an aromatic ring. Use any needed aliphatic or inorganic reagents.

$$H_2N - \bigcirc - SO_2NH - \bigcirc$$

---

[12] This should be done during the laboratory period before the experiment is to be performed. Add potassium hydroxide pellets to the pyridine and store in the laboratory desk until ready to use.

2. Outline the procedure for the purification of aniline in flow diagram form. Indicate the importance of each step in the procedure, and give reasons for doing the steam distillation first with a basic solution and then with an acidic solution. Write equation(s) for reactions which occur when base and then acid are added.

3. In the purification of aniline, sodium hydroxide pellets are used as the drying agent. Why is this compound used rather than magnesium sulfate, calcium chloride or other common drying agents? Would potassium hydroxide be an acceptable substitute for sodium hydroxide?

4. In the preparation of sulfanilamide from *p*-acetamidobenzenesulfonamide, only the acetamido group is hydrolyzed. Provide an explanation for this difference in reactivity of acetamido and sulfonamido groups.

5. Explain the results which were obtained when the solubility of sulfanilamide was determined in 5% hydrochloric acid and in 5% sodium hydroxide. Write equations for any reaction(s) occurring.

6. What would be observed if *p*-acetamidobenzenesulfonamide were subjected to vigorous hydrolysis conditions (*i.e.*, concentrated hydrochloric acid and heat for a long period of time)? Write an equation for the reaction which would occur.

7. With the aid of chemical equations, explain the behavior of sulfathiazole towards acid and base. How do these reagents affect its solubility in water?

8. Outline the workup of sulfathiazole in flow diagram form. Indicate the importance of each step in the procedure, and give reasons for each of them. Write equation(s) for the reactions which occur in the purification process.

SPECTRA OF STARTING MATERIALS AND PRODUCTS

The ir and nmr spectra of aniline are given in Figures 16.3 and 16.4, respectively.

FIGURE 17.7   IR spectrum of nitrobenzene.

FIGURE 17.8   IR spectrum of acetanilide.

FIGURE 17.9   NMR spectrum of acetanilide.

FIGURE 17.10   IR spectrum of *p*-acetamidobenzenesulfonyl chloride.

FIGURE 17.11   IR spectrum of sulfanilamide.

FIGURE 17.12   NMR spectrum of sulfanilamide.

FIGURE 17.13   IR spectrum of 2-aminothiazole.

FIGURE 17.14 NMR spectrum of 2-aminothiazole.

FIGURE 17.15 IR spectrum of sulfathiazole.

17.3 The Synthesis of 1-Bromo-3-chloro-5-iodobenzene

Another interesting multi-step synthesis is the preparation of 1-bromo-3-chloro-5-iodobenzene **(32)**, which can be initiated from benzene. The complete sequence of this eight-step synthesis is outlined in Chart 17.4. Most of the reactions involve electrophilic aromatic substitution. Two important stratagems of aromatic chemistry are incorporated in the synthesis: (1) An amino group is "protected" by conversion to the amide and, after several electrophilic substitution reactions are performed on the molecule, the free amine is regenerated, and (2) an amino group, which was introduced into the molecule initially because of its directing influence and activating properties, is removed in the final step to yield the desired product. The usefulness of the amino group will be indicated as the various steps of the reaction sequence are discussed. It should be noted that the yields of each individual step range from 60 to 96%, which is always

CHART 17.4 Laboratory synthesis of 1-bromo–3-chloro–5-iodobenzene (**32**).

an important consideration in multi-step syntheses. In general, the reaction times are short enough so that one or more steps can be carried out in a single laboratory period.

The synthesis of **32** may be started anywhere along the reaction sequence and, if desired, may be started from benzene. The nitration of benzene, reduction of nitrobenzene and conversion of aniline to acetanilide have already been discussed under Section 17.2 (sulfathiazole), and the experimental details are provided therein. The discussion here will start with the synthesis of 4-bromoacetanilide (**33**) from acetanilide.

4-Bromoacetanilide (33). The first halogen to be introduced into the molecule is bromine. The amino group activates the aromatic ring so greatly that any bromination conditions lead to trisubstitution on aniline, and indeed an aqueous solution of bromine is sufficient to do this [equation (34)]. In order to decrease the reactivity of aniline so that monosubstitution can be realized, aniline is acetylated to give acetanilide. Under mild conditions (bromine in acetic acid), acetanilide can be mono-brominated easily (note that no Lewis acid catalyst, such as $FeBr_3$ is required, for acetanilide is still quite reactive towards bromine).

As the acetamido group, CH_3CONH-, is *ortho,para*-directing, both 2-bromo- and 4-bromoacetanilide are produced; however, under the conditions of the reaction, 95% of the 4-bromo isomer is obtained [equation (35)]. 4-Bromoacetanilide can be separated from the 2-bromo isomer by a single recrystallization of the crude product from methanol.

In general, the separation and purification of *ortho* and *para* isomers from a mixture containing both of them is fairly simple, as illustrated by the ease with which 4-bromoacetanilide is purified. Even though different functional groups may be attached to an aromatic ring, the *para* isomer is more "symmetrical" than is the *ortho* isomer, and one generally observes the former to possess a much higher melting point

$$\text{(34)}$$

$$\text{(35)}$$

18 **33**
95%   5%

than the latter. Also, the *para* isomer is normally less soluble in a given solvent than is the *ortho* isomer, so that fractional crystallization can be used to separate them from one another. For example, in this experiment, the melting point of 4-bromoacetanilide is 167° and that of 2-bromoacetanilide is 99°. The 4-bromo isomer is much less soluble in methanol than is the 2-bromo isomer, so that recrystallization of a mixture of these compounds from methanol gives 4-bromoacetanilide as a solid whereas the 2-bromo isomer remains in solution.

2-Chloro-4-bromoacetanilide (34).

The next halogen to be introduced into the molecule is chlorine. The bromine atom in 4-bromoacetanilide deactivates the ring slightly towards further electrophilic substitution (as compared to acetanilide), but the compound is still reactive enough so that it can be chlorinated without using a catalyst. Chlorine in acetic acid will monochlorinate 4-bromoacetanilide to produce 2-chloro-4-bromoacetanilide (34) [equation (36)]. Although it is possible to use chlorine gas

$$\text{(36)}$$

33 **34**

dissolved in acetic acid, the difficulties and hazards of using this gas with large numbers of students make it desirable to generate chlorine gas *in situ*. One convenient method of *in situ* preparation of chlorine is to allow hydrochloric acid and sodium chlorate (NaClO$_3$) to react; an oxidation-reduction reaction occurs, in which chloride ion is oxidized to chlorine and chlorate ion is reduced to chlorine as shown in equation (37).

OXIDATION: $2\ Cl^{\ominus} = Cl_2 + 2\ e^{\ominus}$ (37a)

REDUCTION: $\underline{12\ H^{\oplus} + 2\ ClO_3{}^{\ominus} + 10\ e^{\ominus} = Cl_2 + 6\ H_2O}$ (37b)

OVERALL: $5\ Cl^{\ominus} + ClO_3{}^{\ominus} + 6\ H^{\oplus} = 3\ Cl_2 + 3\ H_2O$ (37c)

In addition to ease and safety in handling, just the right amount of chlorine gas can be prepared by this method. The reaction between chlorine and 4-bromoacetanilide occurs very rapidly, and little, if any, dichlorination product is obtained. 4-Bromo-2-chloro-acetanilide is purified by fractional crystallization from methanol.

2-Chloro-4-bromoaniline (35).   2-Chloro-4-bromoacetanilide **(34)** can be converted to 2-chloro-4-bromoaniline in excellent yield using acid-catalyzed hydrolysis of the amide [equation (38)]. Concentrated hydrochloric acid in ethanol is employed, with the ethanol being used to increase the solubility of the amide.

$$(38)$$

2-Chloro-4-bromo-6-iodoaniline (36).   The 2-chloro-4-bromoaniline **(35)**, which was obtained in the previous reaction, is highly activated due to the presence of the amino group. Of the positions *ortho* and *para* to the NH_2 group, only the 6-position is un-substituted. Thus, iodine may be added in that position by starting with the free amine and without need for the deactivating effect of the acetamido group. Any electrophilic substitution reaction which might be carried out on **35** would result in monosubstitu-tion at the free 6-position.

Electrophilic substitution reactions involving iodine are generally more difficult to carry out than are chlorination and bromination reactions. One general method uses iodine and potassium carbonate, with the base being present to absorb the hydroiodic acid which forms. Another good method makes use of iodine monochloride, ICl, as the iodinating agent. Of the two halogens present in ICl, chlorine is more *electro-negative* than is iodine. Thus, when the molecule dissociates, the bonding electrons leave with chlorine to give Cl^- and I^+ [equation (39)]. I^+ serves as an electrophile which attacks the aromatic ring and replaces H^+ at the 6-position [equation (40)].

$$I-Cl \xrightleftharpoons{CH_3CO_2H} I^{\oplus} + Cl^{\ominus} \qquad (39)$$

$$(40)$$

It should be pointed out that I^+ is a very weak electrophile compared to Br^+ or Cl^+. One of the main reasons for introducing iodine in the final step of the substitution sequence is that a very activated aromatic ring must be used in order to realize a good yield in any iodination substitution reaction. Thus, bromine and chlorine are introduced with the acetamido group present, since electrophilic substitution with these elements requires a less activated ring than does iodine. Also, with the acetamido group present, the reaction can be controlled to give monosubstitution. Once the chlorine and bromine atoms are in the desired positions, the highly activating free amino group is regenerated and iodination is carried out with ease and without the possibility of disubstitution.

There are no important side reactions in this preparation; unchanged starting material can be separated from the product by fractional crystallization.

1-Bromo-3-chloro-5-iodobenzene (32).   The final step in the synthesis is the removal of the amino group from 2-chloro-4-bromo-6-iodoaniline **(36).** After the amino group has been diazotized with nitrous acid at 0°, the diazo group can be removed by treatment with hypophosphorous acid (H_3PO_2) or by treatment with absolute ethanol. The former is the method most often encountered in textbooks, and indeed it is the best general reagent for replacing the diazo group by a hydrogen. When hypophosphorous acid is used, it is oxidized to phosphorous acid as the diazonium salt is reduced.

The use of ethanol has been found to be successful with certain types of aromatic compounds, in particular those which contain halogen. As this method is easier to use in the laboratory, it will be employed here. Presumably an oxidation–reduction reaction occurs with ethanol as well; in this case ethanol is oxidized to acetaldehyde or acetic acid as the diazo compound is reduced [equation (41)].

$$(41)$$

There is one possible important side reaction which might occur when this procedure is used. The diazonium salt might dissociate to give a carbonium ion and nitrogen. The carbonium ion could then attack ethanol to give an aryl alkyl ether as a side product [equation (42)]. It is well known that this type of reaction occurs with certain types of

$$(42)$$

aromatic amines; for example, o- and m-toluidine (methylanilines) give high yields of the corresponding ethers when diazotized and treated with an alcohol. It appears, however, that ether formation is not important in the present reduction, and if any ether is produced, it is removed in the recrystallization of the product.

In diazotization reactions involving aromatic halo compounds, hydrohalic acids containing a *different* halogen should never be used. There are numerous examples of halogen interchange between the ring and the mineral acid during diazotization and subsequent reduction of the diazo group. This interchange occurs only with the *ortho* and *para* halogens (of which there are three in compound **36**), but not with the *meta* halogens. When a halogen is present in the ring, it is best to produce the nitrous acid from sodium nitrite and *sulfuric acid* as is done here.

EXPERIMENTAL PROCEDURE

It is possible to start the synthesis anywhere along the sequence, but this normally will not extend beyond 4-bromoacetanilide, which is the last compound that is commercially available at a reasonable cost.

In each of the procedures below, the crude product from the previous reaction can be used without further purification. However, methods of purification for each intermediate compound are given, should the student wish to purify the starting materials or to stop somewhere along the reaction sequence. The melting points reported are those for once recrystallized material.

In carrying out a multi-step synthesis, it *may* be necessary to increase or decrease the scale of a subsequent reaction, as dictated by the yield of a previous product. However, do not change reaction *times* unless told to do so by the instructor.

Nitrobenzene, aniline, and acetanilide.   See Section 17.2 for experimental procedures.

4-Bromoacetanilide.   Dissolve 0.06 mole of acetanilide in 30 ml of glacial acetic acid in a 250-ml round-bottomed flask, and add to this mixture slowly and with stirring a solution of 0.061 mole of bromine dissolved in 6 ml of glacial acetic acid. *Use special care when handling bromine; wear gloves and avoid contact with the skin. Consult the instructor before using bromine.* After addition of the bromine-acetic acid solution is complete, stir for several minutes and then add slowly, with stirring, 200 ml of water. On addition of water, a solid will form. Prepare a saturated solution of sodium bisulfite in water, and add just enough of it to discharge the yellow color of the solution. Collect the product by suction filtration; wash the solid well with water and allow it to air-dry.★ The yield of crude product should be about 96%. Recrystallization of the crude product from methanol (about 3–4 ml of methanol per gram of compound) gives 4-bromoacetanilide, mp 171–172°.

2-Chloro-4-bromoacetanilide.

Note: It would be preferable to carry out this reaction in the hood or to invert a funnel over the reaction flask and attach the funnel to the water aspirator.

Suspend 0.05 mole of 4-bromoacetanilide in a mixture of 23 ml of concentrated hydrochloric acid and 28 ml of glacial acetic acid, using a 250-ml flask. Heat the mixture gently on a steam bath until it becomes homogeneous, and then cool the solution to 0°. To the cold mixture, add 0.026 mole of sodium chlorate dissolved in about 7 ml of water. During the addition of the sodium chlorate solution, some chlorine gas is evolved. As the addition is carried out, a yellow precipitate forms and the solution turns yellow. Allow the reaction mixture to stand at room temperature for one hour, and collect the precipitate by suction filtration.★ The material which is obtained is crude 2-chloro-4-bromoacetanilide; the yield should be about 97%. The crude product can be recrystallized from methanol (7–8 ml of methanol per gram of crude product) to give pure 2-chloro-4-bromoacetanilide, mp 153–154°.

2-Chloro-4-bromoaniline.   Mix 0.05 mole of crude 2-chloro-4-bromoacetanilide with 20 ml of 95% ethanol and 13 ml of concentrated hydrochloric acid in a 250-ml Erlenmeyer flask. Heat the mixture on the steam bath for *ca.* 30 min; during the heating, the yellow precipitate should dissolve and be replaced by a white precipitate. At the end of the heating, add 90 ml of hot water. Swirl the flask to dissolve the white solid completely and pour the solution onto 50 g of ice. Add to the resulting mixture 12 ml of 50% sodium hydroxide solution, stirring well during the addition and mixing process. Light brown crystals should precipitate during the addition of the base. Collect these by suction filtration and dry them as thoroughly as possible.★ The yield should be about 91%. Recrystallize the crude product from 30–60° petroleum ether (3–4 ml per gram of product) to give 2-chloro-4-bromoaniline, mp 65–66°.

2-Chloro-4-bromo-6-iodoaniline. Dissolve 0.024 mole of recrystallized 2-chloro-4-bromoaniline in 80 ml of glacial acetic acid, and add about 20 ml of water to the mixture. Prepare a solution of 0.03 mole of technical iodine monochloride in 20 ml of glacial acetic acid in an Erlenmeyer flask and add this solution to the reaction mixture over a period of 8 min. Heat the resulting black mixture on the steam bath until its temperature is 90°, and then add only enough saturated sodium bisulfite solution to turn the color of the mixture bright yellow; note the volume of sodium bisulfite solution added. Dilute the reaction mixture with enough extra water such that the volume of the sodium bisulfite solution *plus* the volume of this added water is about 25 ml. Cool the reaction mixture in an ice bath; 2-chloro-4-bromo-6-iodoaniline will separate as light brown crystals. Collect the solid by suction filtration and wash the crystals with a small amount of 33% acetic acid and then with water.★

It would be advisable to purify the product before going on to the final step of the synthesis. The crude product may be recrystallized from acetic acid-water as follows. Mix the product with glacial acetic acid in the ratio of about 20 ml of glacial acetic acid per gram of product. Heat the mixture on the steam bath and slowly add 5 ml of water per gram of product to the solution as it is heating. On slow cooling, long colorless crystals of 2-chloro-4-bromo-6-iodoaniline will form; filter and dry the pure product. The recovery should be about 80% (based on crude product) and the melting point should be 96–98°.

1-Bromo-3-chloro-5-iodobenzene. Suspend 0.006 mole of 2-chloro-4-bromo-6-iodoaniline in about 10 ml of absolute ethanol in a 250-ml round-bottomed flask. While stirring the mixture, add 4.0 ml of concentrated sulfuric acid dropwise. Equip the flask with a condenser and add 0.01 mole of powdered sodium nitrite in small portions through the condenser. When the addition is complete, heat the mixture on a steam bath for about ten minutes. Add 50 ml of hot water to the flask through the condenser, and steam distil. Collect about 80 ml of distillate (which should be clear and not contain any organic product). The desired product will form as a solid in the condenser, so that care should be taken to insure that the condenser does not plug up completely during the distillation.[13] It is easiest to remove the solid product from the condenser by pouring ether through it; this dissolves and removes the product.★ Dry the ether extracts over anhydrous magnesium sulfate, filter, and distil the ether. (*Use care:* Ether removal is best done in the hood; use a steam bath and be sure that no flames are close by.) Recrystallize the residue from about 10 ml of methanol [14] to give

[13] Should the condenser plug up during the distillation, it can be opened up by running hot water or a slow stream of steam through the condenser jacket.

[14] It may be necessary to heat the filter funnel before filtering the solution, since the solid will crystallize in the stem of a cold funnel before the filtrate reaches the flask. See the discussion in Chapter 2.7 and Figure 2.12.

1-bromo-3-chloro-5-iodobenzene in about 40% yield. The long, nearly colorless needles should melt between 82 and 84°.

EXERCISES

1. Outline synthetic procedures, starting from benzene, for each of the following compounds:

   (a) 1,3,5-tribromobenzene; (b) 2-bromo-4-chloro-6-iodophenol; (c) 2-bromo-4,6-dichloroaniline.

2. In the bromination of acetanilide, using bromine in acetic acid, the major product is 4-bromoacetanilide. Suggest a reason for this. Give another example of an electrophilic substitution reaction which gives the *para* isomer as the predominant product.

3. Give the complete, step-wise mechanism for the reaction which occurs between iodine monochloride and 2-chloro-4-bromoaniline. Suppose that this same reaction were carried out using bromine monochloride, BrCl. What electrophilic substitution reaction might occur when this compound is allowed to react with 2-chloro-4-bromoaniline? Briefly explain.

4. Hydroxylic solvents, such as water or low-molecular-weight alcohols, are often used to purify amides. Briefly explain why these are used rather than hydrocarbon solvents, such as petroleum ether.

REFERENCE

A. Ault and R. Kraig, *Journal of Chemical Education*, **43**, 213 (1966).

SPECTRA OF STARTING MATERIALS AND PRODUCTS

The ir and nmr spectra of aniline are given in Figures 16.3 and 16.4, respectively, and the corresponding spectra of acetanilide in Figures 17.8 and 17.9.

FIGURE 17.16 IR spectrum of 4-bromoacetanilide.

FIGURE 17.17 NMR spectrum of 4-bromoacetanilide.

FIGURE 17.18   IR spectrum of 2-chloro–4-bromoacetanilide.

FIGURE 17.19   NMR spectrum of 2-chloro–4-bromoacetanilide.

FIGURE 17.20   IR spectrum of 2-chloro–4-bromoaniline.

FIGURE 17.21 NMR spectrum of 2-chloro–4-bromoaniline.

FIGURE 17.22 IR spectrum of 2-chloro–4-bromo–6-iodoaniline.

FIGURE 17.23 NMR spectrum of 2-chloro–4-bromo–6-iodoaniline.

FIGURE 17.24   IR spectrum of 1-bromo–3-chloro–5-iodobenzene.

FIGURE 17.25   NMR spectrum of 1-bromo–3-chloro–5-iodobenzene.

Nonbenzenoid aromatic compounds

18.1 Aromaticity

In recent years many fascinating examples of compounds exhibiting chemical properties typical of benzene, yet having no apparent molecular similarity to it, have been discovered. These substances are said to have aromatic character, or "aromaticity," although as yet there is not complete agreement among chemists as to an exact definition of this term.

The problem revolves about which property or properties of benzene are considered to be most typical. The older concept of aromaticity involved resistance toward addition reactions and ease of substitution. This has been updated in terms of mechanisms; both ionic addition and free radical substitution are considered typical of aliphatic compounds, and electrophilic substitution is considered typical of aromatic compounds.

Another criterion that has been applied is *resonance energy*, which may be calculated from heats of combustion or hydrogenation. The resonance energy of an aromatic compound is a measure of the unusual stability of the molecule compared to some aliphatic model and is attributed to the cyclic conjugation of the aromatic compound. Naphthalene (1), phenanthrene (2), and other polycyclic aromatic hydrocarbons and their derivatives are obviously "benzenoid," since they essentially consist of two or more benzene rings fused together. The nitrogen-containing analogs of benzene, such as pyridine (3) and pyrimidine (4), are also usually considered benzenoid.

A different situation is encountered in azulene (5), which contains a cyclic conjugated system in fused five- and seven-membered rings. Azulene, then, must be considered nonbenzenoid. It is an isomer of naphthalene, to which it is converted quantitatively by heating above 350° in the absence of air [equation (1)]. Although less stable than naphthalene, azulene undergoes certain typical electrophilic substitution reactions.[1]

$$\underset{\textbf{5}}{\text{azulene}} \xrightarrow{>350°} \underset{\textbf{1}}{\text{naphthalene}} \qquad (1)$$

The heterocyclic compounds with five-membered rings, such as furan (**6**), pyrrole (**7**), and thiophene (**8**), constitute another class of aromatic compounds. Strictly speaking

$$\underset{\textbf{6}}{\overset{}{\text{O}}} \qquad \underset{\textbf{7}}{\overset{}{\underset{\text{H}}{\text{N}}}} \qquad \underset{\textbf{8}}{\overset{}{\text{S}}}$$

they are not benzenoid, since they do not have six-membered rings. However, they are closely related to benzene and pyridine in that they also have an "aromatic sextet" of electrons. The hetero atoms provide a pair of electrons which enter into conjugation with the four π-electrons of the carbon-carbon bonds, as shown in the resonance structures **6**, **9**, and **10** (there are two additional resonance structures equivalent to **9** and **10**).

$$\underset{\textbf{6}}{\text{structure}} \longleftrightarrow \underset{\textbf{9}}{\text{structure}} \longleftrightarrow \underset{\textbf{10}}{\text{structure}} \text{, etc.}$$

A hydrocarbon system closely related to these heterocyclic compounds is the cyclopentadienyl anion, represented by resonance structures **12**, **13**, and **14**, etc.,

$$\underset{\textbf{11}}{\text{cyclopentadiene}} \xrightarrow{-H^{\oplus}} \underset{\textbf{12}}{\text{anion}} \longleftrightarrow \underset{\textbf{13}}{\text{anion}} \longleftrightarrow \underset{\textbf{14}}{\text{anion}} \text{, etc.}$$

derived from cyclopentadiene (**11**). This compound is unusually acidic for a hydrocarbon ($pK_a = ca.$ 20), presumably owing to stabilization of its anion by delocalization of negative charge as represented by resonance structures **12**, **13** and **14**, etc.

One might infer from these examples that any cyclic conjugated system for which a number of reasonable contributing resonance structures may be written will have aromatic character. That this is not true is demonstrated by the fact that the cyclopentadienyl *cation* (**15**) has never been prepared, even though the same number of contributing resonance structures (**15**, **16**, and **17**, etc.) can be written as for the corre-

---

[1] In connection with this discussion of azulene, see Exercise 6 in Chapter 12.1.

11   **15**   **16**   **17**

sponding anion. Thus, it may be seen that simple *valence bond theory* is inadequate to explain the difference in stability of cyclopentadienyl cation and anion, and certain other aspects of aromaticity.

A *molecular orbital theory* proposed by the German chemist E. Hückel in 1931 has provided an explanation for such problems. Hückel's theory, which was based on quantum mechanics, stated that conjugated cyclic systems in which the number of delocalized π-electrons is $4n + 2$, where $n = 0, 1, 2, 3$, etc., will exhibit special stability, *i.e.*, aromaticity. Benzene, pyridine, furan, and the cyclopentadienyl anion all fit "Hückel's Rule" for $n = 1$; that is, the number of π-electrons is six ($4 \times 1 + 2 = 6$). The cyclopentadienyl *cation* system involves *four* π-electrons, so according to Hückel's Rule, it should not have special stability.

Hückel's Rule also offers an explanation for the fact that cyclooctatetraene **(18)**, with eight π-electrons, and cyclobutadiene **(19)**, with four, are nonaromatic. On the

18   **19**

other hand, cyclopropenyl cations **(20,** $n = 0$), and cycloheptatrienyl cation **(21,** $n = 1$), are predicted to be unusually stable.[2] One of the most significant examples of

20   **21**

the success of prediction based on theory in organic chemistry is the realization of the preparation of these cations and the demonstration of their remarkable stability.

A modern criterion of aromaticity may be described in connection with these cations. A carbonium ion may react with water reversibly to form an equilibrium mixture containing the corresponding alcohol and hydronium ion [equation (2)]. According to

$$R^{\oplus} + 2\ HOH \rightleftarrows ROH + H_3O^{\oplus} \tag{2}$$

$$K_{R^{\oplus}} = \frac{[ROH][H_3O^{\oplus}]}{[R^{\oplus}]} \tag{3}$$

[2] It is interesting to note that the cycloheptatrienyl anion, which has eight π-electrons, is extremely unstable. One can write the same number of equivalent contributing resonance structures for the cycloheptatrienyl cation and anion.

When $\dfrac{[ROH]}{[R\oplus]} = 1$, then $K_R\oplus = [H_3O^{\oplus}]$ (4)

and therefore $pK_R\oplus = \log \dfrac{1}{[H_3O^{\oplus}]} = pH$ (5)

equation (5), the $pK_R\oplus$ of a carbonium ion is equal to the pH of a solution in which the carbonium ion is half converted to the corresponding alcohol. An unconjugated carbonium ion is extremely reactive toward water, so that the $pK_R\oplus$ is a large negative number. The $pK_R\oplus$ of an alkyl carbonium ion, such as the *t*-butyl cation, cannot be measured, and even the triphenylmethyl cation has a $pK_R\oplus$ of -6.63. By contrast, the cycloheptatrienyl cation **(21)** has a $pK_R\oplus$ of $+4.7$ and the cyclopropenyl cation, **(20, R = *n*-C$_3$H$_7$)** has a $pK_R\oplus$ of $+7.2$. The latter cation is thus stable even in neutral water solution; it is the most stable hydrocarbon cation yet known. Since the $pK_R\oplus$ of a cation is a measure of its stability, it may also be used as a criterion of aromaticity in appropriate systems.

Another modern criterion of aromaticity is a *ring current* which is induced in an aromatic compound when it is placed in a magnetic field. An nmr spectrometer makes use of a strong external magnetic field, and evidence for the ring current is provided by the nmr spectrum of an aromatic compound.[3] The ring current attributable to the circulation of π-electrons in benzene results in the deshielding of the six hydrogens (which are *outside* the aromatic ring) so that their nmr signals are shifted downfield (to larger δ values). According to the ring current theory, hydrogen atoms on the *inside* of an aromatic ring should experience a shielding effect, resulting in an upfield shift of their nmr signal. This theory has been confirmed by the nmr spectrum of the interesting hydrocarbon **22**, which is called [18]annulene. This compound, which has 18 π-

22

electrons and thus should be aromatic according to Hückel's Rule ($n = 4$), has 6 hydrogen atoms inside the conjugated ring system and 12 outside. The two nmr signals occur at 1.9 ppm *upfield* from TMS (integration = 6H) and 8.8 ppm *downfield* from TMS (integration = 12H); thus, the inside hydrogen atoms are strongly shielded and the outside hydrogen atoms are deshielded, as expected for a ring current effect.

18.2 Ferrocene

Ferrocene **(23)** was discovered accidentally in 1951. It is an orange, crystalline solid whose structure consists of two cyclopentadienyl anions bonded to a ferrous cation.

[3] Refer to Chapter 4.2, p. 72.

The bonding between the metal and the organic anions involves the π-electrons of the two rings in such a way that all of the carbon atoms are bonded equally to the central ferrous ion. Several other metals form "sandwich-type" compounds with cyclopentadienyl anions, called *metallocenes*. Among the most stable of these, besides ferrocene, are those from the dipositive ions of ruthenium and osmium, since in these compounds, as well as in ferrocene, the metal ion achieves the electronic structure of an inert gas atom.

Ferrocene has the classical aromatic properties of resistance toward reaction with acids and bases, even concentrated sulfuric acid. It is sensitive toward oxidizing acids because the iron is oxidized to the ferric state and the resulting metallocene cation is much less stable (the iron no longer has the krypton electronic structure). Ferrocene does not undergo addition reactions typical of cyclopentadiene, but readily undergoes electrophilic substitutions, such as Friedel-Crafts acylation.

Depending on the catalyst and the conditions, either acetylferrocene **(24)** or 1,1'-diacetylferrocene **(25)** may be produced as the major product of acetylation. The struc-

$$(6)$$

ture of the disubstituted product has been established by degradation studies, using reactions which removed the iron from the organic part of the molecule and which gave no products indicative of the presence of two acetyl groups on a cyclopentadienyl ring. This confirmed the expectation that the second acetyl group would not substitute on the same ring as the first, since the acetyl group would be expected to deactivate the ring toward a second electrophilic attack, by analogy to reactions of benzene derivatives. The fact that only one such diacetylferrocene has been found indicates that the cyclopentadienyl rings are able to rotate about the axis of the bonds to the metal, although the staggered conformation indicated in formula **23** is probably preferred.

The acetylation of ferrocene offers an excellent opportunity for a demonstration of the application of thin-layer chromatography to monitor the course of a reaction, and of column chromatography to effect separations of products from one another and from starting material on a preparative scale. The utility of these methods will be demonstrated in the experiments described below.

18.3   Tropylium Iodide

In contrast to the accidental discovery of ferrocene, the successful synthesis of salts containing the cycloheptatrienyl cation moiety in 1954 was the result of intense competitive research efforts. As mentioned in Section 18.1, Hückel predicted in 1931 that the cycloheptatrienyl, or tropylium, cation should exhibit unusual stability indicative of aromaticity. Many chemists set out on syntheses designed to produce the cation with the intention of proving Hückel's theory to be either correct or incorrect. Ironically, the isolation of a salt containing the cation was delayed by initial failure to anticipate its water-solubility, which might have been expected if Hückel's prediction were correct! However, since no other carbonium ions had previously been found to be stable in water solution,[4] it is not surprising that this possibility was at first overlooked.

In this experiment, a sequence of reactions is used which demonstrates the fact that the tropylium cation is much more stable than the triphenylmethyl cation (refer to Section 18.1 for the comparison in terms of $pK_{R\oplus}$). Triphenylmethyl fluoborate (27) is first prepared from triphenylmethanol [26, equation (7)], and then allowed to react with

$$(C_6H_5)_3COH + HBF_4 \xrightarrow{(CH_3CO)_2O} (C_6H_5)_3C^{\oplus}BF_4^{\ominus} + H_2O \qquad (7)$$
$$\quad\;\; \textbf{26} \qquad\qquad\qquad\qquad\qquad\qquad \textbf{27}$$

(8)

(9)

1,3,5-cycloheptatriene [28, equation (8)]. Hydride ion ($H{:}^{\ominus}$) is transferred from the triene to the triphenylmethyl cation to produce the more stable tropylium cation, which precipitates from ether solution in the form of the fluoborate, 30. Tropylium iodide (31) is much less soluble in water than the bromide (which was the first salt isolated), and can be precipitated by metathetical reaction of tropylium fluoborate with sodium iodide [equation (9)].

EXPERIMENTAL PROCEDURE

A. ACETYLFERROCENE. COLUMN CHROMATOGRAPHY

Place 0.0027 mole of ferrocene and 5 ml of acetic anhydride in a 50-ml Erlenmeyer flask and add 1 ml of 85% phosphoric acid dropwise, with stirring. Attach a calcium chloride drying tube to the flask and heat it for about 10 min on a steam bath. Pour the reaction mixture onto about 20 g of chipped ice (ignoring any tarry material that may cling to the bottom of the flask) and, when the ice has melted, neutralize the solution by the addition of solid sodium bicarbonate. Add the bicarbonate with swirling until bubbles of carbon dioxide are no longer produced when small portions are added.[5] Cool the reaction mixture for about 30 min in an ice-water bath, then add a little solid sodium bicarbonate to the mixture. If reaction occurs, add more bicarbonate until no further bubbling occurs. (See Exercise 1 below.) Now collect the brown solid which has

---

[4] Although the tripropylcyclopropenyl cation is more stable than the tropylium cation (see Section 18.1), it was not prepared until 1962.

[5] The use of pH paper is a more exact way to determine when neutralization is complete. However, if the reaction mixture has darkened so much that the use of pH paper is impractical, the absence of bubbling upon addition of bicarbonate is a satisfactory criterion of neutralization.

precipitated by suction filtration, and wash it with water until the filtrate is pale orange. Dry the product in air.*

Prepare a chromatographic column using a 50-ml buret filled about halfway with acid-washed alumina.[6] Dissolve the dry crude product in a minimum amount of benzene and introduce it onto the alumina column with a capillary pipet. Elute with petroleum ether (bp 60–80°) first, which should bring a band of yellow-orange material down the column much more rapidly than a red-orange material. After the yellow-orange band has been removed from the column with petroleum ether, the red-orange material may be removed more rapidly with petroleum ether–ethyl ether (1:1 by volume).

Evaporate the solvents from the fractions of eluate containing the yellow-orange and red-orange materials and determine the melting point of the crystalline residues. The reported melting point of ferrocene is 173°, of acetylferrocene, 85°. Recrystallize the acetylferrocene from petroleum ether (bp 60–80°), and determine the yield.

Carry out a test with sodium hypoiodite on a small sample of the pure product to furnish evidence of its structure; see Chapter 15.2 for details of the iodoform test.

B. ACETYLFERROCENE AND 1,1′-DIACETYLFERROCENE. THIN-LAYER CHROMATOGRAPHY

1. Phosphoric acid catalyst. Dissolve 0.0054 mole of ferrocene in 20 ml of acetic anhydride in a 125-ml Erlenmeyer flask, fitted with a calcium chloride drying tube, by heating on a steam bath and swirling. Cool the solution to room temperature, and then add dropwise and with swirling 3 ml of 85% phosphoric acid. Reattach the drying tube, note the time, and allow the mixture to stand at room temperature. After two hours,[7] measure out a 5-ml aliquot of the reaction mixture and pour it onto about 20 g of chipped ice. After the ice has melted, neutralize the mixture with solid sodium bicarbonate until effervescence ceases. Extract the resulting mixture with ether, dry the ether extract with calcium chloride, and evaporate the ether.* Dissolve the solid residue in a small amount of benzene and use the benzene solution to spot a thin-layer chromatographic (tlc) plate.[8] Dissolve some ferrocene in benzene and place a spot alongside the spot from the reaction mixture. Use a mixture of benzene and absolute ethanol (30:1 by volume) to develop the tlc plate, and determine the R_f values of ferrocene and acetylferrocene.

Allow the remainder of the reaction mixture to stand with the drying tube attached and, after periods of one or two days and/or a week, measure out 5-ml aliquots and pour them onto 20 g of chipped ice and repeat the work-up and tlc procedure described for the two-hour reaction mixture. Compare the thin-layer chromatograms obtained from the aliquots of the reaction mixture after longer reaction times. Determine the R_f for 1,1′-diacetylferrocene, which should be found as a product from the longer reaction periods.

2. Aluminum chloride catalyst. Dissolve 1.0 g of ferrocene in *ca.* 15 ml of dichloromethane. Calculate amounts of acetyl chloride and powdered *anhydrous* aluminum

---

[6] See Chapter 3.5 for details of preparing a chromatographic column. The column should be used the same day it is prepared, so it may be convenient to allow the crude acetylferrocene to dry in the desk until the next laboratory period, when the column will be prepared and used.

[7] This period may be used for the preparation of tlc plates (see reference B.6 in Chapter 3), or if these are furnished ready for use, part 2 of this experiment (using aluminum chloride catalyst) may be initiated.

[8] Commercial silica-gel plates (without fluorescent indicator) may be used conveniently. For details of tlc technique, see Chapter 3.6.

chloride ($AlCl_3$) which will provide 1.5 moles of each of these compounds per mole of ferrocene. Weigh out the correct amounts of these reagents within ±0.02 g and dissolve both immediately in 25 ml of dichloromethane. *Caution:* Neither of these reagents should be exposed to moist air any longer than necessary. Do not allow either of them to come in contact with the skin. The vapor of acetyl chloride is irritating and harmful to breathe.

Add the solution of ferrocene to the solution of acetyl chloride and aluminum chloride, swirl to mix, note the time, and allow the mixture to stand in a flask protected by a calcium chloride drying tube. After intervals of 1-, 15-, 30-, and 60-min, measure out a 5-ml aliquot of the reaction mixture and pour it onto about 15 g of chipped ice. Neutralize the aliquot with solid sodium bicarbonate until effervescence ceases, add about 10 ml of dichloromethane, and separate the layers. Wash the organic solution with distilled water and dry it over calcium chloride.* Evaporate the solvent from the extract and dissolve the solid residue in a small amount of benzene.* Carry out tlc analysis of the benzene solution as described in part 1 above. Determine R_f values for ferrocene, acetylferrocene, and 1,1'-diacetylferrocene.

As an interesting extension of this experiment, some of the class members may use different molar ratios of ferrocene, acetyl chloride, and aluminum chloride, and the effect on the relative rates of production of acetylferrocene and 1,1'-diacetylferrocene may be determined. Other suggested molar ratios are ferrocene:acetyl chloride:aluminum chloride = 1:1:1 and 1:2:2.

C. TROPYLIUM IODIDE

In a 150-ml beaker, place 0.069 mole of acetic anhydride and add 2.2 g of 48% fluoboric acid dropwise with stirring and cooling. *Caution: Fluoboric acid is a highly corrosive liquid: do not allow it to come in contact with the skin.* Now add 0.015 mole of triphenylmethanol to this mixture and stir until all the solid dissolves to form a bright orange solution. Pour the solution into 200 ml of anhydrous ether in a 400-ml beaker, stir for a few minutes and then permit the precipitate of triphenylmethyl fluoborate to settle. Pour a few milliliters of ether carefully down the side of the beaker; if any turbidity appears, the precipitation was incomplete, and more ether should be added.

The next step is to collect the triphenylmethyl fluoborate on a filter. Since it is sensitive to moisture in the air, be prepared to dissolve it *immediately* in 10 ml of *anhydrous* nitromethane by rapidly carrying out the operations described below. Use a 4-cm Büchner funnel and a well-fitting paper to collect the triphenylmethyl fluoborate by suction filtration. As soon as the filtrate has run through the funnel, release the vacuum and insert the funnel stem into a clean, dry Erlenmeyer flask. Pour about 7 or 8 ml of anhydrous nitromethane onto the triphenylmethyl fluoborate on the filter paper and stir with a glass rod until solution is complete. Lift the paper slightly with the stirring rod so that the solution runs through the funnel and into the flask below. Add another 2 or 3 ml of anhydrous nitromethane to wash the filter paper and funnel, and allow the wash liquid to drain into the flask with the main part of the solution.

To the nitromethane solution of triphenylmethyl fluoborate, add 0.010 mole of 1,3,5-cycloheptatriene. After stirring for two or three minutes, pour the solution, with stirring, into 200 ml of anhydrous ether. Allow the white precipitate of tropylium fluoborate to settle and test for completeness of precipitation by adding a little ether as before.

Collect the precipitate by suction filtration, transfer it to a beaker, and then dissolve it in a *minimum amount* of hot water. To the warm solution, add dropwise about 5 ml of a saturated solution of sodium iodide; an immediate precipitation of red crystals of tropylium iodide should occur. Collect the crystals on a filter and wash them with a little *ice-cold* water. Air-dry these crystals and determine the yield, based on 1,3,5-cycloheptatriene.

EXERCISES

1. Why does more sodium bicarbonate react after the reaction mixture from ferrocene, acetic anhydride and phosphoric acid has been neutralized once and then allowed to stand 30 minutes? Hint: It is not necessary to wait and neutralize a second time when acetyl chloride is used as the acetylating agent.
2. Judging from the procedure for the column chromatography of crude acetylferrocene, (a) which is more strongly adsorbed on alumina, ferrocene or acetylferrocene? (b) In which solvent, petroleum ether or ethyl ether, is acetylferrocene more soluble? Why?
3. Ferrocene cannot be nitrated successfully by the usual mixed nitric acid-sulfuric acid medium. Why not?
4. If ferrocene were locked in the conformation shown in **23**, how many isomers of 1,1′-diacetylferrocene would there be? It may be helpful to draw the structure as it would appear from a position above the molecule as shown in **23**, *i.e.*, in line with the axis of the bonds between the rings and the iron.
5. In part C, what is the function of the acetic anhydride that is added to the 48% fluoboric acid? Why does the mixture require cooling?
6. What would happen to triphenylmethyl fluoborate if it were allowed to stand in moist air for several hours? What *weight* of water is required to react with 0.015 mole of triphenylmethyl fluoborate?
7. Apparently $C_{10}H_{10}$ ([10]annulene) and $C_{14}H_{14}$ ([14]annulene) are much less stable than [18]annulene, even though these compounds fit Hückel's Rule for $n = 2$ and $n = 3$. Suggest an explanation, considering possible geometries of [10]- and [14]annulene.

SPECTRA OF STARTING MATERIALS AND PRODUCTS

For the nmr spectrum of triphenylmethanol, see Figure 13.6.

FIGURE 18.1   IR spectrum of ferrocene.

FIGURE 18.2   NMR spectrum of ferrocene.

FIGURE 18.3   IR spectrum of acetylferrocene.

FIGURE 18.4   NMR spectrum of acetylferrocene.

FIGURE 18.5 IR spectrum of 1,3,5-cycloheptatriene.

FIGURE 18.6 NMR spectrum of 1,3,5-cycloheptatriene.

Optical activity
Carbohydrates

19.1 Introduction

Two types of isomerism, geometric and structural (or skeletal), have been encountered in previous chapters. Geometric isomerism, which occurs in alkenes for example, is a subdivision of the more general type of isomerism known as *stereoisomerism*. Stereoisomers can be defined as isomers having the same structural formula, but different relative orientations of their atoms in space.

The general phenomenon of stereoisomerism also encompasses another specific type of isomerism, *optical isomerism*. A compound is said to display optical isomerism if and only if its molecules are not superimposable (structurally identical) with their *mirror images*. Two optical isomers whose *only* difference is that one is the nonsuperimposable mirror image of the other constitute an *enantiomeric* pair, and such isomers are referred to as *enantiomers*. Figure 19.1 shows two examples: (a) 2-Chlorobutane displays optical

FIGURE 19.1 Optical isomerism.

isomerism; (b) 2-chloropropane does not. The two structures shown in (a) bear an enantiomeric relationship to one another.

The most commonly encountered type of organic compound displaying optical isomerism is one that contains a carbon atom tetrahedrally bonded to four *different* groups. This type of carbon atom is called an *asymmetric carbon atom*. The presence of one asymmetric carbon atom bestows asymmetry to an entire molecule because, in such a case, the molecule cannot be superimposable with its mirror image. The asymmetry of 2-chlorobutane in Figure 19.1(a) is due to the presence of an asymmetric carbon atom.

Molecules may be dissymmetric [1] (display optical isomerism) without an asymmetric atom. Examples of this type of molecule are shown in Figure 19.2. The allenes exhibit optical isomerism when the groups *a* and *b* are different; the mirror image structures shown are nonsuperimposable. This latter point deserves further emphasis: *the best test for optical isomerism is the nonsuperimposability of mirror images.* Figure 19.2(b)

FIGURE 19.2   Molecular dissymmetry. (a) the allenes; (b) *trans*-cyclooctene.

shows the results of this test as applied to *trans*-cyclooctene: There are two stereo-isomeric forms of this compound which constitute an enantiomeric pair.

A molecule may have more than one asymmetric carbon atom; when this is the case, the number of conceivable stereoisomers for the compound is greater than two. Of course, when only one asymmetric atom is present the number of isomers is only two, these isomers being mirror images of one another.

The structures shown in Figure 19.3 are the four possible stereoisomers of 2-chloro-

FIGURE 19.3   Enantiomers and diastereomers.

3-phenylbutane, a compound that contains two asymmetric carbon atoms. Among these four isomers, there are two enantiomeric pairs; one member of each pair is the nonsuperimposable mirror image of the other member of the same pair. Note, however, that neither member of one pair is a mirror image of either member of the other pair; stereoisomers that are not mirror images of one another are called *diastereomers* (diastereomeric isomers). Therefore, either structure of a pair is enantiomeric to the other member of the pair, but is diastereomeric to either member of the other pair.

The *maximum* number of stereoisomers for a compound may be represented by the following formula, wherein *n* represents the number of asymmetric atoms:

$$\text{Number of stereoisomers} = 2^n \tag{1}$$

[1] To be strictly correct, molecules that have nonsuperimposable mirror images should be termed dissymmetric. Asymmetric molecules constitute a sub-set of the larger set of dissymmetric molecules; the most commonly encountered type of asymmetric molecule is one containing asymmetric atoms.

The number of stereoisomers actually encountered for a compound may be less than indicated by the formula, but not greater. Thus with some compounds, even though asymmetric atoms are present, at least one of the stereoisomers may be superimposable with, and thus identical to, its mirror image. In such a case, that stereoisomer does not have an enantiomeric partner, and the total number of possible stereoisomers is reduced by one. Such stereoisomers are termed *meso;* an example is shown in Figure 19.4.

mirror
superimposable

FIGURE 19.4   *meso*-2,3-Dihydroxybutane.

With very few exceptions, asymmetric compounds have a rotational effect on the common plane of polarized light [2] as it passes through the substance. The phenomenon can be explained as follows. Consider that an asymmetric molecule has an asymmetric distribution of electron density. As the polarized light passes through the molecule, its electric component will necessarily be affected in an asymmetric manner by the electrons of the molecule. The result is that the plane of the polarized light is twisted about the axis of propagation. Any substance having this effect on polarized light is termed as *optically active.* A symmetric molecule (one that is superimposable with its mirror image) will be optically inactive, since the light, as it passes through the substance, will on the average [3] encounter a symmetric distribution of electrons and will thus be unaffected.

In a symmetric environment, enantiomers are identical with regard to all physical constants and chemical properties, with one exception; although each member of the enantiomeric pair rotates the plane of polarized light by angles of the same magnitude, they do so in opposite directions. By convention, when looking through the sample *into* the beam of light, a rotation which is observed to be clockwise, or to the right, is termed a *positive* rotation, and a counterclockwise rotation is termed a *negative* rotation. Should the sample under consideration be a mixture of exactly equal amounts of the two enantiomers, the rotation of one of the components will be exactly offset by the equal but opposite rotation of the other. The net rotation will be observed as zero. Such a mixture is called a *racemic modification.* In order that there be any net observed rotation in a mixture of enantiomers, one member of the pair must be present in amounts greater than the other.

---

[2] Electromagnetic radiation (including light) is composed of propagated electric and magnetic fields. From a wave-mechanical point of view, a ray of light is composed of two plane waves, one electric and one magnetic, oriented at right angles to one another. Considering for simplicity only the electric waves, ordinary light is completely unordered, with these waves oriented at all possible angles in the plane perpendicular to the direction of propagation. In plane-polarized light, each ray has the plane of its electric wave oriented parallel to the corresponding plane of all other rays.

[3] It must be realized that such an experiment is macroscopic rather than microscopic, for even with a very small sample of the substance under consideration, a very large number of molecules are present. Any measurement must necessarily be an average observation of the effect on all rays within the polarized light by all molecules through which they pass.

It should be noted that enantiomers have identical physical and chemical properties in a symmetric environment; a racemic modification cannot be separated into its constituent isomers by any of the methods of separation and purification that have been presented. There are methods of achieving this type of separation, although these techniques will not be discussed here.[4] On the other hand, diastereomeric isomers may vary widely in their chemical and physical properties. Therefore, they may be separated by the usual physical methods (Chapters 2 and 3).

The observed angle of rotation depends on a number of factors: The nature of the compound, its concentration (in solution) or density (neat liquid), the length of the sample through which the light must pass (the path length), temperature, solvent, and wavelength of light. The concentration, or density, and path length of the sample are important considerations because they determine the average number of optically active molecules through which a beam of light will pass. The magnitude of rotation is the result of a cumulative effect, and if the light, on the average, passes through a greater number of active molecules, the observed magnitude of rotation will be greater. The dependence is essentially linear with respect to both concentration and path length.

It is certainly desirable to present the optical rotation as a physical constant of the active compound. Therefore, it is common practice to report the *specific rotation* rather than the observed angle of rotation. The specific rotation is calculated by correcting the observed rotation to *unit* concentration and path length with the following equation:

$$[\alpha] = \frac{\alpha}{l \times c} \text{ or } \frac{\alpha}{l \times d} \tag{2}$$

$[\alpha]$ = specific rotation (degrees)
α = observed rotation (degrees)
l = path length (decimeters)
c = concentration (g/ml of solvent, solution)
d = density (g/ml, neat)

In order to specify the other variables upon which the rotation depends, the temperature and wavelength (mμ) employed are presented as superscript and subscript, respectively, on the symbol for specific rotation, and the solvent used is denoted in parentheses following the numerical value and sign of the specific rotation, *e.g.*, $[\alpha]_{490}^{25°} + 23.4°$ (CH_3OH). Usually, a sodium lamp emitting light at 589 mμ, the sodium D line, is used as a light source, and in this case it is common to use D in place of the numerical value of the wavelength in the expression, *e.g.*, $[\alpha]_D^{25°} - 15.2°$ (H_2O).

Various types of commercial polarimetric apparatus are available. Some, such as the Rudolph polarimeter, are manually operated and require direct readings by the operator, whereas other, newer varieties are automatic and utilize photoelectric cells for measurements of greater accuracy and precision. The fundamental attributes of these instruments are similar however, and may be amply understood within the context of the following discussion.

A very simply constructed student polarimeter has been described by Shaw.[5] The polarimeter is constructed from a mailing tube in each end of which a piece of Polaroid film has been permanently affixed [Figure 19.5(a)]. The tube is such that the two ends may be rotated relative to one another. The sample is held within a glass tube placed within the mailing tube [Figure 19.5(b)]. Light passing into the polarimeter at

---

[4] Discussions of the general approaches that may be employed to separate enantiomers may be found in most modern textbooks of organic chemistry.

[5] William H. R. Shaw, "A Mailing-Tube Polarimeter," *Journal of Chemical Education*, **32**, 10 (1955).

FIGURE 19.5   Construction of the Shaw polarimeter. (a) The mailing tube and its accessories; (b) The sample tube.

one end will be polarized by the Polaroid film at that end (the *polarizer*); as the polarized light passes on through the sample tube containing a solution of an optically active substance, the plane of the polarized light will be rotated through an angle. The angle of rotation is determined by rotating the section of the tube through which the observer is viewing until the light intensity observed is the same as that observed when the sample tube contains only pure solvent. The number of degrees through which this

FIGURE 19.6   Schematic illustration of optical rotation ($\alpha = ca. -45°$).

analyzer section of the tube must be turned is then equivalent to the optical rotation, α, of the sample (see Figure 19.6). This can be converted to the specific rotation by application of equation (2).

19.2   *Mutarotation of D-Glucose*

As a typical example of the utility of optical rotation measurements, a polarimeter will be used to determine the rate of mutarotation of α-D-glucose, a carbohydrate [6].

[6] The "D" of D-glucose and D-fructose is a symbol used to designate the relative configuration of these sugars to that of D-glyceraldehyde, the standard of configuration for carbohydrates. The sign of optical rotation of a carbohydrate is independent of its configuration. See any modern organic textbook for further explanation.

D-Glucose exists in two diastereomeric forms, α-D-glucose (1) and β-D-glucose (2), which are in equilibrium with one another in aqueous solution via the intermediacy of an open chain structure (3, Figure 19.7).

FIGURE 19.7 Solution equilibrium of D-glucose.

Note that the open chain structure has four asymmetric carbon atoms, whereas when hemiacetal formation (cyclization) occurs between the aldehyde group and the C-5 hydroxyl group, an additional asymmetric center is formed at C-1. Since there are two possible configurations about this new center, there must be two possible isomeric hemiacetals: 1 (the α-form) and 2 (the β-form). These two isomers are disastereomers, and can be separated from one another by rather specific techniques of crystallization, as is done in the experiment.

The α-form has a specific rotation of $[\alpha]_D^{25°} + 112°$. When it is placed in water solution the specific rotation gradually changes until it reaches a constant $[\alpha]_D^{25°} + 52.7°$. The β-form, in water solution, undergoes a change of specific rotation from $[\alpha]_D^{25°} + 19°$ to $[\alpha]_D^{25°} + 52.7°$. Quite obviously, $+52.7°$ represents the specific rotation of an equilibrium mixture of 1 and 2. The equilibration of optical isomers which gives rise to an equilibrium rotation is called *mutarotation*.

By following the rate of change of optical rotation of α- or β-D-glucose, information concerning the *rate* of mutarotation can be obtained. For a reaction of the type shown in equation (3),

$$A \underset{k_2}{\overset{k_1}{\rightleftarrows}} B \tag{3}$$

the following expression can be written [7]:

$$2.303 \log \frac{A_e - A_o}{A_e - A_t} = (k_1 + k_2)t \tag{4}$$

where A_e is the concentration of A at equilibrium, A_o is the concentration of A at time $t = 0$, and A_t is the concentration at time t during the equilibration process. Note that the term $(A_e - A_o)$ represents the total extent of reaction from the beginning to the final establishment of equilibrium, that is, the total change in concentration of A, whereas the term $(A_e - A_t)$ represents the extent of reaction remaining at time t. We may substitute for these terms any other terms that also represent the ratio of the total extent of reaction and the extent remaining at time t. Any terms utilized must vary

[7] The interested student can find the derivation of this expression in A. A. Frost and R. G. Pearson, *Kinetics and Mechanism*, 2nd Ed., John Wiley and Sons, Inc., New York, 1961, p. 186.

linearly during the course of reaction, as does concentration. Earlier discussion indicated that optical rotation is linearly related to concentration; thus we may substitute the optical rotation of the solution measured at the beginning, end, and at time t for the concentrations in equation (3) to give equation (5):

$$2.303 \log \frac{\alpha_e - \alpha_o}{\alpha_e - \alpha_t} = (k_1 + k_2)t \tag{5}$$

Thus we see that measurement of the optical rotation of the initial solution of α-D-glucose, of the rotation of the same solution at equilibrium, and of the rotations at a series of times in between gives the data needed for calculation of $(k_1 + k_2)$. This is accomplished by plotting the left-hand term against t; the slope of the straight line obtained is $(k_1 + k_2)$. It is not possible to obtain k_1 and k_2 individually unless the value of the equilibrium constant is also known (see Experimental Procedure).

EXPERIMENTAL PROCEDURE

Preparation of α-D-glucose. α-D-Glucose is the form obtained when D-glucose is crystallized slowly at room temperature from a mixture of acetic acid and water. Dissolve 50 g of anhydrous D-glucose (dextrose) in 25 ml of distilled water contained in a 500-ml Erlenmeyer flask. Heat the mixture on a steam bath to effect complete solution. The solution will be quite viscous and should be continuously stirred during this step. When solution is complete (*no* crystals remaining), remove the syrup from the steam bath, add 100 ml of *glacial* acetic acid which has been previously cooled in an ice bath to 0°, and swirl the flask until the mixture becomes homogeneous. Stopper the flask with a cork and allow it to remain undisturbed in the desk until the next laboratory period.

Collect the α-D-glucose by suction filtration, wash the crystals thoroughly with 50–60 ml of 95% ethanol, followed by 50–60 ml of absolute ethanol. Either air-dry the crystals or dry them in an oven at *ca.* 80° for 1–2 hr.

Rate of mutarotation. Before coming to class, prepare in the laboratory record book a chart for collecting the following information during the performance of this experiment: temperature, concentration, path length of sample tube, time of mixing of solution, calibration factor for Shaw polarimeter (if used), [8] and a series of measurements of time and optical rotation.

Obtain from the instructor a polarimeter and sample tube. Note the scale attached to the polarizer section of the mailing tube. Count the number of divisions on that scale about the circumference of the tube and divide into 360° in order to determine the calibration factor.

Carefully fill the sample tube with distilled water and put the end-piece in place in such a way that no air bubbles remain trapped within. Place the sample tube in the polarimeter and, viewing through the analyzer section, point the assembly at the daytime sky. Turn the analyzer end of the polarimeter until the observed field turns deep purple and place a light pencil mark on the polarizer scale in alignment with the zero

[8] The directions given here apply to the Shaw polarimeter. Although most undergraduate laboratories do not have more elaborate polarimeters available, some do; should the instructor wish such an instrument to be used, he will issue instructions in its use.

pointer on the analyzer section. This mark will serve as a zero point for later determinations of rotation. Empty and carefully dry the sample tube and end-pieces.

Weigh 10–15 g of α-D-glucose accurately to the nearest 0.05 g and transfer the sample *completely* to a dry 250-ml volumetric flask. Fill the flask to within a few milliliters of the volumetric mark with distilled water; tightly stopper the flask and shake the contents to effect complete solution. The fine solid should dissolve completely within about 30 sec. When the solid is approximately one-half dissolved, note the time to the nearest minute and record it. As soon as the solid is dissolved, carefully fill the flask to the mark with distilled water using a dropper. Again stopper and shake the flask until the solution is homogeneous (*ca.* 5–10 sec). Working carefully but rapidly, fill the polarimeter sample tube with this solution and insert the end-piece leaving no air bubbles. Place a thermometer in the volumetric flask and, as soon as there is time, read and record the temperature. Insert the sample tube in the polarimeter, put the analyzer in place, setting the pointer at the zero point. Point the polarimeter at the daytime sky and turn the analyzer until the field turns deep purple.[9] Since α-D-glucose has a positive rotation, the analyzer will need to be turned to the right (clockwise). Note the time to the nearest minute and the number of divisions [10] the pointer is displaced from the zero point; record these data. During the next 6 or 7 min take a similar reading once a minute, recording the data obtained. Then take additional readings every 3 or 4 min for the next 40–50 min. Remove the sample tube from the polarimeter and measure and record the path length of the solution through which the polarized light passes.

Approximately 4.5 to 5 hr will be required for the rotation to drop to within experimental error of the equilibrium value. If it is agreeable with the instructor, place the sample tube in the desk until the next laboratory period, or, if not, empty the solution from the sample tube back into the volumetric flask, stopper the flask [11] and place it in the desk until the next meeting. The instructor should have placed identifying numbers on each of the polarimeters available. Note which polarimeter was used in this experiment and record the number in the notebook.

At the next laboratory period and using *the same* polarimeter as before, refill the sample tube with the equilibrated solution and take a final reading, which will be α_e.

Treatment of data.

1. Define the first recorded time at which a measurement was taken as time $t = 0$. The observed rotation at that time is then α_o. For each reading following the defined zero point, determine and record the elapsed time Δt (from $t = 0$) in minutes.

2. Perform the calculations needed for the graphical plot in the next operation.

3. On a piece of graph paper, plot 2.303 log $(\alpha_e - \alpha_o)/(\alpha_e - \alpha_t)$ (vertically) *vs.* elapsed time Δt (horizontally). Determine the slope of the best straight line that can be drawn through these points; [12] the value of the slope (in min^{-1}) is equivalent to the value $(k_1 + k_2)$.

4. From the following equation,[13] calculate the value of $(k_1 + k_2)$ to have been ex-

---

[9] Due to reflections within the tube, the field will appear as a "bulls-eye target." Pay particular attention to the color within the "eye." Since further readings will be taken and it is desired that they be consistent, note carefully the color shade at which the reading has been taken, and for all subsequent readings adjust the polarimeter to the same shade.

[10] If the scale has been constructed from millimeter-lined graph paper, estimation of the displacement to 0.2 division should be possible.

[11] Evaporation must be prevented so there will be no concentration changes.

[12] See footnote 13 in Chapter 11 (p. 180).

pected at the temperature at which the kinetic determination was made. Compare this with the value obtained.[14]

$$\log (k_1 + k_2) = 11.0198 - 3873/T \ (°K) \tag{6}$$

5. On a separate piece of graph paper, plot α_t (vertically) *vs.* Δt (horizontally). A curve should be obtained. Draw the best curved line through these points and extrapolate the curve from time $t = 0$ back to the time of initial mixing. Using the extrapolated value of α obtained at that time, calculate the initial specific rotation, using equation (2), of the α-D-glucose used in the experiment. Remember, pure α-D-glucose has a specific rotation of $[\alpha]_D^{25°} +112°$.

6. Using the value of the equilibrium constant, K_e, for the equilibrium between α- and β-D-glucose obtained by other workers, and the value of $(k_1 + k_2)$ obtained in this experiment, calculate the individual values of k_1 and k_2 [see equation (7)].

$$K_e = 1.762 = \frac{k_1}{k_2} \tag{7}$$

EXERCISES

1. Would any different experimental results have been expected if one had started with β-D-glucose rather than with α-D-glucose? Explain.
2. From the specific rotation values of $+112°$ for α-D-glucose, $+19°$ for β-D-glucose, and $+52.7°$ for the equilibrium mixture, show by calculation that the value for the equilibrium constant given in equation (7) is correct.
3. Write a mechanism for the isomerization of α- to β-D-glucose.
4. An experimentally observed rotation appearing as $60°$ could just as correctly be interpreted as a rotation of $240°$, $420°$, $600°$, $-120°$, $-300°$, etc. How could it be determined which rotation is the correct one?

19.3 The Hydrolysis of Sucrose

Sucrose, a familiar foodstuff, is a carbohydrate having the structure **4**. As shown in the structure, the sucrose molecule is formed by a linkage between the hemiacetal form of α-D-glucose and one of the hemiketal forms of D-fructose (**5**).[6] The observation that

4

5

---

[13] C. S. Hudson and J. K. Dale, *Journal of the American Chemical Society*, **39**, 320 (1917).

[14] From the authors' experience with this experiment, the experimental and calculated values of $(k_1 + k_2)$ should compare within a factor of two.

sucrose does not undergo mutarotation (Section 19.2) is evidence that the linkage is through the glycosidic hydroxyl groups of each sugar.

Sucrose, $[\alpha]_D^{25°}$ +66.5°, undergoes acid-catalyzed hydrolysis to give a mixture of D-glucose and D-fructose [equation (8)]. Under the conditions of the hydrolysis, the

$$\text{Sucrose} \xrightarrow{\text{H}_3\text{O}^{\oplus}} \text{D-Glucose} + \text{D-Fructose} \qquad (8)$$

glucose undergoes rapid mutarotation to give the equilibrium mixture of α- and β-forms, $[\alpha]_D^{25°}$ +52.7°, whereas the fructose is formed as an equilibrium mixture of the isomers shown in Figure 19.8 and has $[\alpha]_D^{25°}$ −92°. Isomers **6** and **7**, five-membered

FIGURE 19.8 Solution equilibria of D-fructose.

ring hemiketals, predominate in the equilibrium mixture. The observation that the sign of optical rotation changes upon hydrolysis of sucrose has led to the name *invert sugar* for the product mixture. The enzyme *invertase* accomplishes the same chemical result as does the acid-catalyzed hydrolysis of sucrose.

Formation of osazones. Most carbohydrates, particularly when impure, crystallize from solution only with difficulty, tending instead to form syrups; this frequently makes their direct characterization difficult. Many carbohydrates react with phenylhydrazine to form bright yellow, crystalline derivatives called *osazones*. These derivatives may usually be identified readily both by their temperatures of decomposition (or melting points) and their crystalline forms.

Not all sugars form osazones, for only those that have an aldehydo or keto carbonyl group, either free or in equilibrium with a hemiacetal or hemiketal, will react with phenylhydrazine. For example, α- and β-D-glucose (**1** and **2**) in water solution are in dynamic equilibrium with the ring-opened form **3**, which, because of its aldehydo group, will react with phenylhydrazine (see Chapter 15.1).

Equation (9) shows the formation of glucosazone (**8**), the osazone derivative of glucose. Note that three equivalents of phenylhydrazine are required and that two of these reagent molecules are incorporated into the osazone structure. Additional products include aniline and ammonia.

$$
\begin{array}{c}
\text{CHO} \\
\text{H—C—OH} \\
\text{HO—C—H} \\
\text{H—C—OH} \\
\text{H—C—OH} \\
\text{CH}_2\text{OH}
\end{array}
\;+\;3\;C_6H_5NHNH_2\;\xrightarrow{\;H_3O^{\oplus}\;}\;
\begin{array}{c}
\text{CH=NNHC}_6\text{H}_5 \\
\text{C=NNHC}_6\text{H}_5 \\
\text{HO—C—H} \\
\text{H—C—OH} \\
\text{H—C—OH} \\
\text{CH}_2\text{OH}
\end{array}
\;+\;C_6H_5NH_2\;+\;NH_3 \qquad (9)
$$

$$\textbf{8}$$

The accepted mechanism for the reaction is presented in Figure 19.9. Following the formation of a phenylhydrazone, **9**, an internal oxidation-reduction reaction is accomplished by tautomeric migration of two hydrogens from C-2 to the hydrazone moiety to give **10**. The newly formed carbonyl group condenses with a second equivalent of phenylhydrazine to give **11** which undergoes subsequent tautomerization to **12**. Following a 1,4-elimination of aniline which produces either **13** or **14**, a third equivalent of phenylhydrazine condenses with the imine group to give the osazone, **15**, and ammonia. Although it may appear that **15** should suffer further reaction by intramolecular oxidation-reduction between the secondary alcohol group at C-3 and the hydrazone group at C-2, the reaction stops at this point so that only two phenylhydrazine units are introduced. The formation of the intramolecular hydrogen bond shown in **15** has been established as the cause of limitation of the reaction to the first two carbons of the chain.[15]

$$
\begin{array}{c}
\text{CHO} \\
\text{HCOH} \\
\text{HCOH}
\end{array}
\xrightarrow{C_6H_5NHNH_2}
\begin{array}{c}
\text{CH=NNHC}_6\text{H}_5 \\
\text{HCOH} \\
\text{HCOH}
\end{array}
\longrightarrow
\begin{array}{c}
\text{CH}_2\text{NHNHC}_6\text{H}_5 \\
\text{C=O} \\
\text{HCOH}
\end{array}
\xrightarrow{C_6H_5NHNH_2}
\begin{array}{c}
\text{CH}_2\text{NHNHC}_6\text{H}_5 \\
\text{C=NNHC}_6\text{H}_5 \\
\text{HCOH}
\end{array}
$$

$$\quad\textbf{9}\qquad\qquad\textbf{10}\qquad\qquad\textbf{11}$$

$$
\textbf{11}\longrightarrow
\begin{array}{c}
\text{CH—NHNHC}_6\text{H}_5 \\
\text{C—NHNHC}_6\text{H}_5 \\
\text{HCOH}
\end{array}
\xrightarrow{-\,C_6H_5NH_2}
\begin{array}{c}
\text{CH=NH} \\
\text{C=NNHC}_6\text{H}_5 \\
\text{HCOH}
\end{array}
\;\text{or}\;
\begin{array}{c}
\text{CH=NNHC}_6\text{H}_5 \\
\text{C=NH} \\
\text{HCOH}
\end{array}
$$

$$\qquad\qquad\textbf{12}\qquad\qquad\qquad\textbf{13}\qquad\qquad\textbf{14}$$

$$
\textbf{13 or 14}\xrightarrow{C_6H_5NHNH_2}
\quad + \;NH_3
$$

$$\textbf{15}$$

FIGURE 19.9 Mechanism of osazone formation.

[15] When 1-methylphenylhydrazine $[H_2NN(CH_3)C_6H_5]$ is used in place of phenylhydrazine, the reaction proceeds down the chain readily at least as far as C-5. Note that in this reagent, the N-H proton involved in the hydrogen bonding in **15** has been replaced with a methyl group. O. L. Chapman, W. J. Welstead, Jr., T. J. Murphy and R. W. King, *Journal of the American Chemical Society*, **86**, 732, 4968 (1964).

The experiments below involve the hydrolysis of sucrose, analysis of the product mixture by osazone formation, and observation of the optical rotation changes occurring upon hydrolysis.

EXPERIMENTAL PROCEDURE

A. HYDROLYSIS OF SUCROSE

Place 10–15 g of pure sucrose, whose weight has been determined as accurately as possible, in a 250-ml round-bottomed flask. Add *ca.* 100 ml of water, swirl the contents of the flask to effect solution and then add three or four drops of concentrated hydrochloric acid to the mixture. Heat the solution at reflux for 2 hr.

At the end of the period of reflux, cool the reaction mixture to room temperature and carefully transfer *all* of the solution to a 250-ml volumetric flask. Use small amounts of water to rinse the round-bottomed flask and add the rinses to the rest of the solution. Dilute the mixture to a volume of 250 ml. With the Shaw polarimeter, whose use is described in Section 19.2, determine the specific rotation of the product mixture from the hydrolysis of sucrose. Save at least 20 ml of this solution for later use.

B. SPECIFIC ROTATION OF SOME PURE SUGARS

Prepare aqueous solutions [16] of accurately known concentrations (*ca.* 0.1 g/ml) of sucrose, anhydrous D-glucose (dextrose, the equilibrium mixture of α- and β-forms), and D-fructose [levulose]. Using the Shaw polarimeter, determine the specific rotations of each of these sugars.

C. FORMATION OF OSAZONES

Using the general procedure described below, attempt preparation of osazone derivatives of D-glucose, D-fructose and sucrose.

Heat a large beaker of water to boiling. In a large test tube, dissolve 1 g of sugar in 20 ml of water and add to this solution *either* 3 ml of glacial acetic acid, 3 g of sodium acetate, and 2 ml of phenylhydrazine [17] *or* 3 g of sodium acetate and 3 g of phenylhydrazine hydrochloride. Stir the solution thoroughly, place the test tube in the beaker of boiling water, and discontinue heating. Allow the test tube to remain in the cooling water bath for 30 min. In all cases where osazone derivatives will be produced, a precipitate should have formed after 5–10 min.

After *ca.* 30 min, remove the test tube, cool the contents to room temperature, and collect the precipitate by suction filtration. Recrystallize a small sample of the osazone from ethanol-water, and determine the temperature of decomposition (melting point) of the dry, pure osazone. Determine the temperature of decomposition of a *mixture* of the osazone derivatives of glucose and fructose.

Prepare an osazone derivative of the product of hydrolysis of sucrose. Substitute 20 ml of the solution which was diluted for rotation measurements, part A, for the 1 g of sugar and 20 ml of water used in the above procedure. Determine the temperatures of decomposition of this osazone *and* of a mixture of this osazone with the osazone of fructose.

---

[16] The total volume of solution needed will depend on the size of the sample tubes used in the Shaw polarimeters.

[17] Phenylhydrazine is toxic and care should be exercised in its handling.

EXERCISES

1. Offer an explanation for the change in sign of the optical rotation of sucrose following hydrolysis. Calculate the specific rotation of *invert* sugar from the known equilibrium rotations of D-glucose and D-fructose. How does this number compare with that determined experimentally?

2. In what way is the specific rotation of invert sugar analogous to the specific rotation of a racemic mixture?

3. The results of the experiments performed offer evidence concerning the structure of D-fructosazone. Indicate what evidence is offered and provide the structure of this compound.

4. Suggest a mechanism for the internal oxidation-reduction reaction which occurs in the formation of osazones, *i.e.*, **9** → **10**, Figure 19.9.

5. Explain why sucrose does not form an osazone under the conditions of the experiment performed. If the conditions for osazone preparation were altered by heating the reaction mixture at reflux for a period of time, would you expect an osazone to form? Why or why not?

20

Identification of organic compounds

Systematic procedures for the identification of organic compounds were developed much later than those for inorganic compounds and elements. The first successful scheme of organic qualitative analysis was developed by Professor Oliver Kamm and culminated in his textbook published in 1922. This scheme is the one on which most textbooks are still based,[1] and we shall refer to it, including the modifications and modernizations that have been made, as *classical qualitative organic analysis*.

As we have mentioned earlier (Chapters 3 and 4), in recent years the development of instrumental methods of separation and analysis (particularly chromatographic and spectroscopic techniques) has revolutionized the laboratory practice of organic chemistry. However, the interest in classical qualitative organic analysis remains high because it is recognized by teachers and students alike as the most effective as well as the most interesting means of teaching fundamental organic chemistry. For this reason, in this book we have retained an adequate outline of the classical scheme and have also provided an introduction to some of the more *modern spectroscopic methods of identification and structure determination*.

20.1 Separation of Mixtures of Organic Compounds

In this chapter we shall be concerned with the identification of a pure organic compound. However, it must be recognized that when a chemist is faced with the problem of identifying an organic compound, it is seldom pure, but is usually mixed with by-products and starting materials. Modern methods of separation, particularly chromatographic procedures, make the isolation of a pure compound easier than it used to be, but one must not lose sight of the importance of classical techniques of separation.[2] These have been utilized in many of the experimental procedures in the preceding chapters of this book.

The common basis of the procedures most often used to separate mixtures of organic compounds is the difference in *polarity* which exists or may be induced in the com-

---

[1] See references 1 and 2 at the end of this chapter.

[2] Methods of separation and purification of organic compounds are treated in detail in Chapters 2 and 3.

ponents of the mixture. This difference in polarity is exploited in nearly all of the separation techniques, including distillation, recrystallization, extraction, and chromatography. The greatest differences in polarity, which make for the simplest separations, are those which exist between salts and non-polar organic compounds. Whenever one or more of the components of a mixture can be converted to a salt, it can be separated easily and efficiently from the nonpolar components by extraction or distillation.

We shall not attempt in this chapter to give additional general or specific directions for separating mixtures. If you are given an unknown mixture to separate into its components before identifying them individually, review Chapters 2 and 3, and study the experimental procedures of the preparative experiments of the book. Examine each procedure carefully to determine what impurities are removed from the desired product and what separation techniques are utilized to accomplish this in each step of the "work-up." These same techniques may be employed in separating the components of an unknown mixture.

20.2 Classical Qualitative Organic Analysis

The system consists of six fundamental steps, which are usually best carried out in the sequence listed below.
1. **Preliminary examination** of physical and chemical characteristics
2. Determination of **physical constants**
3. **Elementary analysis**
4. **Solubility tests,** including acid-base reactions
5. **Classification tests;** functional reactivity other than acid-base reactions
6. **Preparation of derivatives**

It is a tribute to the effectiveness of the system that, although the number of organic compounds is many thousands of times the number of common inorganic ions, it is usually possible to identify a known organic compound with more certainty than an inorganic compound. However, with the exception of a few guidelines, there is no rigid regimen of "cook-book" directions to be followed; a student must rely on his own judgment and initiative in choosing a course of attack on his unknown.

Discussions of Theory and Directions for Experimental Procedures

1. The preliminary examination. The preliminary examination may provide more information with less effort than any other part of the identification procedure, if it is carried out intelligently! The simple observation that the unknown is a *crystalline* solid, for example, eliminates from consideration a major fraction of all organic compounds, since most of them are liquids at room temperature. The *color* is also informative; most pure organic compounds are white (or colorless). A brown color is most often characteristic of small amounts of impurities; *e.g.*, aromatic amines and phenols quickly become discolored by the formation of trace amounts of highly colored air-oxidation products. Color in a pure organic compound is usually attributable to conjugated double bonds.

The *odor* of many organic compounds is highly distinctive, particularly among those of lower molecular-weight. A conscious effort should be made to learn and recognize the odors which are characteristic of several classes of compounds, such as the alcohols, esters, ketones, aliphatic and aromatic hydrocarbons. The odors of certain compounds demand respect, if not admiration, even when they are encountered in small amounts and at considerable distance; for example, the unpleasant odors of thiols (mercaptans), isonitriles, and higher carboxylic acids and diamines cannot be described definitively, but they are recognizable, once encountered. The more pleasant odors may also be recognized more easily than described. *Be cautious* in smelling unknowns since some compounds are not only disagreeable but also irritating to the mucous membranes. Large amounts of organic vapors should never be inhaled since many compounds are toxic.

The *ignition test* is a highly informative procedure. Heat a small amount (one drop of a liquid or *ca.* 50 mg of a solid) gently on a small spatula or crucible cover, at first above or to the side of a microburner flame. Make a note as to whether a solid melts at low temperature or only upon heating more strongly. Observe the flammability and the nature of any flame. A yellow, sooty flame is indicative of an aromatic or highly unsaturated aliphatic compound; a yellow but nonsooty flame is characteristic of aliphatic hydrocarbons. Oxygen content in a substance makes its flame more colorless (or blue); extensive oxygen content lowers or prevents flammability, as does halogen content. The unmistakable odor of sulfur dioxide indicates the presence of sulfur in the compound.

If a white, nonvolatile residue is left after ignition, add a drop of water and test the solution with litmus or pH paper; a sodium (or other metal) salt is indicated by an alkaline test.

2. Physical constants. If the unknown is a solid, determine its melting point by the capillary tube method (Chapter 1.4). If the melting range is more than 2°, recrystallize the sample. (Chapter 2.7).

If the unknown is a liquid, determine its *boiling point* by the micro boiling-point procedure (Chapter 1.5). If the boiling point is indefinite or non-reproducible, or if the unknown sample is discolored or inhomogeneous, distil it (Chapter 2) and determine the boiling point. Other physical constants which are useful for liquids, particularly in the case of hydrocarbons, ethers, and other less reactive compounds, are the *refractive index* and the *density* (Chapter 1.6 and 1.7, respectively). Consult your instructor about the advisability of making these measurements; full directions are given in references 1 and 2 at the end of this chapter.

3. Elementary analysis. A knowledge of the elements other than carbon and hydrogen present in an unknown organic compound is an enormous advantage in identification. The most commonly occurring elements are oxygen, nitrogen, sulfur and the halogens. There are no simple tests for oxygen, and the other elements are most commonly held by covalent bonds, so that they do not respond directly to the usual ionic tests. However, if the organic unknown is fused with molten sodium, most compounds react so that N, S, and X are converted to the ions CN^-, $S^=$, CNS^-, and X^-. After the excess sodium is carefully decomposed, the aqueous solution containing these anions is analyzed by conventional methods of inorganic analysis.

Sodium fusion. Support a small, soft-glass test tube in a vertical position in a clamp with an asbestos liner (or none—*no rubber*). Place a clean cube of sodium (about 3 to 4 mm on an edge) in the tube and heat it gently with a micro burner flame until the sodium melts, and the vapors rise about 1 cm. Then remove the flame and drop a small amount of the organic compound (about 5 to 10 mg) directly onto the molten sodium.

Caution: Safety glasses must be worn and the face should be kept away from the mouth of the tube, as some compounds react violently or explosively with the molten sodium. Heat the bottom of the tube again, then remove the flame and add a second 5- to 10-mg portion of the organic compound. Now heat the bottom of the tube until it is a dull red color. Next remove the flame and allow the tube to cool. Add about 1 ml of ethanol to decompose the excess sodium. After the reaction has subsided, apply gentle heat to boil the ethanol and place an inverted funnel connected to a vacuum line (or aspirator) over the mouth of the test tube. When the ethanol has been removed, heat the bottom of the test tube again and then drop the hot tube into a small beaker containing about 15 ml of distilled water (*caution*).[3] Break up the tube further with a stirring rod, heat the solution to boiling and filter. The filtrate is used for the following tests.

   a. Sulfur.   Acidify a 1- or 2-ml sample of the filtrate with acetic acid and add a few drops of 5% lead acetate solution. A black precipitate of lead sulfide indicates the presence of sulfur in the original organic compound.

   b. Nitrogen.   Adjust the pH of 1 ml of the filtrate to 13 (an indicator paper such as Hydrion E may be used). Add two drops each of a saturated solution of ferrous ammonium sulfate and of 30% potassium fluoride solution. Boil the mixture gently for about 30 sec and then carefully acidify it by the dropwise addition of 30% sulfuric acid until the precipitate of iron hydroxide just dissolves. An excess of acid should be avoided. Appearance of the deep blue precipitate of potassium ferric ferrocyanide (Prussian blue) indicates the presence of nitrogen. If the presence of a blue precipitate is uncertain and the result is only what appears to be a green or blue solution, filter it; a blue color remaining on the filter paper is a weak but positive test for nitrogen. Nitrogen present in an organic molecule is converted by sodium fusion into cyanide ion.

   c. Halogens.   Acidify about 2 ml of the filtrate by dropwise addition of nitric acid (test with litmus) and boil the solution gently for several minutes to expel any hydrogen sulfide or cyanide that may be present. Cool the solution and add several drops of 5% silver nitrate. A heavy precipitate of silver halide indicates the presence of chlorine, bromine, or iodine in the original organic compound. (A faint turbidity may be due to the presence of traces of halogen impurities in the original compound, the reagents, or even the glass of the sodium-fusion test tube.) Tentative identification of the particular halogen may be made on the basis of color, silver chloride is white, silver bromide is pale yellow and silver iodide is yellow, but as all three halides darken rapidly in light, positive identification must be made by standard inorganic qualitative procedures.[1,4]

   4. Solubility tests.   Solubility tests, including acid-base reactions, were a fundamental part of the original Kamm system and still may be used to give indicative information; however, it should be recognized that definite assignment of an unknown to a formal solubility class is rather arbitrary because of the large number of compounds that exhibit borderline behavior, no matter where the dividing lines are drawn. The classification scheme is outlined in Figure 20.1.

   Make the tests with the solvents or reagents in left-to-right sequence, *i.e.*, determine the solubility in water first. In considering water-solubility, for our purposes a com-

---

[3] Alternatively, a Pyrex test tube may be used for the fusion. After the tube has cooled slightly, water may be added to the tube and the mixture stirred and then poured out into a small beaker. Rinse the tube with more water, combining the rinse with the main solution. The total amount of water should be about 15 ml. Boil the mixture, filter, and use the filtrate as described.

[4] See reference 3 at the end of this chapter.

FIGURE 20.1 Classification by means of solubility and acid-base reactions, with typical examples of each class.

pound is said to be soluble if it dissolves to the extent of 3 g in 100 ml of water or, more practically, 100 mg in 3 ml of water. If an unknown dissolves to this extent in water, do not test it in sodium hydroxide solution or any of the other reagents, but test its aqueous solution with litmus paper and determine its subclassification as shown at the top right of Figure 20.1.

If an unknown is insoluble in water, test it in 5% NaOH solution. In this, and in the other reagent solutions, the definition of solubility is not 3%, but any greater solubility in the acid or base than in water,[5] which reflects the presence of an acidic or basic functional group in the water-insoluble unknown compound. Unknowns that are insoluble in water but soluble in dilute base are also tested for solubility in dilute (5%) hydrochloric acid, since both acidic and basic groups might be present in the molecule. However, there is no point in testing compounds that are soluble in dilute HC1, or neutral

[5] This may be determined as follows: after shaking the unknown with dilute acid, for example, decant the liquid or filter it from undissolved sample and make the filtrate basic; the formation of a precipitate or turbidity is indicative of significant solubility of the sample in acid.

water-insoluble compounds containing **N** or **S**, in concentrated H_2SO_4, since they will invariably dissolve or react with it.[6] Any detectable reaction such as evolution of a gas or formation of a precipitate is considered "solubility" in concentrated H_2SO_4.

The borderline for water-solubility of monofunctional organic compounds, by our definition, is most commonly at or near the member of the homologous series containing five carbon atoms. Thus, butanoic acid is soluble, pentanoic acid is borderline, and hexanoic acid is insoluble; 1-butanol is soluble, 1-pentanol is borderline, and 1-hexanol insoluble. The relationship between molecular surface area and solubility is pointed up, however, by the observation that 2-methyl-2-butanol is clearly soluble (12.5 g per 100 ml of water).

We strongly recommend that at this stage of the investigation each student present a *preliminary report* to his instructor of the results of his determination of physical constants, elementary analysis and solubility tests. The instructor should advise the student of any errors (in the case of melting and boiling points a satisfactory limit of accuracy is $\pm 5°$) so as to prevent serious loss of time on "wild goose chases."

5. Classification tests. Under this heading are grouped all the reactions, other than acid-base reactions, which are indicative of other functional groups and allow assignment of the unknown to a structural class, such as alkene, alkyne, alcohol, aldehyde, etc. Several reactions of this sort have been encountered previously in this book, and the remainder of these classification tests will be presented later. The structural classes that will be considered are:

A. NEUTRAL COMPOUNDS
  Carbonyl Compounds (aldehydes and ketones), p. 334
  Unsaturated Compounds (alkenes), p. 335
  Alkyl Halides, p. 336
  Aryl Halides, p. 337
  Alcohols, p. 339
  Esters, p. 349
  Amides, p. 352
  Nitriles, p. 351
  Nitro Compounds, p. 348
  Aromatic Hydrocarbons, p. 337
B. ACIDIC COMPOUNDS
  Carboxylic Acids, p. 344
  Phenols, p. 342
C. BASIC COMPOUNDS
  Amines, p. 345

6. Preparation of derivatives. In the classical approach, it is often useful to convert the suspected compound into another compound (a derivative) which is a solid. The melting points of two such solid derivatives, along with the melting point or boiling point of the unknown compound, often serve to identify the unknown completely. Some methods for preparing derivatives have been given previously, and in addition, other representative and dependable procedures for preparing solid derivatives of the other types of compounds will be presented here. Short tables of some selected compounds, along with their melting or boiling points and melting points of several solid derivatives, are also presented.

[6] This is the reason for the dashed arrow in Figure 20.1.

Each of the classes of compounds tabulated above will be considered in the following manner. *Classification tests* for the functional group characteristic of the type of compound will be given (or in those cases where these tests have already been presented, reference will be made to suitable places in the text), and these will be followed by brief descriptions of methods of *preparation of several representative solid derivatives*. Reference will be made to the tables of compounds and derivatives which follow.

As mentioned in connection with the outline of the six fundamental steps, in general it is wise to carry these out in the order listed. On the other hand, there is nothing necessarily sacred about the order of the first four steps. It usually makes little difference whether one carries out an elementary analysis before doing solubility tests, or *vice versa*. For example, if one found a water-insoluble compound to be soluble in dilute hydrochloric acid, an amine is a possibility, and an immediate test for nitrogen is indicated. Conversely, if a positive nitrogen test had been found first, a solubility test for an amine would be a logical next step.

There is one warning that should be emphasized strongly, however. Do not go directly to the preparation of a derivative on the basis of a "hunch" as to the class of compound to which your unknown belongs, or even as to its specific identity, but rather make certain of the type of functional group present by obtaining one or more positive classification tests before attempting the preparation of any derivative. "Jumping the gun"—going directly from a melting point or odor to a derivative is the most common error of beginners in qualitative organic analysis. For example, it is very frustrating to try to make a ketone derivative when the unknown is actually an ester.

In deciding which of the classification tests to try first, one should use the information gained in the first four steps. To continue with the example used above, if the unknown has been found to contain nitrogen and to be soluble in dilute acidic solution, a classification test for an amine should be applied first. In the absence of clear indication of the type of compound, the most reliable classification tests should be applied first. Among the most reliable, certainly, are those for aldehydes and ketones.

ALDEHYDES AND KETONES (Tables 20.1 and 20.2)

A. CLASSIFICATION TESTS

1. 2,4-Dinitrophenylhydrazine. This reagent will give a positive test for either an aldehyde or a ketone. Experimental details for this test are given in Chapter 15.1.

2. Chromic acid. This reagent serves as a clear-cut method for distinguishing between aldehydes and ketones. The reactions are discussed in detail in Chapter 14.2, and the experimental procedures are given there.

3. Iodoform test using sodium hypoiodite (NaOI). This test is used to detect the presence of methyl ketones (compounds having a terminal –COCH₃ group), as compared to a ketone which does not contain this specific functional group. Hydroxymethyl compounds (–CHOHCH₃) also give iodoform with NaOI, so that it is always necessary to consider this alternative before making a final decision; the problem is solved by determining that a keto group is present or absent. This reaction is discussed in Chapter 15.2, and experimental procedures accompany that discussion.

4. Tollens' test. Another method for distinguishing between aldehydes and ketones (an alternative to the chromic anhydride test) is the Tollens' test. A positive test determines the presence of an aldehyde group, whereas no reaction occurs with ketones.

Tollens' reagent consists of silver ammonia complex, $Ag(NH_3)_2^+$, in an ammonia solution. This reagent, on reaction with an aliphatic or aromatic aldehyde, is reduced, whereas the aldehyde is oxidized to the corresponding carboxylic acid; the silver is reduced from the +1 state to elemental silver, and the silver often is deposited as a silver mirror on the side of the test tube. Thus, the formation of the silver mirror is considered a positive test. The equation for the reaction which occurs is given below.

$$RCHO + 2\ Ag(NH_3)_2 + 2\ HO^\ominus \longrightarrow 2\ \underline{Ag} + RCO_2^\ominus NH_4^\oplus + H_2O + 3\ NH_3 \quad (1)$$

Similar tests for aldehydes make use of Fehling's and Benedict's reagents which contain complex salts (tartrate and citrate, respectively) of cupric ion as the oxidizing agents. With these reagents, a positive test is the formation of a brick red precipitate of cuprous oxide (Cu_2O) which forms when Cu^{++} is reduced to Cu^+ by the aldehyde. These two tests are more useful in distinguishing between aliphatic and aromatic aldehydes, since the aliphatic compounds give a fast reaction. We shall not discuss further or use either of these tests.

PROCEDURE

Prepare the Tollens' reagent by placing 2 ml of a 5% solution of silver nitrate and a drop of 10% sodium hydroxide in a test tube. Add a 2% solution of ammonia drop by drop to this mixture, with shaking, until the brown precipitate of silver oxide just dissolves. Avoid an excess of ammonia. Prepare the reagent just before it is to be used, and do not store the solution since it decomposes on standing and yields an explosive precipitate. Add 50–100 mg of the unknown compound to the reagent; the formation of a silver mirror or a black precipitate constitutes a positive test.[7] If no reaction takes place at room temperature, warm the solution slightly in a beaker of warm water.

B. DERIVATIVES

Two very suitable solid derivatives of aldehydes and ketones are the *2,4-dinitrophenylhydrazones* and the *semicarbazones*. Oximes are also sometimes useful. Methods for preparing and purifying each of these are given in Chapter 15.1; see Tables 20.1 and 20.2 for lists of possible compounds and their derivatives.

UNSATURATED COMPOUNDS

CLASSIFICATION TESTS

Two common types of unsaturated compounds are alkenes and alkynes, characterized by the carbon-carbon double and triple bond, respectively, as the functional group. There are no simple direct ways to prepare solid derivatives of unsaturated aliphatic compounds having no other functional groups, but it is often useful to detect the presence of these two functional groups. The two common qualitative tests for unsaturation are the reaction of the compounds with *bromine in carbon tetrachloride* and with *potassium permanganate*. In both cases, a positive test is denoted by decoloration of the reagent. Discussion and experimental procedures for these two tests may be found in Chapter 6.1.

[7] If the test tube is not scrupulously clean, the silver will not deposit on the wall as a mirror, but will appear as a black precipitate.

ALKYL HALIDES

Qualitative tests for alkyl halides are useful in deciding whether the compound in question is a primary, secondary, or tertiary halide. In general, it is quite difficult to prepare solid derivatives of alkyl halides, so that we shall limit this discussion to the two qualitative tests: (1) the reaction with *alcoholic silver nitrate* solution; and (2) the reaction with *sodium iodide in acetone*.

1. Alcoholic silver nitrate. If a compound is known to contain a halogen (bromine, chlorine or iodine), information concerning its environment may be obtained from observation of its reaction with alcoholic silver nitrate. The overall reaction is shown in equation (2). Such a reaction will be of the $S_N 1$ type. As has been asserted earlier

$$RX + AgNO_3 \xrightarrow{\text{ethanol}} \underline{AgX} + RONO_2 \tag{2}$$

(Chapter 11.2), tertiary halides are more reactive than secondary halides, which are in turn more reactive than primary halides, in an $S_N 1$ reaction. Differing rates of silver halide precipitation would be expected from halogen in each of these environments, *viz.*, primary < secondary < tertiary. These differences are best determined by testing authentic samples of primary, secondary and tertiary halides with silver nitrate in separate test tubes and observing the results.

Alkyl bromides and iodides react more rapidly than chlorides, and the latter may require warming to produce a reaction in a reasonable period of time.

Aryl halides are unreactive towards the test reagent as are, in general, any vinyl or alkynyl halides. Allylic halides, even when primary, show reactivities as great as or greater than tertiary halides due to resonance stabilization of the resulting allyl cation.

PROCEDURE

Add 1 drop of the alkyl halide to 2 ml of a 2% solution of silver nitrate in ethanol. If no reaction is observed within 5 min at room temperature, warm the mixture in a beaker of boiling water and observe any change. Note the color of any precipitates; silver chloride is white, silver bromide is pale yellow, and silver iodide is yellow. If there is any precipitate, add several drops of 5% nitric acid solution to it and note any changes; the silver halides are insoluble in acid. To determine expected reactivities, test primary, secondary, and tertiary halides in this manner. If possible, use alkyl iodides, bromides, and chlorides so that differences in halogen reactivity can also be observed.

2. Sodium iodide in acetone. Another method for distinguishing between primary, secondary, and tertiary halides makes use of sodium iodide dissolved in acetone. This test complements the alcoholic silver nitrate test, and when these two tests are used together, it is possible to determine fairly accurately the gross structure of the attached alkyl group.

The test depends on the fact that both sodium chloride and sodium bromide are not very soluble in acetone, whereas sodium iodide is. The reactions which occur are:

$$RCl + NaI \xrightarrow{\text{acetone}} RI + \underline{NaCl} \tag{3}$$

$$RBr + NaI \xrightarrow{\text{acetone}} RI + \underline{NaBr} \tag{4}$$

This reaction is an $S_N 2$ substitution in which iodide ion is the nucleophile; the order of reactivity is primary > secondary > tertiary.

With this test reagent, primary bromides give a precipitate of sodium bromide in about 3 min at room temperature, whereas the primary and secondary chlorides must be heated to about 50° before reaction occurs. Secondary and tertiary bromides react at 50°, but the tertiary chlorides fail to react in a reasonable time. It should be noted that this test is necessarily limited to bromides and chlorides.

PROCEDURE

Place 1 ml of the sodium iodide–acetone test solution in a test tube and add 2 drops of the chloro or bromo compound. If the compound is a solid, dissolve about 50 mg of it in a minimum volume of acetone, and add this solution to the reagent. Shake the test tube and allow it to stand for 3 min at room temperature. Note whether a precipitate forms; if no change occurs after 3 min, warm the mixture in a beaker of water at 50°. After 6 min of heating, cool to room temperature and note whether any precipitate forms.

Carry out this reaction with a series of primary, secondary and tertiary halides, both chlorides and bromides. Note in all cases the differences in reactivity as evidenced by the rate of formation of sodium bromide or chloride.

ARYL HALIDES AND AROMATIC HYDROCARBONS
(Tables 20.3 and 20.4)

A. CLASSIFICATION TEST

This test for the presence of an aromatic ring should be performed only on compounds which have been shown to be insoluble in concentrated sulfuric acid (see solubility tests). The test involves the reaction between an aromatic compound and chloroform in the presence of anhydrous aluminum chloride catalyst. The colors produced in this type of reaction are often quite characteristic for certain aromatic compounds, whereas aliphatic compounds give little or no color with this test. Some typical examples are tabulated below.

Type of Compound	Color
Benzene and homologs	Orange to Red
Aryl Halides	Orange to Red
Naphthalene	Blue
Biphenyl	Purple

Often these colors change with time and ultimately yield brown-colored solutions. Carbon tetrachloride may be used in place of chloroform; it yields similar colors.

The test is based upon a series of Friedel-Crafts alkylation reactions; for benzene, the ultimate product is triphenylmethane [equation (5)].

$$3 \ C_6H_6 + CHCl_3 \xrightarrow{\text{AlCl}_3} (C_6H_5)_3CH + 3 \ HCl \qquad (5)$$

The colors arise due to formation of species such as triphenylmethyl cations $(C_6H_5)_3C^+$, which remain in the solution as $AlCl_4^-$ salts; ions of this sort are highly colored due to the extensive delocalization of charge which is possible throughout the three aromatic rings.

The test is significant if positive, but a negative test does not rule out an aromatic structure; some compounds are so unreactive that they do not readily undergo Friedel-Crafts reactions (Chapter 7.1).

PROCEDURE

Heat about 100 mg of *anhydrous* aluminum chloride in a Pyrex test tube held almost horizontally until the material has sublimed to 3 or 4 cm above the bottom of the tube. Allow the tube to cool until it is almost comfortable to touch and then add down the side of the tube about 20 mg of a solid or 1 drop of a liquid unknown, followed by 2 or 3 drops of chloroform. The appearance of a bright color, ranging from red to blue, where the sample and chloroform come in contact with the aluminum chloride, is a positive indication of an aromatic ring.

Positive tests for aryl halides are difficult to obtain directly, and some of the best evidence for their presence involves indirect methods. Elemental analysis will indicate the presence of halogen. If *both* the silver nitrate and sodium iodide–acetone tests are negative, then the compound is most likely a vinyl or aromatic halide, both of which are very unreactive towards silver nitrate and sodium iodide. Distinction between a vinyl and an aromatic halide can be made by means of the aluminum chloride–chloroform test.

B. DERIVATIVES

Two types of derivatives can be used to characterize aromatic hydrocarbons and aryl halides. These are prepared by (1) nitration and (2) side-chain oxidation. The second method involves oxidation of a side chain to a carboxylic acid group. Since carboxylic acids are often solids, they themselves serve as suitable derivatives. The mechanism and equations for this oxidation reaction have been discussed (Chapter 16.1). The two common methods of oxidation use either chromic acid or potassium permanganate; although the experimental details of a permanganate oxidation are given in Chapter 16, procedures for doing both types of oxidations will be given here since much smaller amounts of compound normally must be used.

1. Nitration.   Some of the best solid derivatives of aryl halides are mono- and dinitration products. Two general procedures can be used for nitration. Whenever nitrating a compound, whether it is known or unknown, use care since many of these compounds react vigorously under typical nitration conditions.

Method A: This method yields *m*-dinitrobenzene from benzene or nitrobenzene and the *p*-nitro derivative from chloro- or bromobenzene, benzyl chloride, or toluene. Dinitro derivatives are obtained from phenol, acetanilide, naphthalene, and biphenyl. Add about 1 g of the compound to 4 ml of concentrated sulfuric acid. Add 4 ml of concentrated nitric acid dropwise to this mixture; shake after each addition. Carry out the reaction in a large test tube or small Erlenmeyer flask and heat at 45° for *ca.* 5 min, using a beaker of water to supply the heat. Then pour the reaction mixture onto 25 g of ice, and collect the precipitate on a filter. The solid may be recrystallized from aqueous ethanol if needed.

Method B: This method is the best one to use for nitrating halogenated benzenes, since dinitration occurs to give compounds which have higher melting points and are easier to purify than are mononitration products obtained from Method A. The xylenes, mesitylene, and pseudocumene yield trinitro compounds. Follow the procedure used for Method A, except that 4 ml of *fuming* nitric acid should be used in place of the *concentrated* nitric acid. Warm the mixture using a steam cone for 10 min, and work up as above. If little or no nitration occurs, substitute *fuming* sulfuric acid for the *concentrated* sulfuric acid.

2. Side-chain oxidation. a. Permanganate method. Add 1 g of the compound to a solution prepared from 80 ml of water and 4 g of potassium permanganate. Add 1 ml of 10% sodium hydroxide solution and heat the mixture at reflux until the purple color characteristic of the permanganate has disappeared; this will normally take from 30 min to 3 hr. At the end of the reflux period, cool the mixture and carefully acidify it with dilute sulfuric acid. Now heat the mixture for an additional 30 min and cool it again; remove excess brown manganese dioxide (if any) by addition of sodium bisulfite solution. The bisulfite serves to reduce the manganese dioxide to manganous ion, which is water soluble. Collect the solid acid which remains by suction filtration. Recrystallize the acid from benzene or aqueous ethanol. If little or no solid acid is formed, this may be due to the fact that the acid is somewhat water-soluble. In this case, extract the aqueous layer with chloroform, ether, or dichloromethane, dry the organic extracts and remove the organic solvent by means of a steam bath in the hood. Recrystallize the acid which remains. In this particular method, the presence of base during the oxidation often means that some silicic acid will form on acidification; thus, purification before determining the melting point is necessary.

b. Chromic acid method. Dissolve 7 g of sodium dichromate in 15 ml of water, and add 2 to 3 g of the compound to be oxidized. Add 10 ml of concentrated sulfuric acid to the mixture with mixing and cooling. Attach a reflux condenser to the flask and heat gently until a reaction ensues; as soon as the reaction begins, remove the flame and cool the mixture if necessary. After spontaneous boiling subsides, heat the mixture at reflux for two hours. Pour the reaction mixture into 25 ml of water, and collect the precipitate by filtration. Transfer the solid to a flask and add 20 ml of 5% sulfuric acid, and then warm the flask on a steam cone with stirring. Cool the mixture, collect the precipitate and wash it with about 20 ml of cold water. Dissolve the residue in 20 ml of 5% sodium hydroxide solution and filter the solution. Add the filtrate, with stirring, to 25 ml of 10% sulfuric acid. Collect the new precipitate, wash with water, and recrystallize from either benzene or aqueous ethanol.

Lists of possible aromatic hydrocarbons and halides appear in Tables 20.3 and 20.4, along with derivatives.

ALCOHOLS (Table 20.5)

A. CLASSIFICATION TESTS

The tests for the presence of a hydroxy group not only detect the presence of the group, but may also give an indication as to whether it occupies a primary, secondary, or tertiary position.

1. Chromic acid in acetone can be used to detect the presence of a hydroxy group, provided that it has been shown previously that the molecule does not contain an aldehyde function. The reactions and experimental procedures for this test are given in Chapter 14.2. It has been pointed out that chromic acid does not distinguish between primary and secondary alcohols, for primary and secondary alcohols *both* give a positive test whereas tertiary alcohols do not.

2. The Lucas test can be used to distinguish between primary, secondary, and tertiary alcohols. The reagent used is a mixture of concentrated hydrochloric acid and zinc chloride, which on reaction with alcohols converts them to the corresponding alkyl chlorides. With this reagent, primary alcohols give no appreciable reaction, secondary alcohols react more rapidly, and tertiary alcohols react very rapidly. A positive test depends on the fact that the alcohol is soluble in the reagent, whereas the alkyl chloride is not; thus, the formation of a second layer or an emulsion comprises a positive test. The solubility of the alcohol in the reagent places limitations on the utility of the test, and in general, only monofunctional alcohols with six or less carbon atoms, as well as

PRIMARY: $RCH_2OH + HCl \xrightarrow{ZnCl_2}$ No reaction   (6)

SECONDARY: $R_2CHOH + HCl \xrightarrow{ZnCl_2} R_2CHCl + H_2O$ (7)

TERTIARY: $R_3COH + HCl \xrightarrow{ZnCl_2} R_3CCl + H_2O$ (8)

polyfunctional alcohols, can be used. Note the similarity of this reaction with the nucleophilic displacement reactions between alcohols and hydrohalic acids which have been discussed in Chapter 11.2. In the Lucas test that the presence of zinc chloride, which is a Lewis acid, greatly increases the reactivity of alcohols toward hydrochloric acid.

PROCEDURE

Add 10 ml of the hydrochloric acid–zinc chloride reagent (Lucas reagent) to about 1 ml of the compound in a test tube. Stopper the tube and shake; allow the mixture to stand at room temperature. Try this test with a primary, a secondary, and a tertiary alcohol and note the time which is required for the formation of the alkyl chloride, which will appear either as a second layer or as an emulsion.

3. The ceric nitrate reagent can also be used as a qualitative test for alcohols. Although this reagent has been used primarily for phenols, it does give a positive test with alcohols. Discussion about and experimental procedures for this test are given under Phenols, p. 342.

B. DERIVATIVES

Two common derivatives of alcohols are the urethans and the benzoate esters; the former are best for primary and secondary alcohols, whereas the latter are useful for all types of alcohols.

1. Urethans. When an alcohol is allowed to react with an aryl substituted isocyanate, $ArN{=}C{=}O$, addition of the alcohol occurs to give a urethan [equation (9)].

$$ArN{=}C{=}O + R'OH \longrightarrow ArNH{-}\overset{\overset{\displaystyle O}{\|}}{C}{-}OR' \tag{9}$$

Some commonly used isocyanates are α-naphthyl, p-nitrophenyl, and phenyl iso-cyanate. A major side reaction which occurs is that of water with the isocyanate; water hydrolyzes the isocyanate to an amine, and the amine reacts with more isocyanate to give a disubstituted urea [equations (10) and (11)]. Since the ureas are high melting

$$ArN{=}C{=}O + H_2O \rightarrow (ArNH{-}COOH) \rightarrow ArNH_2 + CO_2 \tag{10}$$

a carbamic acid
(unstable)

$$ArN{=}C{=}O + ArNH_2 \longrightarrow ArNH{-}\overset{\overset{\displaystyle O}{\|}}{C}{-}NHAr \tag{11}$$

a disubstituted urea

amides, owing to their symmetry, their presence makes purification of the desired urethan quite difficult. In using this procedure, take precautions to insure that the alcohol is anhydrous. This procedure works best for alcohols which are water-insoluble and can therefore be obtained easily in anhydrous form.

This type of derivative can also be useful for phenols; the procedure given here has been generalized so that it can be used for alcohols and phenols. Other derivatives of phenols will be given below.

PROCEDURE

Place 1 g of the *anhydrous* alcohol or phenol in a test tube and add 0.5 ml of phenyl isocyanate or α-naphthyl isocyanate. (If a phenol is used, add 2 or 3 drops of pyridine as a catalyst.) If the reaction is not spontaneous, warm the test tube over a steam bath for 5 min, being careful to exclude condensed steam from the inside of the test tube. Cool the mixture in a beaker of ice and scratch the sides of the test tube to in-duce crystallization. Purify the crude solid so obtained by recrystallizing it from carbon tetrachloride or petroleum ether.

2. 3,5-Dinitrobenzoates.   The reaction between 3,5-dinitrobenzoyl chloride and an alcohol gives the corresponding ester. This method is useful for primary, secondary, and tertiary alcohols, especially those which are water-soluble and which are likely to contain traces of water.

$$\text{ROH} + \text{Cl}{-}\overset{\overset{\displaystyle O}{\|}}{C}{-}\!\!\!\bigcirc\!\!\!\overset{NO_2}{\underset{NO_2}{}} \longrightarrow \text{RO}{-}\overset{\overset{\displaystyle O}{\|}}{C}{-}\!\!\!\bigcirc\!\!\!\overset{NO_2}{\underset{NO_2}{}} + \text{HCl} \tag{12}$$

PROCEDURE

Mix 2 ml of the alcohol with about 0.5 g of 3,5-dinitrobenzoyl chloride in a test tube, and boil the mixture gently for 5 min. Add *ca.* 10 ml of distilled water and cool the solution in an ice bath to obtain a solid product. Collect the crude product, wash it with 10 ml of 2% sodium carbonate solution, and recrystallize the product using aqueous alcohol. A minimum volume of recrystallization solvent should be used; adjust the solvent composition by adding just enough water to the alcohol so that the product dissolves in the hot solution but yields crystals when cooled.

A list of alcohols and their derivatives is given in Table 20.5.

PHENOLS (Table 20.6)

A. CLASSIFICATION TEST

There are several tests which can be used to detect the presence of a phenolic hydroxy group: (1) Bromine water, (2) ceric nitrate reagent, and (3) ferric chloride solution. In addition to these, solubility tests give a preliminary indication of a phenol since phenols are soluble in 5% sodium hydroxide solution but generally insoluble in 5% sodium bicarbonate solution.

1. Bromine water. Phenols, in general, are highly reactive toward electrophilic substitution, and consequently are brominated readily by bromine water. The rate of bromination is much greater in water than in carbon tetrachloride solution. The water, being more polar than carbon tetrachloride, increases the ionization of bromine and thus enhances the ionic bromination mechanism. Although hydrogen bromide is liberated, it is not observed when using water as the solvent. Phenols are so reactive

$$\text{(phenol)} + 3\,Br_2 \longrightarrow \text{(2,4,6-tribromophenol)} + 3\,HBr \tag{13}$$

that all unsubstituted positions *ortho* and *para* to the hydroxy group are brominated. The brominated compounds so formed are often solids and can also be used as derivatives (see below). Aniline and substituted anilines are also very reactive toward bromine and react analogously; however, solubility tests can be used to distinguish between anilines and phenols.

PROCEDURE

Prepare a 1% aqueous solution of the unknown, and add bromine water dropwise to this solution; continue addition until the bromine color remains. Note how much bromine water was used, and try this experiment on phenol and aniline for comparison purposes.

2. Ceric nitrate reagent. Alcohols and phenols are capable of replacing nitrate ions in complex cerate anions, resulting in a change from a yellow to a red solution:

$$(NH_4^{\oplus})_2 Ce(NO_3)_6^{\ominus} + ROH \rightarrow (NH_4^{\oplus})_2 Ce(OR)(NO_3)_5^{\ominus} + HNO_3 \tag{14}$$

$$\text{yellow} \qquad\qquad\qquad\qquad \text{red}$$

Alcohols and phenols having no more than ten carbon atoms give a positive test. Alcohols give a red solution. Phenols give a brown to greenish-brown precipitate in aqueous solution; in dioxane a red to brown solution is produced. Aromatic amines may be oxidized by the reagent and give a color indicating a positive test.

PROCEDURE

Dissolve about 20 mg of a solid or 1 drop of a liquid unknown in 1–2 ml of water and add 0.5 ml of the ceric ammonium nitrate reagent; shake and note the color. If the unknown in insoluble in water, dissolve it in 1 ml of dioxane before adding 0.5 ml of the reagent.

3. Ferric chloride test. Most phenols and enols react with ferric chloride to give colored complexes. The colors vary, depending not only on the nature of the phenol or enol, but also on the solvent, concentration, and time of observation. Some phenols which do not give coloration in aqueous or alcoholic solution do so in chloroform solution, especially after addition of a drop of pyridine The nature of the colored complexes is still uncertain; they may be ferric phenoxide salts that absorb visible light to give an excited state in which electrons are delocalized over both the iron atoms and the conjugated organic system. The production of a color is typical of phenols and enols; however, many of them do *not* give colors, so that a negative ferric chloride test must not be taken as significant without supporting information (*e.g.*, the ceric nitrate and bromine water tests).

PROCEDURE

Dissolve 30 to 50 mg of the unknown compound in 1–2 ml of water (or a mixture of water and ethanol if the compound is not water-soluble) and add several drops of a 2.5% aqueous solution of ferric chloride. Most phenols produce red, blue, purple, or green colorations; enols give red, violet, or tan colorations.

B. Derivatives

Two useful solid derivatives of phenols are α-naphthyl urethan and bromo derivatives. The preparation of urethans has been discussed previously in this chapter (Alcohols, page 339) and whereas any substituted urethan could be prepared, the majority of the derivatives reported are the α-naphthyl urethans. They are suggested as the urethans of choice.

PROCEDURE

The bromination of phenols may be carried out as follows. Prepare a brominating solution by dissolving 15 g of potassium hydroxide in 100 ml of water and adding 10 g of bromine. Dissolve 1 g of the phenol in water, ethanol, acetone, or dioxane, and add the brominating solution to it dropwise. Add just enough of the bromine solution to impart a yellow bromine color to the solution. Add about 50 ml of water to the mixture. Shake the mixture vigorously and remove the solid by filtration. Wash the solid with a dilute solution of sodium bisulfite to remove excess bromine and recrystallize it from ethanol or aqueous ethanol.

A listing of phenols and their derivatives is given in Table 20.6.

CARBOXYLIC ACIDS (Table 20.7)

A. CLASSIFICATION TEST

One of the best qualitative tests for the carboxylic acid group is solubility in basic solutions. Carboxylic acids are soluble in both 5% sodium hydroxide solution and in 5% sodium bicarbonate solution, from which they can be regenerated by addition of acid. Solubility properties have been discussed previously in this chapter.

Information of a quantitative nature can be obtained about carboxylic acids, this being the *equivalent weight* or *neutralization equivalent*. This value can be obtained by titrating a known weight of the acid with a known volume of standardized base solution. Further discussion of the equivalent weight is given in Chapter 16.1.

PROCEDURE

Dissolve an accurately weighed sample (about 0.2 grams) of the acid in 50 to 100 ml of water or ethanol or a mixture of the two. It may be necessary to warm the mixture to dissolve the compound completely. Titrate the solution with a previously standardized sodium hydroxide solution having a concentration of about 0.1 N. Use phenolphthalein as the indicator, and from this data calculate the equivalent weight as described in Chapter 16.1.

B. DERIVATIVES

Three good solid derivatives of carboxylic acids are: (1) amides, (2) anilides, and (3) *p*-toluidides. Each of these three derivatives is prepared from the corresponding acid chloride by treatment of the latter with ammonia, aniline, or *p*-toluidine. The amides are generally less satisfactory than the other two, for they tend to be more soluble in water and as a result, harder to isolate. The acid chlorides are most conveniently prepared from the acid, or its salt, and thionyl chloride.

$$RCOOH \text{ (or } RCOO^{\ominus} Na^{\oplus}) + SOCl_2 \rightarrow RCOCl + SO_2 + HCl \text{ (or NaCl)} \quad (15)$$

AMIDES: $$RCOCl + 2 NH_3 \xrightarrow{\text{cold}} RCONH_2 + NH_4^{\oplus} Cl^{\ominus} \quad (16)$$

ANILIDES: $$RCOCl + 2 C_6H_5NH_2 \rightarrow RCONHC_6H_5 + C_6H_5NH_3^{\oplus} Cl^{\ominus} \quad (17)$$

TOLUIDIDES: $$RCOCl + 2 \text{ } p\text{-}CH_3C_6H_4NH_2 \rightarrow RCONHC_6H_4CH_3$$
$$+ p\text{-}CH_3C_6H_4NH_3^{\oplus} Cl^{\ominus} \quad (18)$$

PROCEDURE

The acid chloride may be prepared by heating, under reflux, 5 ml of thionyl chloride and 1 g of the acid or its sodium salt. Heat for 15–30 min, and use care to protect the apparatus from moisture; keep the vapors from filling the laboratory [see Chapter 16 (Experimental Procedure) for additional comments about using thionyl chloride].

1. Amides. Pour the mixture containing the acid chloride and unchanged thionyl chloride into 15 ml of ice-cold, concentrated ammonia; be very careful, for a vigorous reaction occurs. Collect the precipitated amide on a filter and recrystallize from water or aqueous ethanol. Determine the melting point of the product.

2. Anilides and *p*-toluidides. Prepare the acid chloride as before, using 1 g of acid or its salt and 2 ml of thionyl chloride. Heat the mixture at gentle reflux for 30 min, and then cool it. To the cold mixture, add a solution of 1 to 2 g of the appropriate amine (aniline or *p*-toluidine) dissolved in 30 ml of benzene. Heat the mixture on a steam bath for *ca.* 2 min, and then decant the benzene solution into a separatory funnel. Wash the organic layer with 2 ml of water, 5 ml of 5% hydrochloric acid, 5 ml of 5% sodium hydroxide solution, and finally with 2 ml of water. Evaporate the benzene and recrystallize the amide from either water or aqueous ethanol.

A list of carboxylic acids and their derivatives is given in Table 20.7.

AMINES (Tables 20.8 and 20.9)

A. CLASSIFICATION TEST

Two common qualitative tests for amines are the *Hinsberg test* and the *nitrous acid reaction*. In addition to the initial solubility tests, which indicate the presence of an amine, these tests allow one to classify an amine as either primary, secondary, or tertiary.

1. The Hinsberg test.   The reaction between primary or secondary amines and benzenesulfonyl chloride in the presence of sodium hydroxide solution yields the corresponding substituted benzenesulfonamide, whereas tertiary amines do not react with this reagent. The equations below depict the reactions which occur. The reaction is

PRIMARY:

$$RNH_2 + \langle\bigcirc\rangle-SO_2Cl \xrightarrow{NaOH} \langle\bigcirc\rangle-SO_2NHR + NaCl + H_2O \qquad (19)$$

Insoluble in water

excess HCl ⇅ excess NaOH

$$\langle\bigcirc\rangle-SO_2\overset{\ominus}{N}RNa^{\oplus} + H_2O \qquad (20)$$

Soluble in water

SECONDARY:

$$R_2NH + \langle\bigcirc\rangle-SO_2Cl \xrightarrow{NaOH} \langle\bigcirc\rangle-SO_2NR_2 + NaCl + H_2O \qquad (21)$$

Insoluble in water

excess NaOH ↓

No reaction

TERTIARY: $R_3N + \langle\bigcirc\rangle-SO_2Cl \xrightarrow{NaOH}$ No reaction $\xrightarrow{HCl} R_3NH^{\oplus}Cl^{\ominus}$ (22)
   (R_3N)   Soluble

normally carried out in excess base. If the amine is primary, the sulfonamide first formed has an acidic hydrogen which is removed by base to yield the soluble sodium salt. On addition of acid, the water-insoluble sulfonamide is regenerated. If the amine is secondary, the sulfonamide which is produced is insoluble in both base and acid. If the amine is tertiary, it does not react, and is recovered as unchanged amine. It will be insoluble in base but will be soluble in acid. These three possible combinations of acid and base solubility of the product allow for unique distinction between primary, secondary, and tertiary amines.

However, the Hinsberg test must be carried out carefully and the results interpreted intelligently. A carelessly executed test is worse than none attempted, in that it may be misleading. If there is any doubt regarding the interpretation of the results, carry out the test on known compounds and compare the results with those of the unknown.

PROCEDURE

Mix 5 ml of 10% sodium hydroxide, 0.3 ml of the amine, and 0.4 ml of benzene-sulfonyl chloride (*Caution:* It is a lachrymator) in a test tube. Stopper the tube and shake the mixture vigorously. If no reaction appears to occur, heat the reaction mixture on a steam bath. Some secondary amines react slowly. After making sure that the solution is still basic (if not, add more base until it is), cool the solution in an ice bath and remove by filtration or decantation any residue which remains.

Observe whether the residue is a solid or liquid and whether it is more or less dense than the aqueous solution. Test the solubility of the residue in water, in acid, and in base. Solubility of the residue in acid indicates that it is a tertiary amine. The reason for testing solubility in water is that some sulfonamide salts of primary amines are insoluble in base but soluble in water; thus solubility in water is indicative of a primary amine.

Acidify the water solution to see if a precipitate forms. Acidify the filtrate. If a precipitate forms, attempt to get it to crystallize by cooling the mixture in an ice bath and by scratching the sides of the tube. The formation of a precipitate here is indicative of a primary amine.

For purposes of comparison, it is suggested that the student try these tests with a known primary, secondary, and tertiary amine.

2. Nitrous acid test. This test should always be used to confirm and extend the benzenesulfonyl chloride test.

PROCEDURE

Dissolve about 50 mg of a solid or two drops of a liquid amine in 2 ml of 2 N hydrochloric acid and cool the solution to 0–5° in an ice bath. Add about 5 drops of a cold 20% aqueous solution of sodium nitrite.

a. Immediate evolution of a *colorless* gas (nitrogen) is indicative of a primary aliphatic amine.

$$RNH_2 \xrightarrow[\text{HCl}]{\text{NaNO}_2} [RN_2^{\oplus} \, Cl^{\ominus}] \xrightarrow{\text{H}_2\text{O}} N_2 + ROH + RCl + \text{alkenes} \qquad (22)$$

b. Primary aromatic amines give diazonium salts which are stable at low temperatures. If vigorous nitrogen evolution was not observed, add a few drops of the cold reaction mixture to a cold solution of 50 mg of β-naphthol in 2 ml of 2 N sodium hydroxide. The formation of an orange to red azo dye is indicative of a primary aromatic amine.

$$\text{(23)}$$

c. If no nitrogen was evolved, but an insoluble yellow or orange liquid separated, the unknown amine is indicated to be a secondary amine, either aliphatic or aromatic.

$$R_2NH + HONO \longrightarrow R_2N-N{=}O + HOH \tag{24}$$

d. If no appreciable reaction was observed, the unknown is indicated to be a tertiary aliphatic amine, whose hydrochloric acid salt is insoluble in acidic solution.

$$R_3N + HCl \rightleftharpoons R_3NH^{\oplus}Cl^{\ominus} \tag{25}$$

A yellow coloration may result from reaction of an N,N-dialkylaniline with nitrous acid; the p-nitroso derivative is green, but, in acidic solution, its salt is yellow.

$$R_2N{-}\langle\bigcirc\rangle + HONO + HCl \longrightarrow R_2\overset{\oplus}{N}H{-}\langle\bigcirc\rangle{-}N{=}O + H_2O \tag{26}$$
$$Cl^{\ominus}$$

B. Derivatives

Suitable derivatives of primary and secondary amines are the benzamides and benzenesulfonamides.

$$RNH_2 \text{ (or } R_2NH) + C_6H_5COCl \xrightarrow{\text{Pyridine}} C_6H_5\overset{O}{\overset{\|}{C}}{-}NHR \text{ (or } C_6H_5\overset{O}{\overset{\|}{C}}{-}NR_2) \tag{27}$$

$$RNH_2 \text{ (or } R_2NH) + C_6H_5SO_2Cl \xrightarrow{\text{NaOH}} C_6H_5SO_2{-}NHR \text{ (or } C_6H_5SO_2NR_2) \tag{28}$$

PROCEDURE

The method for preparing the benzenesulfonamides has been discussed under the Hinsberg test. The derivatives can be prepared using that method, but sufficient amounts of material should be used so that the final product can be purified by recrystallization from 95% ethanol.

The benzamides can be prepared as follows. Dissolve 0.5 g of the amine in 5 ml of dry pyridine and 10 ml of dry benzene. Add, dropwise, 0.5 ml of benzoyl chloride. Heat the resulting mixture in a water bath for 30 min at *ca.* 60–70°, and then pour the solution into 100 ml of water. Separate the benzene layer and wash the aqueous layer once with 10 ml of benzene. Combine the organic extracts and wash them with water followed by 5% sodium carbonate solution. After drying over magnesium sulfate and removing the drying agent by filtration, distil the mixture to remove most of the benzene. When the volume is 3 to 4 ml, add 20 ml of hexane to the mixture. Remove the solid which is formed by filtration and wash it with a small volume of hexane. Recrystallize the product from one of the following solvents: cyclohexane–hexane mixtures, cyclohexane–ethyl acetate mixtures, ethanol, or aqueous ethanol.

Although the above methods are satisfactory with primary and secondary amines, tertiary amines do not undergo the same reactions. In general, one must take advantage of the fact that tertiary amines do form salts. Two useful crystalline salts are the ones from methyl iodide and picric acid.

$$R_3N: + CH_3I \longrightarrow [R_3\overset{\oplus}{N}\!\!-\!\!CH_3]\, I^{\ominus} \tag{29}$$

$$\tag{30}$$

PROCEDURE

The methyl iodide derivative may be prepared as follows: Mix 0.5 g of the amine with 0.5 ml of methyl iodide and warm the test tube with a water bath for several minutes. Cool the test tube in an ice-water bath; the tube may be scratched with a rod to help induce crystallization. Purify the product by recrystallization from absolute ethanol or methanol or from ethyl acetate.

The picric acid derivative (often called the picrate) may be prepared by mixing 0.3 to 0.5 g of the compound with 10 ml of 95% ethanol. If solution is not complete, remove the excess solid by filtration. Add 10 ml of a saturated solution of picric acid in 95% ethanol to the mixture and heat the mixture to boiling. Cool the solution slowly and remove the yellow crystals of the picrate salt by filtration. Recrystallize the salt from ethanol.

A list of primary and secondary amines and their derivatives can be found in Table 20.8, whereas tertiary amines and derivatives appear in Table 20.9.

NITRO COMPOUNDS (Table 20.10)

A. CLASSIFICATION TEST

Ferrous hydroxide test. Organic compounds that are oxidizing agents will oxidize ferrous hydroxide (blue) to ferric hydroxide (brown). The most common organic compounds that function in this way are the *nitro compounds*, both aliphatic and aromatic, which are in turn reduced to amines in the reaction. Other less common types of com-

$$RNO_2 + 6 \underset{\text{blue}}{\underline{Fe(OH)_2}} + 4\,H_2O \longrightarrow RNH_2 + 6 \underset{\text{brown}}{\underline{Fe(OH)_3}} \tag{31}$$

pounds which give the same test are nitroso compounds, hydroxylamines, alkyl nitrates, alkyl nitrites, and quinones.

PROCEDURE

In a 10 × 75-mm or smaller test tube,[8] mix about 20 mg of a solid or 1 drop of a liquid unknown with 1.5 ml of freshly prepared 5% ferrous ammonium sulfate solution. Add 1 drop of 6 *N* sulfuric acid and 1 ml of 2 *N* potassium hydroxide in methanol. Stopper the tube immediately and shake it. A positive test is indicated by the blue precipitate turning rust-brown within one minute. (A slight darkening or greenish coloration of the blue precipitate should not be considered a positive test.)

---

[8] A small test tube is used in order to lessen the exposure of the reagent to air, which may bring about some oxidation.

B. DERIVATIVES

Two different types of derivatives of nitro compounds can be prepared. Aromatic nitro compounds can be di- and trinitrated with nitric acid and sulfuric acid. Discussion of and procedures for nitration have been given under "Aryl Halides," and the student is referred there for additional information.

The other method for preparation of a derivative can be utilized for both aliphatic and aromatic amines. This involves the reduction of the nitro compound to the corresponding primary amine, followed by conversion of the amine to a benzamide or benzenesulfonamide, as described under "Amines." The reduction is most often carried out with tin and hydrochloric acid.

$$\text{RNO}_2 \text{ or } \text{ArNO}_2 \xrightarrow[\text{HCl}]{\text{Sn}} \text{RNH}_2 \text{ or } \text{ArNH}_2 \tag{32}$$

PROCEDURE

Carry out the reduction of the nitro compound by combining 1 g of the compound and 2 g of granulated tin in a small flask. Attach a reflux condenser and add, in small portions, 20 ml of 10% hydrochloric acid. Shake after each addition. After addition is complete, warm the mixture using a steam bath for 10 min. If the nitro compound is insoluble, add 5 ml of 95% ethanol to increase its solubility. Decant the warm, homogeneous solution into 10 ml of water, and add enough 40% sodium hydroxide solution so that the tin hydroxide completely dissolves. Extract the basic solution with several 10-ml portions of ether. Dry the ether solution over potassium hydroxide pellets and remove the ether by distillation.

The residue contains the primary amine. Convert it to one of the derivatives described under "Amines."

A list of some nitro compounds and their derivatives appears in Table 20.10.

ESTERS (Table 20.11)

A. CLASSIFICATION TESTS

A test for the presence of the ester group involves the use of hydroxylamine and ferric chloride. The former converts the ester to a hydroxamic acid, which then complexes with Fe(III) to give a colored species.

$$\underset{\substack{\\ }}{\text{R}-\overset{\overset{\text{O}}{\|}}{\text{C}}-\text{OR}'} + \text{H}_2\text{NOH} \longrightarrow \underset{\substack{\text{hydroxamic}\\\text{acid}}}{\text{R}-\overset{\overset{\text{O}}{\|}}{\text{C}}-\text{NHOH}} + \text{R}'\text{OH} \tag{33}$$

$$3\ \text{R}-\overset{\overset{\text{O}}{\|}}{\text{C}}-\text{NHOH} + \text{FeCl}_3 \longrightarrow \left[\underset{\substack{\\ }}{\text{R}-\text{C}} \underset{\underset{\text{H}}{\overset{|}{\text{N}-\text{O}}}}{\overset{\overset{\text{O}}{\diagdown}}{\diagup}} \text{Fe} \right]_3 + 3\ \text{HCl} \tag{34}$$

colored

All carboxylic acid esters (including polyesters and lactones) give magenta colors which vary in intensity depending on structural features in the molecule. Acid chlorides and anhydrides also give positive tests. Formic acid produces a red color, but other free acids give negative tests. Primary or secondary aliphatic nitro compounds give a positive test because ferric chloride reacts with the *aci* form which is present in basic solution. Most imides give positive tests. Some amides, but not all, give light magenta coloration, whereas most nitriles give a negative test. (A modification of the procedure below will be given later which will yield a positive test for amides and nitriles.)

PROCEDURE

Before performing the final test, it is necessary to run a preliminary (or "blank") test since some compounds will give a positive test even though they do not contain an ester linkage.

1. Preliminary test. Mix 1 ml of 95% ethanol and 50–100 mg of the compound to be tested, and add 1 ml. of 1 *M* hydrochloric acid. Note the color which is produced when 1 drop of 5% aqueous ferric chloride solution is added. If the color is orange, red, blue, or violet, the following test for the ester group does not apply and cannot be used.

2. Final test. Mix 40–50 mg of the unknown, 1 ml of 0.5 *M* hydroxylamine hydrochloride in 95% ethanol, and 0.2 ml. of 6 *M* sodium hydroxide. Heat the mixture to boiling, and after cooling it slightly, add 2 ml. of 1 *M* hydrochloric acid. If the solution is cloudy, add more (about 2 ml) 95% ethanol. Add 1 drop of 5% ferric chloride and observe the color. Add more ferric chloride solution if the color does not persist, and continue to add it until it does. Compare the color which you obtain here with that from the preliminary test. If the color is burgundy or magenta, as compared to the yellow color in the preliminary experiment, the presence of an ester group is indicated.

B. QUANTITATIVE TEST

Once an ester group has been detected, it is possible to obtain information about its equivalent weight. This value, termed the *saponification equivalent*, has already been discussed in Chapter 16.2; the principles and equations involved in this technique are presented there.

PROCEDURE

Dissolve approximately 3 g of potassium hydroxide in 60 ml of 95% ethanol. Allow the small amount of insoluble material to settle to the bottom, and fill a 50-ml buret with the clear solution by decantation. Measure exactly 25.0 ml of the alcoholic solution into each of two 125-ml Erlenmeyer flasks. Weigh *accurately* into one of the flasks a 0.3–0.4-g sample of pure, dry ester; the other basic solution will be used as a "blank." Fit each of the flasks with a reflux condenser by means of a rubber stopper which has been washed with dilute sodium hydroxide solution and rinsed with distilled water.

Heat the solutions in both flasks at a gentle reflux for 1 hr. When the flasks have cooled, loosen the stoppers in the flasks, and rinse each condenser and stopper with about 10 ml of distilled water, catching the rinse water in the flask. Add phenolphthalein and titrate each of the solutions with standardized hydrochloric acid which is approximately 0.5 *N*.

The difference in the volumes of hydrochloric acid required to neutralize the base in the flask containing the sample and in the "blank" flask represents the amount of potassium hydroxide which reacted with the ester. The volume difference (in milliliters) multiplied by the normality of the hydrochloric acid equals the number of *milli*equivalents of potassium hydroxide consumed. Using the titration data, calculate the saponification equivalent of the unknown ester.

Should the ester not completely saponify in the allotted time, heat under reflux for longer periods (2–4 hr). In some cases, higher temperatures may be required; is so, diethylene glycol must be used as a solvent. Consult the instructor for details about this method.

C. Derivatives

In order to characterize an ester completely, it is necessary to prepare solid derivatives of both the acid and the alcohol components. The problem here is to isolate both of these components in pure form so that suitable derivatives can be prepared. One such way is to carry out the ester hydrolysis in a high-boiling solvent. If the alcohol is low-boiling, it can be distilled from the reaction mixture and characterized. The acid which remains in the mixture can be isolated also. Derivatives of acids and alcohols have already been discussed.

$$\text{RCOOR}' + \text{HO}^\ominus \rightarrow \text{RCOO}^\ominus + \text{R}'\text{OH} \tag{35}$$

PROCEDURE

In a small reaction vessel mix 3 ml of diethylene glycol, 0.6 g (2 pellets) of potassium hydroxide, and 10 drops of water. Heat until the solution is homogeneous, and cool to room temperature again. Add 1 ml of the ester and equip the apparatus with a condenser. Heat to boiling again with swirling and, after the ester layer dissolves (3–5 min), recool the solution. Equip for a simple distillation and heat the flask strongly so that the alcohol distils. The distillate, which is fairly pure and dry, can be used for the preparation of a solid derivative. This method is generally quite useful, and all but high-boiling alcohols can be removed by direct distillation.

The residue which remains after distillation contains the salt of the carboxylic acid. Add 10 ml of water to the residue and mix thoroughly. Acidify the solution with 6 N sulfuric acid. If a solid acid is expected, allow the mixture to stand until crystals of the acid form, and collect them by filtration. If crystals do not form, extract the aqueous acidic solution with ether or dichloromethane, dry the organic solution, and evaporate the solvent. Use the residual acid to prepare a derivative.

A list of esters and their boiling or melting points is given in Table 20.11. Alcohols and carboxylic acids and their derivatives are given in Tables 20.5 and 20.7, respectively.

NITRILES (Table 20.12)

A. Classification tests

A qualitative test which may be used for nitriles is similar to that for esters. Common nitriles (as well as amides—see below) give a colored solution on treatment with hydroxylamine and ferric chloride:

$$\text{R—C}{\equiv}\text{N} + \text{H}_2\text{NOH} \longrightarrow \text{R—}\overset{\displaystyle \text{NH}}{\underset{}{\text{C}}}\text{—NHOH} \xrightarrow{\text{FeCl}_3} \left[\text{R—C} \diagdown \text{Fe} \right]_3 + 3\,\text{HCl} \tag{36}$$

PROCEDURE

Prepare a mixture consisting of 2 ml of 1 *M* hydroxylamine hydrochloride in propylene glycol, 30–50 mg of the compound which has been dissolved in a minimum amount of propylene glycol, and 1 ml of 1 *N* potassium hydroxide. Heat the mixture to boiling for 2 min and cool to room temperature; add *ca.* 0.5–1.0 ml of a 5% alcoholic ferric chloride solution. A red to violet color is a positive test. Yellow colors are negative, and brown colors and precipitations are neither positive nor negative.

B. DERIVATIVE

On hydrolysis, in either acidic or basic solution, nitriles are ultimately converted to the corresponding carboxylic acids. Using methods given previously, it is then possible to prepare a derivative of the acid.

$$\text{Basic Hydrolysis: } RCN + NaOH \xrightarrow{H_2O} RCOO^{\ominus}Na^{\oplus} + NH_3 \qquad (37)$$

$$\text{Acidic Hydrolysis: } RCN \xrightarrow[H_2SO_4]{H_2O} RCONH_2 \xrightarrow[H_2SO_4]{H_2O} RCOOH + NH_4^{\oplus} \qquad (38)$$

PROCEDURE

1. Basic hydrolysis. Mix 10 ml of 10% sodium hydroxide solution and 1 g of the nitrile. Heat the mixture to boiling and note the odor of ammonia. After the mixture is homogeneous, cool it and make it acidic to litmus. If the acid is a solid, collect the crystals by filtration. If it is a liquid, extract the acidic solution with ether; after drying the ether solution, remove the ether by distillation. The residue that remains is the acid. Prepare a suitable derivative of the acid using procedures given previously.

2. Acidic hydrolysis. Treat 1 g of the nitrile with 10 ml of concentrated sulfuric acid or concentrated hydrochloric acid, and warm the mixture to 50° for *ca.* 30 min. Dilute the mixture with water (*Caution:* Add the mixture slowly to water if sulfuric acid has been used) and heat the mixture at gentle reflux for 30 min to 2 hr. The organic layer will be the acid. Cool the mixture and either collect the crystals by filtration or extract the liquid with ether. Prepare derivatives of the acid.

A list of some nitriles is given in Table 20.12; carboxylic acids and their derivatives are given in Table 20.7.

AMIDES (Table 20.13)

A. CLASSIFICATION TEST

A qualitative test for an amide group is the same as that given for a nitrile (see above).

$$\underset{\substack{\| \\ O}}{R-C}-NH_2 + H_2NOH \longrightarrow \underset{\substack{\| \\ O}}{R-C}-NHOH \xrightarrow{FeCl_3} \left[R-C \underset{\substack{| \\ H}}{\overset{O}{\underset{N-O}{\diagdown}}} \right]_3 Fe + 3\,HCl \qquad (39)$$

PROCEDURE

Follow exactly the procedure given for nitriles. The colors which are observed with amides are the same as those with nitriles.

B. DERIVATIVES

Like nitriles, amides must be hydrolyzed (acidic or basic) to give an amine and a carboxylic acid. In the case of unsubstituted amides, ammonia is liberated, but with substituted amides, a substituted amine is obtained. In those cases, it is necessary to classify the amine as being primary or secondary, and to prepare derivatives of both the acid and the amine.

$$RCONR'_2 \xrightarrow[\text{H}_2\text{O}]{\text{H}^\oplus \text{ or HO}^\ominus} RCOOH + HNR'_2 \tag{40}$$
$$(R' = \text{alkyl, aryl, or H})$$

PROCEDURE

1. Basic hydrolysis. Carry out the procedure described for the hydrolysis of nitriles. Distil the ammonia or volatile amine from the alkaline solution into a container of dilute hydrochloric acid. Neutralize this acidic solution, carry out the Hinsberg test, and prepare a derivative of the amine. If the amine is not volatile, it may be extracted from the aqueous layer with ether, the solution dried over potassium hydroxide pellets, and the ether removed to give the amine. After the amine has been obtained, either by distillation or extraction from the hydrolysis mixture, make the alkaline solution acidic and isolate the acid (either by filtration if a solid or by extraction if a liquid). Characterize the acid by preparing a suitable solid derivative.

2. Acidic hydrolysis. Carry out the hydrolysis using the method described for nitriles. In this case, the free acid is liberated and can be removed by filtration or extraction with ether. Prepare a derivative of the acid. Make the acidic hydrolysis mixture alkaline to liberate the amine. Collect the amine by distillation or extraction, characterize it by the Hinsberg test and make a derivative.

A list of amides is given in Table 20.13. Lists of amines and acids and their derivatives are given in Tables 20.7, 20.8, and 20.9.

20.3   Modern Spectroscopic Methods of Analysis

One major limitation to the classical system of qualitative organic analysis which has not been mentioned but which is inherent to the system is that only *known compounds* can be identified. The research chemist is constantly faced with the task of identifying new compounds. Although much information about the *type* of a compound may be derived from the classical system, until recently the complete identification of an unknown organic structure required a combination of degradation and synthesis which was usually a lengthy and laborious task.

The advent of modern spectroscopy has changed this picture dramatically, because not only known compounds but also new and unknown compounds may be identified quickly and certainly by a combination of spectroscopic methods, such as those described in Chapter 4. A number of examples might be cited of the structural identification in a matter of weeks or months of molecules of a complexity greater than those of other compounds which defied the lifework of several of the great 19th and early 20th century organic chemists. One of the significant early examples of this was the application of infrared spectroscopy and X-ray diffraction to the determination of the structure of the penicillin-G molecule during World War II.

$$\begin{array}{c} CH_3 \\ | \\ CH_3-C \qquad CH-CO_2H \\ | \qquad\qquad | \\ S \qquad\quad N \\ \ \ \ \ CH \quad C=O \\ \ \ \ \ CH \\ | \\ C_6H_5CH_2-CO-NH \end{array}$$

Penicillin-G

Modern procedure for identification of organic compounds usually involves a combination of classical and spectroscopic methods. Ideally a student should be introduced to applications of spectroscopy in organic chemistry through use of the instruments which produce the spectra. However, some of these instruments are very expensive, so that this is not feasible in most instances. The next best alternative is for students to have access to the spectra of typical known compounds for study, and then to be provided with spectra of "unknowns" to identify.

In the preceding chapters of this book over 90 ir spectra and over 60 nmr spectra of starting materials and products have been presented. A careful study of these spectra, aided by the material of Chapter 4 and supplemented by the instructor, should enable a student to make meaningful use of ir and nmr spectra in the identification of unknown organic compounds. The spectral data may serve as a substitute for, or to confirm, many of the "wet" classification tests, *e.g.*, a strong ir absorption in the 1690–1760 cm^{-1} region is certainly as indicative of a carbonyl group as formation of a 2,4-dinitrophenylhydrazone, and nmr absorptions in the 6–8.5 δ region are a more reliable indication of an aromatic compound than a color test with CHCl$_3$ and AlCl$_3$. In addition, the quantitative aspects of nmr spectroscopy make it possible to identify an unknown compound with a unique structure even more certainly than by the preparation of several known derivatives.

However, it must be emphasized that in spectroscopic analysis, just as in classical qualitative organic analysis, *careful interpretation of the data* and a certain amount of *chemical intuition* or common sense must be exercised. The student should be cautioned not to "go overboard" in his enthusiasm with modern spectroscopy. Although there are some problems which can be resolved quickly and almost uniquely by spectroscopy, there remain some problems which can be resolved just as simply, and much more economically, by classical qualitative organic analytical procedures.

The instructor will indicate how he wishes any spectroscopic data provided to be utilized. In some instances he may wish that it be used in place of, or to supplement, classification tests on an actual unknown sample issued, although a derivative may still be required. In other instances, he may allow identification of the unknown on the basis of physical constants that have been determined on a sample, plus the spectroscopic data provided, but no derivative will be required. Or he may issue some "paper unknowns" which consist of ir and nmr spectra,[9] with some additional information such as elements present, molecular weight, etc., but no actual sample will be issued.

REFERENCES

1. R. L. Shriner, R. C. Fuson, and D. Y. Curtin, *The Systematic Identification of Organic Compounds*, 5th Ed., John Wiley & Sons, Inc., New York, 1964.

2. N. D. Cheronis, J. B. Entrikin, and E. M. Hodnett, *Semimicro Qualitative Organic Analysis*, 3rd Ed., Interscience Publishers, New York, 1965.

[9] In some cases ultraviolet and mass spectral data might also be provided.

3. T. R. Hogness, W. C. Johnson, and A. R. Armstrong, *Qualitative Analysis and Chemical Equilibrium*, 5th Ed., Holt, Rinehart & Winston, Inc., New York, 1966.

4. D. J. Pasto and C. R. Johnson, "*Organic Structure Determination,*" Prentice-Hall, Inc., Englewood Cliffs, N. J., 1969.

20.4   Tables of Derivatives

The following abbreviated tables of common organic compounds and their usual derivatives are arranged according to classes of compounds. The melting and boiling points listed represent the highest points in the range actually observed.

TABLE 20.1   ALDEHYDES

| | | | Derivatives | |
| | | | Semi-carbazone | 2,4-Dinitrophenyl-hydrazone |
Name of Compound	BP	MP		
Propionaldehyde	49°		154°	156°
Glyoxal	50		270	328
Acrolein	52		171	165
Isobutyraldehyde	64		126	182
n-Butyraldehyde	75		106	123
Trimethylacetaldehyde	75		190	209
2-Methylbutanal	93		103	120
Chloral	98		90d	131
Pentanal	103			107
Crotonaldehyde	104		199	190
Hexanal	131		106	104
Heptanal	156		109	108
Furfural	161		202	230(214)
3-Cyclohexenecarboxaldehyde	165		155	
Benzaldehyde	179		222	237
Phenylacetaldehyde	194		156	121
Salicylaldehyde	197		231	248
o-Tolualdehyde	200		212	195
p-Tolualdehyde	204		215	234
Citronellal	207		92	78
o-Chlorobenzaldehyde	208		225	214
p-Chlorobenzaldehyde	214	47°	230	270
o-Anisaldehyde	246	39	215	254
p-Anisaldehyde	248		210	254d
Cinnamaldehyde	252		215	255d
p-Bromobenzaldehyde		57	228	
Vanillin		81	230	271d
p-Nitrobenzaldehyde		106	221	320d

TABLE 20.2 KETONES

Name of Compound	BP	MP	Derivatives Semi-carbazone	2,4-Dinitrophenyl-hydrazone
Acetone	56°		190°	126°
Butanone	80		136	117
3-Methyl-2-butanone	94		113	120
3-Pentanone	102		139	156
2-Pentanone	102		112	144
Pinacolone	106		158	125
4-Methyl-2-pentanone	117		132	95
Chloroacetone	119		164d	125
2, 4-Dimethyl-3-pentanone	125		160	95
2-Hexanone	129		122	110
4-Methyl-3-penten-2-one	130		164	203
Cyclopentanone	131		203	146
2,4-Pentanedione	139		122(mono); 209(di)	209
4-Heptanone	145		133	75
5-Methyl-2-hexanone	145		147	95
3-Heptanone	148		103(152)	
2-Heptanone	151		127	89
Cyclohexanone	156		167	162
6-Methyl-3-heptanone	160		132	
5-Methyl-3-heptanone	160		102	
4-Hydroxy-2-pentanone	166			203
2,6-Dimethyl-4-heptanone	169		122	92
2-Octanone	173		123	58
Acetophenone	203	20°	199	240
l-Menthone	209		189	146
3,5,5-Trimethylcyclohexene-3-one	215		200	
Benzyl methyl ketone	216	27	198	156
Propiophenone	218	20	174	191
p-Methylacetophenone	226	28	205	258
2-Undecanone	228		122	63
Dibenzyl ketone	330	34	146	100
p-Methoxyacetophenone	258	38	198	220
Benzophenone		48	167	239
p-Bromoacetophenone		51	208	235
Desoxybenzoin	320	60	148	204
Benzalacetophenone	348	58(62)	168	245
Benzoin		133	206d	245
p-Hydroxybenzophenone		135	194	242
d,l-Camphor		178	248	164

TABLE 20.3 AROMATIC HYDROCARBONS

Name of Compound	BP	MP	Derivatives		Oxidation Product
			Nitration Product		
			Position	MP	
Benzene	80°		1,3	89°	
Toluene	111		2,4	70	Benzoic Acid
Ethylbenzene	136		2,4,6	37	Benzoic Acid
p-Xylene	138		2,3,5	13	Terephthalic Acid
m-Xylene	139		2,4	83	Isophthalic Acid
o-Xylene	142		4,5	71	Phthalic Acid
Isopropylbenzene	153		2,4,6	109	Benzoic Acid
t-Butylbenzene	169		2,4	62	Benzoic Acid
m-Diethylbenzene	181		2,4,6	62	Isophthalic Acid
Tetralin	206		5,7	95	Phthalic Acid
Cyclohexylbenzene	237		4	58	Benzoic Acid
Diphenylmethane		25°	2,4,2′,4′	172	
Dibenzyl		53	4,4′	180	
Biphenyl		70	4,4′	237	
Naphthalene		80	1	61	
Triphenylmethane		92	4,4′,4″	206	
Acenaphthene		96	5	101	
Fluorene		115	2,7	199	

TABLE 20.4 ARYL HALIDES

Name of Compound	BP	MP	Nitration Product Position	Nitration Product MP	Sidechain Oxidation Product	Sidechain Oxidation MP
Chlorobenzene	132°		2,4	52°		
Bromobenzene	156		2,4	75		
o-Chlorotoluene	159		3,5	63	o-Chlorobenzoic acid	140°
m-Chlorotoluene	162		4,6	91	m-Chlorobenzoic acid	158
p-Chlorotoluene	162		2	38	p-Chlorobenzoic acid	242
o-Dichlorobenzene	179		4,5	110		
o-Bromotoluene	182		3,5	82	o-Bromobenzoic acid	147
2,4-Dichlorotoluene	200		3,5	104	2,4-Dichloro-benzoic acid	160
α-Chloronaphthalene	259		4,5	180		
p-Bromotoluene	184	28°	2	47	p-Bromobenzoic acid	251
p-Dichlorobenzene		53	2	54		
β-Chloronaphthalene		56	1,8	175		
p-Dibromobenzene		89	2,5	84		

TABLE 20.5   ALCOHOLS

Name of Compound	BP	MP	Derivatives 3,5-Dinitro-benzoate	Phenylurethan
Methanol	65°		108°	47°
Ethanol	78		93	52
Isopropyl Alcohol	82		123	88
t-Butyl Alcohol	83		142	136
n-Propyl Alcohol	97		74	51
sec-Butyl Alcohol	99		76	65
t-Pentyl Alcohol	102		117	42
Isobutyl Alcohol	108		87	86
3-Pentanol	116		97	48
n-Butyl Alcohol	118		64	57
1-Methoxy-2-propanol	119		85	
2-Pentanol	119		61	
3-Methyl-1-butanol	132		61	57
1-Pentanol	138		46	46
Cyclopentanol	141		115	132
2-Methyl-1-pentanol	148		51	
1-Hexanol	156		58	42
Cyclohexanol	161		113	82
Furfuryl Alcohol	172		81	45
2,6-Dimethyl-4-heptanol	173			62
1-Heptanol	177		47	68
2-Octanol	179		32	114
Ethylene Glycol	198		169	157
Methylphenylcarbinol	203		95	94
Benzyl Alcohol	206		113	78
β-Phenethyl Alcohol	220		108	80
Cinnamyl Alcohol		33°	121	90
1-Tetradecanol		39	67	74
(−)-Menthol		42		111
1-Hexadecanol		50	66	73
Neopentyl Alcohol		53		144
1-Octadecanol		60	66	80
Benzoin		133		165
(−)-Cholesterol		148		168

TABLE 20.6   PHENOLS

Name of Compound	BP	MP	α-Naphthyl-urethan	Bromo Derivatives	
o-Chlorophenol	175°	7°	120°		
Phenol	180	42	133	Tribromo	95°
o-Cresol	190	31	142	Dibromo	56
o-Bromophenol	195		129	Tribromo	95
p-Cresol	202	36	146	Dibromo	49
m-Cresol	202	3	128	Tribromo	84
2-Methoxyphenol	205	32	118	4,5,6-Tribromo	116
p-Chlorophenol	217	43	166		
o-Nitrophenol		45	113	4,6-Dibromo	117
p-Ethylphenol		47	128		
p-Bromophenol		63	169	Tribromo	95
3,5-Dichlorophenol		68		Tribromo	189
3,5-Dimethylphenol		68		Tribromo	166
2,4,6-Trimethylphenol		69		Dibromo	158
2,5-Dimethylphenol		74	173	Tribromo	178
α-Naphthol		94	152	2,4-Dibromo	105
m-Nitrophenol		97	167	Dibromo	91
p-t-Butylphenol		100	110	Bromo	50
				Dibromo	67
o-Hydroxyphenol		105	175	Tetrabromo	193
m-Hydroxyphenol		110		Tribromo	112
p-Nitrophenol		114	151	2,6-Dibromo	142
p-Hydroxybenzaldehyde		115		3,5-Dibromo	181
β-Naphthol		122	157	Bromo	84
2,3-Dihydroxyphenol		133	230 (tri)	Dibromo	158

TABLE 20.7   ACIDS

Name of Compound	BP	MP	Derivatives		
			Anilide	p-Toluidide	Amide
Acetic	118°		114°	147°	
Propionic	140		106	126	
Pivalic	164	35°	129	120	154°
Chloroacetic	185	63	134	120	118
α-Chloropropionic	186		92	124	
Dichloroacetic	189		118	153	
Cyclohexylcarboxylic		30	146		186
β-Phenylpropionic		48	92	135	82
(Hydrocinnamic)					
Trichloroacetic		57	94	113	141
Phenylacetic		76	118	136	154
Glycolic		80	97	143	120
o-Methoxybenzoic		100	131		128
o-Toluic		104	125	144	142
m-Toluic		111	126	118	97
Benzoic		122	163	158	128
Cinnamic		133	153	168	147
meso-Tartaric		140			190
o-Chlorobenzoic		142	118	131	139
m-Nitrobenzoic		146	155		142
o-Bromobenzoic		150	141		155
Benzilic		150	175		154
Salicylic		158	136		139
m-Chlorobenzoic		158	122		134
p-Toluic		178	145	160	158
Anisic		184	169	186	162
Phthalic		208d	169	201(di)	149
p-Hydroxybenzoic		215	202	204	162
p-Nitrobenzoic		241	217	204	201
p-Chlorobenzoic		242	194		179

TABLE 20.8 PRIMARY AND SECONDARY AMINES

Name of Compound	BP	MP	Derivatives Benzamide	Derivatives Benzene-sulfonamide
Isopropylamine	33°		100°	26°
n-Propylamine	49		84	36
sec-Butylamine	63		76	70
n-Butylamine	77		42	
Piperidine	106		48	93
Di-n-propylamine	110			51
n-Hexylamine	129		40	96
Morpholine	130		75	118
Cyclohexylamine	134		149	89
N-Methylcyclohexylamine	147		86	
Aniline	183		163	112
Benzylamine	184		105	88
N-Methylaniline	196		63	79
o-Toluidine	199		146	124
m-Toluidine	203		125	95
N-Ethylaniline	205		60	
o-Ethylaniline	216		147	
p-Ethylaniline	216		151	
o-Anisidine	225		60	89
m-Chloroaniline	230		120	121
Tetrahydroisoquinoline	250	20	75	67
Dibenzylamine	300		112	68
p-Toluidine	200	45	158	120
α-Naphthylamine		50	160	167
2,5-Dichloroaniline		50	120	
Diphenylamine	302	54	180	124
p-Anisidine	240	58	154	95
2-Aminopyridine		60		165(di)
N-Phenyl-α-naphthylamine		62	152	
p-Bromoaniline		66	204	134
p-Chloroaniline	232°	70	192	122
o-Nitroaniline		71	110	104
N-Phenyl-β-naphthylamine		108	148	
m-Nitroaniline	284	114	155	136
Anthranilic Acid		147	182	214
p-Nitroaniline		147	199	139
p-Nitro-N-methylaniline		152	112	121
Carbazole	351	246	98	

TABLE 20.9 TERTIARY AMINES

Name of Compound	BP	MP	Derivatives	
			Picric Acid	Methyl Iodide
Triethylamine	89°		173°	
Pyridine	116		167	117°
2-Picoline	129		169	230
3-Picoline	143		150	92
4-Picoline	143		167	152
Tri-*n*-propylamine	156		116	208
2,4-Lutidine	159		183	113
2,4,6-Collidine	172		156	
2-Methyl-4-ethylpyridine	173		123	
N,N-Dimethylaniline	193		163	228*d*
Tri-*n*-butylamine	211		106	186
N,N-Diethylaniline	218		142	102
Quinoline	239		203	72(133)
Isoquinoline	243		222	159

TABLE 20.10 AROMATIC NITRO COMPOUNDS

Name of Compound	BP	MP	Acyl Derivatives of Reduced Amine		Other Derivatives
			Benzene-sulfonamide	Benzamide	
Nitrobenzene	210°		112°	160°	m-Dinitrobenzene 90°
o-Nitrotoluene	224		124	147	2,4-Dinitrotoluene 70
m-Nitrotoluene	231	16°	95	125	m-Nitrobenzoic Acid 140
o-Nitroanisole	265		89		2,4,6-Trinitroanisole 68
o-Chloronitrobenzene	246	32	129	99	2,4-Dinitrochlorobenzene 52(50)
o-Bromonitrobenzene	261	43		116	2,4-Dinitrobromobenzene 72
2,4-Dinitrochlorobenzene		52		178	2,4-Dinitrophenol 114
p-Nitroanisole	258	54	95	154	2,4-Dinitroanisole 89
β-Nitronaphthalene		78	136	162	
p-Chloronitrobenzene	242	83	121	192	p-Nitrophenol 114
Picryl chloride		83	211		Picric Acid 122
m-Dinitrobenzene	302	90	194	240	m-Nitroaniline 114
4-Nitrobiphenyl		114		230	Acetamide 171
p-Bromonitrobenzene	259	126	136	204	p-Nitrophenol 114

TABLE 20.11   ESTERS

Name of Compound	BP	MP	Name of Compound	BP	MP
Methyl acetate	57°		n-Propyl benzoate	230°	
Ethyl acetate	77		Methyl m-chlorobenzoate	231	
Isopropyl acetate	91		Ethyl salicylate	234	
tert-Butyl acetate	98		n-Butyl benzoate	249	
n-Butyl acetate	126		Methyl p-toluate		33°
n-Amyl acetate	149		Methyl cinnamate		36
Cyclohexyl acetate	175		Benzyl cinnamate		39
n-Heptyl acetate	192		Phenyl salicylate		42
Phenyl acetate	197		Methyl p-chlorobenzoate		44
o-Cresyl acetate	208		Ethyl m-nitrobenzoate		47
m-Cresyl acetate	212		α-Naphthyl acetate		49
p-Cresyl acetate	213		Ethyl p-nitrobenzoate		56
Ethyl benzoate	213		Phenyl benzoate		69
Methyl o-toluate	215		β-Naphthyl acetate		71
Benzyl acetate	217		p-Cresyl benzoate		71
Isopropyl benzoate	218		Methyl m-nitrobenzoate		78
Ethyl phenylacetate	229				

TABLE 20.12 NITRILES

Name of Compound	BP	MP	Name of Compound	BP	MP
Acetonitrile	81°		m-Chlorobenzonitrile		41°
Propionitrile	97		o-Chlorobenzonitrile	232°	47
Chloroacetonitrile	127		p-Methoxybenzonitrile		62
Benzonitrile	191		p-Chlorobenzonitrile		92
o-Tolunitrile	205		o-Nitrobenzonitrile		110
m-Tolunitrile	212		m-Nitrobenzonitrile		118
p-Tolunitrile	217	27°	p-Nitrobenzonitrile		147
Phenylacetonitrile	234				

TABLE 20.13 AMIDES

Name of Compound	MP	Name of Compound	MP
N-n-Propylacetanilide	50°	o-Bromobenzanilide	141
m-Acetotoluide	66	Salicylamide	142
Propionamide	81	o-Toluamide	142
Acetamide	82	m-Nitrobenzamide	143
o-Nitroacetanilide	92	p-Toluanilide	145
n-Butyranilide	95	Cinnamanilide	151
Trichloroacetanilide	97	m-Nitrobenzanilide	154
N-Methylacetanilide	102		
Acetanilide	114	o-Bromobenzamide	155
Dichloroacetanilide	118	m-Nitroacetanilide	155
o-Toluanilide	125	p-Toluamide	160
m-Toluanilide	126	Benzanilide	163
p-Methoxyacetanilide	127	p-Bromoacetanilide	167
Benzamide	130	p-Chloroacetanilide	179
m-Chlorobenzamide	134	p-Chlorobenzanilide	194
m-Bromobenzanilide	136	p-Nitrobenzamide	201

Appendix I
Drying agents (Desiccants)

The important procedure of drying either a reagent, solvent or product will be encountered at some stage of nearly every reaction performed in the organic chemistry laboratory. The techniques of drying solids and liquids and some of the drying agents which are commonly used will be described in this appendix.

Solids. It is important to dry solid organic compounds because water and organic solvents not only may affect melting points and quantitative elemental analyses, but also may cause difficulties if the wet solid is used in a reaction whose success depends on the absence of small amounts of water or other liquids. If the solid has been recrystallized from a volatile organic solvent, it can usually be dried to an extent satisfactory for most purposes by air-drying at room temperature. The process can be accelerated by spreading the solid out on a piece of filter paper or on a clay plate, either of which will serve to absorb traces of water or any excess solvent present. Another useful method to enhance the rate of drying is to collect the solid on a Büchner funnel held in a filter flask, to press the solid as dry as possible with a clean cork, and then to pull air through the filter cake by means of the aspirator (water pump).

If the solid compound is hygroscopic or has been recrystallized from water or a high-boiling solvent, it is usually necessary to dry the substance at atmospheric pressure or under vacuum in an oven operating at a temperature below the melting or decomposition point of the solid. For air-sensitive solids, drying must be done either in an inert atmosphere, such as that provided by nitrogen, or under vacuum. For samples which are to be submitted for quantitative elemental analysis, the solid is normally dried to constant weight by heating under vacuum.

It is often convenient to store dried organic solids in desiccators containing desiccants such as silica gel, phosphorus pentoxide, calcium chloride, etc., although tightly stoppered bottles make good storage vessels. Of course, it is also possible to use a desiccator to dry an organic solid wet with water. A solid that is contaminated with a hydrocarbon solvent can be dried by placing it in a desiccator containing a block of paraffin, which serves to absorb the hydrocarbon.

Liquids. The drying of organic liquids is particularly important because their most common method of purification is by distillation. Any water present in an organic compound being distilled may react with the compound at or below the temperature of the distillation or it may co- or steam distil with the liquid and thereby contaminate the distillate.[1]

Two of the more important general requirements for a drying agent are that neither it nor its hydrolysis product will react chemically with the organic liquid which is to be dried and that it can be easily and *completely* separated from the dried liquid. Another important consideration for a drying agent is that it be efficient in its action so that most or all of the water will be removed by the desiccant.

The more commonly used drying agents and some of their properties are listed in the accompanying table. Of the agents listed, calcium chloride, sodium sulfate, and magnesium sulfate will generally suffice for the needs of the basic course in organic chemistry laboratory.

It should be noted that the desiccants given in the table function in one of two ways: Either the drying agent interacts reversibly with water, by the process of adsorption or absorption [equation (1)], or it reacts irreversibly, with water serving as an acid or a base. In the case of reversible hydration, a certain amount of water will remain in equilibrium with the hydrated drying agent; the lower the amount of water left at equilibrium, the greater is the *efficiency* of the desiccant. Thus, magnesium sulfate is found to be more efficient in its drying power than is sodium sulfate but less efficient than calcium sulfate. Desiccants which remove water by irreversible chemical reaction clearly have very high efficiencies, but are generally more expensive than the other types of drying agents.

$$\text{Drying Agent} + H_2O \rightleftarrows (\text{Drying Agent}) \cdot x\, H_2O \qquad (1)$$

With those drying agents that operate by formation of hydrates, it is imperative that the drying agent be *completely* removed by filtration or decantation *before* distillation of the dried liquid, since most of the hydrates decompose with loss of water at temperatures above 30–40°. For those desiccants, such as calcium hydride, sodium, and phosphorus pentoxide, which react vigorously with water and for those like calcium sulfate, which have a low capacity for removing water, a preliminary drying using a less reactive and efficient desiccant with a high capacity is normally required. When a drying agent which reacts with water to evolve hydrogen is used, appropriate precautions must be taken to vent the hydrogen and thereby prevent buildup of this highly flammable gas.

It is poor technique to use an unnecessarily large quantity of drying agent when drying a liquid since the desiccant may adsorb or absorb the organic liquid along with water. Also, mechanical losses upon filtration or decantation of the dried solution may become significant. Of course, the amount of drying agent required depends upon the quantity of water present and upon the capacity of the desiccant. In general, a portion of drying agent that covers the bottom of the vessel in which the liquid is contained should suffice. If additional desiccant is required, more can be added. Swirling the container of liquid and desiccant enhances the rate of drying when desiccants such as calcium chloride and magnesium sulfate are being used, since it hastens the establishment of the equilibrium for hydration.

---

[1] The drying of some organic solvents makes profitable use of the codistillation of water with the solvent in a process termed azeotropic distillation. For example, benzene can be dried in this manner. Solids can also sometimes be dried by dissolving them in a suitable solvent, removing any water by azeotropic distillation, and then recovering the solid by removal of the solvent.

Drying Agent	Acid-Base Properties	Product(s) with water	Comments [2]
$CaCl_2$	Neutral	$CaCl_2 \cdot H_2O$ $CaCl_2 \cdot 2H_2O$ $CaCl_2 \cdot 6H_2O$	High capacity and fast action; not highly efficient; good preliminary drying agent; readily separated from dried solution because $CaCl_2$ is available as large granules; cannot be used to dry either alcohols and amines (because of compound formation) or phenols, esters, and acids [because drying agent contains some $Ca(OH)_2$]; hexahydrate decomposes (loses water) above 30°.
Na_2SO_4	Neutral	$Na_2SO_4 \cdot 7H_2O$ $Na_2SO_4 \cdot 10H_2O$	Inexpensive; high capacity; slow action and low efficiency; good general preliminary drying agent; physical form is that of a powder so filtration required for removal of drying agent from dried solution; decahydrate decomposes above 33°.
$MgSO_4$	Neutral	$MgSO_4 \cdot H_2O$ $MgSO_4 \cdot 7H_2O$	Faster and more efficient than Na_2SO_4; good general drying agent; requires filtration for removal of drying agent from dried solution; heptahydrate decomposes above 48°.
$CaSO_4$	Neutral	$CaSO_4 \cdot \frac{1}{2}H_2O$	Low capacity but extremely fast action and high efficiency; preliminary drying of solution with drying agent of higher capacity recommended; hemihydrate can be dehydrated by heating at 235° for two to three hours.
$CuSO_4$	Neutral	$CuSO_4 \cdot H_2O$ $CuSO_4 \cdot 3H_2O$ $CuSO_4 \cdot 5H_2O$	More efficient than $MgSO_4$ or Na_2SO_4 but more expensive than either.
K_2CO_3	Basic	$K_2CO_3 \cdot 1\frac{1}{2}H_2O$ $K_2CO_3 \cdot 2H_2O$	Fair efficiency and capacity; good for esters, nitriles and ketones; cannot be used with acidic organic compounds.
H_2SO_4	Acidic	$H_3O^{\oplus}HSO_4^{\ominus}$	Good for alkyl halides and aliphatic hydrocarbons; cannot be used with even such weak bases as alkenes and ethers; high efficiency.
P_2O_5	Acidic	HPO_3 $H_4P_2O_7$ H_3PO_4	See comments under H_2SO_4; also good for ethers, aryl halides and aromatic hydrocarbons; extremely high efficiency; preliminary drying of solution recommended; dried solution can be distilled from drying agent.

[2] *Capacity*, as used in this table, refers to the amount of water which can be removed by a given weight of drying agent; *efficiency* refers to the amount of water, if any, in equilibrium with the hydrated desiccant.

Drying Agent	Acid-Base Properties	Product(s) with water	Comments [2]
CaH	Basic	$H_2 + Ca(OH)_2$	High efficiency but somewhat slow action; good for basic, neutral or weakly acidic compounds; cannot be used for base-sensitive substances; preliminary drying of solution recommended; dried solution can be distilled from drying agent.
Na	Basic	$H_2 + NaOH$	Good efficiency but slow action; cannot be used on compounds sensitive to alkali metals or to base; care must be exercised in destroying excess drying agent; preliminary drying required; dried solution can be distilled from drying agent.
BaO or CaO	Basic	$Ba(OH)_2$ or $Ca(OH)_2$	Slow action but high efficiency; good for alcohols and amines; cannot be used with compounds sensitive to base; dried solution can be distilled from drying agent.
KOH or NaOH	Basic	Solution	Rapid and efficient but use limited almost exclusively to drying of amines.
Molecular Sieve #3A or #4A [3]	Neutral	Water strongly adsorbed	Rapid and highly efficient; preliminary drying recommended; dried solution can be distilled from drying agent if desired. Molecular Sieve is the trade name for aluminosilicates whose crystal structure contains a network of pores of uniform diameter. The pore size of sieves #3A and 4A is such that only water and other small molecules such as ammonia can pass into the sieve; water is strongly adsorbed as water of hydration; hydrated sieves can be reactivated by heating at 300–320° under vacuum or at atmospheric pressure.

[3] The numbers refer to the nominal pore size, in Ångstrom units, of the sieve.

Appendix II
The laboratory notebook

A notebook used in the laboratory is to be a *complete* record of the experimental work. An 8×10-inch bound book is normally satisfactory for this purpose. The criterion used in judging what should be in the notebook is that the record should be so thorough and so well organized that anyone who reads the experiment can understand it and can see exactly what has been done, and thus could repeat it in precisely the same way the original work was done, if necessary.

All data are to be recorded in the notebook *at the time that they are obtained.* There is no reason for recording anything on odd pieces of paper to be transcribed into the notebook later. Neatness is desirable, but it is less important than having a notebook that is complete. Copying over experimental data costs time, cannot possibly improve them and, at times, may even worsen the data because errors may be made in copying. The notes made at the time of performance necessarily constitute the primary record.

In setting up the notebook at the beginning of the laboratory work, the following general structure should be used:

1. Leave room at the beginning of the notebook for a Table of Contents and keep it up-to-date.
2. Number the pages, if they are not already numbered.
3. Start every new experiment on a fresh page. Make all entries *in ink*.
4. It is not necessary to copy out the details of experimental procedure when you make no variation whatsoever from it. However, every experiment should have some *reference* (so that if the procedure is a standard one, it can be checked). In essence, your notebook should be a log of your laboratory operations; dates, times, and other pertinent conditions should be entered regularly.

It is particularly important that any *variations* from standard procedure and the reasons for them be noted; if this is not done, it is not possible to reconstruct later just what was done and why it was done. Nothing should ever be deleted from the notebook; merely draw a line through something *considered* wrong and amend it appropriately. The reasons which seem adequate at a particular time may later be found to be inadequate, and a result rejected at an early stage because it seemed impossible may eventually turn out to be the only worthwhile outcome of a particular experiment.

The notebook should not only contain a complete record of any observations but should also reflect precisely what was *concluded* from these observations and the *reasons* for these conclusions. Often one finds in laboratory work that data have been interpreted incorrectly, and only a careful and complete record of the experimental observations will allow one to discover the error in interpretation.

In organic chemistry, two distinctly different "types" of experiments are normally carried out in the laboratory. The first of these is the "preparative-type" of experiment, in which one compound is converted into another. The second of these is the "investigative-type" of experiment, in which one studies physical properties such as boiling point or melting point, or chemical properties, such as qualitative tests for various functional groups in a molecule. These two types of experiments require slightly different types of notebook write-up. As an aid in preparing an acceptable notebook, suggested formats for both preparative- and investigative-type experiments are presented below. However, the amount of advance preparation and the notebook form that is to be used will probably be discussed by the instructor. The formats given here are illustrative only.

A. NOTEBOOK FORMAT FOR PREPARATIVE-TYPE EXPERIMENTS

The important information that should be included in describing preparative-type experiments is the following:

I. Introduction.   Prepare a brief statement regarding the work to be done.

II. Main reaction(s) and mechanisms.   Give the main reaction(s) leading to the preparation of the desired compound. Where possible, include the mechanisms of the reactions.

III. Table of reagents and products.   List, in tabular form, the molecular weights of each reagent and product, and calculate and enter the number of moles of each reagent to be used. Use the chemical equation(s) for the main reactions to derive the theoretical molar ratio of reagents and products, and enter these values in the list.

Determine which reagent, if any, is used in less than the theoretical molar ratio required; if more than one such reagent is found, find which of these deviates most from the molar ratio needed. This will be the *limiting reagent* in the reaction, *i.e.*, the reagent which will determine the maximum amount of product that can be formed. After the limiting reagent has been consumed, the desired reaction will cease no matter how much of the other reagents remain.

IV. Yield data.   Using the table of reagents and products, calculate the maximum theoretical yield of product that one can expect to obtain from the starting materials. The theoretical yield of product (in moles) can be calculated by determining the moles of product formed per mole of limiting reagent used:

Theoretical yield (in moles) = (theoretical molar ratio of product to limiting reagent) × (moles of limiting reagent actually used)

Multiplying this theoretical yield by the molecular weight of the product gives the expected yield (in grams).

Determine the actual yield of product, in grams, that is obtained in the experiment. Use the actual yield and the theoretical yield to determine the percent yield, which is:

$$\% \text{ yield} = \frac{\text{actual yield (grams)}}{\text{theoretical yield (grams)}} \times 100$$

V. Observed properties of the product. Enter the physical properties of the product that is obtained from the preparation. Important data include melting point or boiling point, color, crystalline form, etc. It is important to record these properties accurately.

VI. Side reactions. List all possible side reactions that may occur in the reaction. To help determine what these might be, it is often necessary to consult additional sources, such as lecture notes or the lecture textbook.

VII. Other methods of preparation. List alternate methods of preparation that could be used to prepare this same compound. Consult a textbook, and comment briefly on the advantages of the other methods as opposed to the one used. Suggest a reason why the present method of preparation is preferable, if it is.

VIII. Method of purification. A very important phase of synthetic organic chemistry is the purification of the desired product. Because of the many and varied reactions that organic compounds undergo, frequently the most difficult step in the preparation of a compound is not its actual formation but its isolation in pure form from side products and unchanged starting materials. In order to accomplish this, list all the compounds that could possibly be present on the basis of a consideration of the main and side reactions. Devise a purification scheme, using a flow sheet, to show how various experimental procedures eliminate the undesired substances and yield the pure product. By examining a purification scheme in this manner, the purpose of each step in the procedure becomes clear, and one may be able to predict what impurities, if any, may contaminate the product.

IX. Answers to exercises.

This type of format is shown in Figure 1, using an experiment involving the chlorination of cyclohexane (Chapter 5.1, page 84) as an illustration. The figure represents an open notebook; it should be noted that a new experiment is normally started on the left-hand page.

B. NOTEBOOK FORMAT FOR INVESTIGATIVE-TYPE EXPERIMENTS

A format that can be used for investigative-type experiments is similar to that shown in Figure 1. The description of a new experiment should normally start on the left-hand page, and each new experiment should contain a title and reference. The important information which should be included in the notebook is:

I. Introduction. Give a brief introduction to the experiment and state the purpose(s) of it. This should take up no more than half of the page.

II. Experiments and results. Give a brief statement (one or two lines) of each experiment that is to be performed, and leave sufficient room to enter the results as they are obtained. Do *not* recopy the experimental details from the text— just identify each experiment that is to be done. Continue this section as far as needed.

III. Conclusions. After completing the assigned experiments, state briefly the conclusions which have been reached on the basis of the results. If the experiment was the identification of an unknown, summarize the results of the findings here.

IV. Answers to exercises.

Figure 1. Sample for preparative-type experiment

Notebook Page No. 3.

CHLORINATION OF CYCLOHEXANE

Reference: Chapter 5.1 in Roberts, Gilbert, Rodewald and Wingrove, page 84

I. INTRODUCTION.

Chlorocyclohexane is to be prepared from cyclohexane using sulfuryl chloride as reagent.

II. MAIN REACTION(S) AND MECHANISM(S).

$$C_6H_{12} + SO_2Cl_2 \rightarrow C_6H_{11}Cl + HCl + SO_2$$

(Mechanism intentionally omitted)

Notebook Page No. 4.

III. TABLE OF REAGENTS AND PRODUCTS

Compound	M.W.	Wt. Used, g.	Moles Used	Ratio of Moles: Theory	Used	Other Data
Cyclohexane	84	33.6	0.4	1	2	Density = 0.779 bp 81°
Sulfuryl chloride	135	27.0	0.2	1	1	Density = 1.667 bp 69°
Chlorocyclohexane	118.5			1		Density = 1.016 colorless bp 142.5°
Benzoyl peroxide	—	0.1	—	(Initiator)		

Limiting reagent: *Sulfuryl chloride*

IV. YIELD DATA.

(To determine the *theoretical yield*, it is necessary first to determine the *limiting reagent*. The equation for the main reaction shows that 1 mole of cyclohexane reacts with 1 mole of sulfuryl chloride. Observe, however, that in this experiment 0.4 mole of cyclohexane is allowed to react with 0.2 mole of sulfuryl chloride, so that sulfuryl chloride is the *limiting reagent*. Since the theoretical molar ratio of chlorocyclohexane to sulfuryl chloride is 1:1, the number of moles of chlorocyclohexane produced will equal the number of moles of sulfuryl chloride used.)

Theoretical yield of chlorocyclohexane = (moles $C_6H_{11}Cl$) (M.W. $C_6H_{11}Cl$) = (0.2 mole) (118.5 g/mole) = 23.7 g.

If, in an actual experiment, 15.0 g of pure chlorocyclohexane was obtained, the percent yield would be:

$$\% \text{ Yield} = \frac{15.0 \text{ g}}{23.7 \text{ g}} \times 100 = 63\%$$

Figure 1. Sample for preparative-type experiment, continued

Notebook Page No. 5.

V. OBSERVED PROPERTIES OF THE PRODUCT.
bp 138–140°; colorless liquid; insoluble in water.

VI. SIDE REACTIONS.

$C_6H_{11}Cl + SO_2Cl_2 \rightarrow C_6H_{10}Cl_2 + SO_2 + HCl$
+ other polychlorinated cyclohexyl compounds.

VII. OTHER METHODS OF PREPARATION.

$C_6H_{10} + HCl \rightarrow C_6H_{11}{-}Cl$
$C_6H_{11}{-}OH + HCl \rightarrow C_6H_{11}{-}Cl + H_2O$
$C_6H_{11}{-}OH + SOCl_2 \rightarrow C_6H_{11}{-}Cl + HCl + SO_2$

Notebook Page No. 6.

VIII. METHOD OF PURIFICATION.

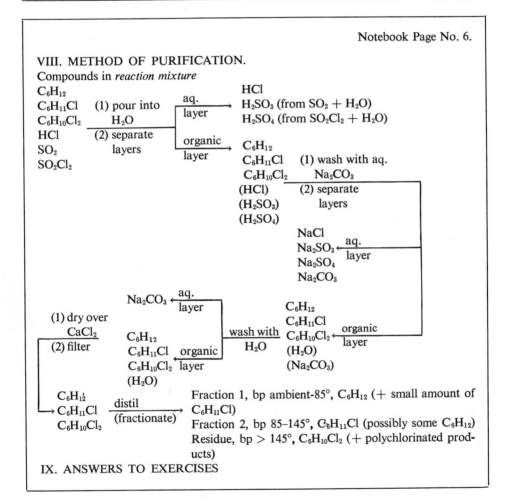

IX. ANSWERS TO EXERCISES

Appendix III

List of nmr and ir spectra

Compound	NMR (Figure No.)	IR (Figure No.)
p-Acetamidobenzenesulfonyl chloride		17.10
Acetanilide	17.9	17.8
Acetophenone	15.20	15.19
Acetylferrocene	18.4	18.3
Adipic acid		14.5
2-Aminothiazole	17.14	17.13
Aniline	16.4	16.3
Anthracene		12.4
trans-Benzalacetophenone	15.22	15.21
Benzaldehyde	14.7	14.6
Benzamide	16.2	16.1
Benzanilide	16.6	16.5
Benzoic acid	13.2	13.1
Benzoic anhydride		16.7
Benzonitrile		16.8
p-Benzoquinone		9.4
Benzyl alcohol		14.8
Benzyl chloride		15.12
N-Benzyl-m-nitroaniline	15.11	15.10
N-Benzylidene-m-nitroaniline		15.9
4-Bromoacetanilide	17.17	17.16

Compound	NMR *(Figure No.)*	IR *(Figure No.)*
Bromobenzene		13.5
1-Bromo-3-chloro-5-iodobenzene	17.25	17.24
n-Butyl alcohol	11.12	11.11
n-Butylbenzene		7.1(a)
sec-Butylbenzene		7.1(c)
t-Butylbenzene	5.14	5.13
		7.1(d)
Butylbenzenes, reaction mixture	7.4 (a), (b)	7.3 (a), (b) and (c)
2-Chloro-4-bromoacetanilide	17.19	17.18
2-Chloro-4-bromoaniline	17.21	17.20
2-Chloro-4-bromo-6-iodoaniline	17.23	17.22
1-Chlorobutane	5.6	5.5
Chlorocyclohexane		5.8
2-Chloro-2-methylbutane	11.4	11.3
2-Chloronaphthalene		12.3
Cinnamaldehyde	15.2	15.1
N-Cinnamylaniline	15.8	15.7
N-Cinnamylideneaniline		15.6
N-Cinnamylidene-*m*-nitroaniline		15.3
N-Cinnamyl-*m*-nitroaniline	15.5	15.4
1,3,5-Cycloheptatriene	18.6	18.5
Cyclohexane		5.7
cis-Cyclohexane-1,2-dicarboxylic acid		6.23
Cyclohexanol	6.15	6.14
Cyclohexanone	10.6	10.5
Cyclohexene	6.17	6.16
4-Cyclohexene-*cis*-1,2-dicarboxylic acid		6.22
4-Cyclohexene-*cis*-1,2-dicarboxylic anhydride	9.8	9.7
1,3-Cyclopentadiene	9.2	9.1
1,x-Dichlorobutanes (mixture)	5.4	
2,3-Dimethylbutane		4.3(a)
Dimethyl fumarate	6.21	6.20
Dimethyl maleate		6.19
Dimethyl maleate and dimethyl fumarate, mixture	6.18	
trans, *trans*-1,4-Diphenyl-1,3-butadiene	15.18	15.17
Ethylbenzene	5.10	5.9
Ferrocene	18.2	18.1

Compound	NMR (Figure No.)	IR (Figure No.)
2-Furaldehyde	10.4	10.3
Indene	12.2	12.1
Isobutyl alcohol	4.8(b)	4.8(a)
	14.2	14.1
Isobutylbenzene		7.1(b)
sec-Butyl-p-xylene ($C_{12}H_{18}$)	4.9(b)	4.9(a)
Isopropylbenzene	5.12	5.11
Maleic anhydride		9.3
Methyl benzoate	13.4	13.3
2-Methylbutane		4.3(b)
2-Methyl-1-butene	11.8	11.7
2-Methyl-2-butene	11.6	11.5
Methylbutenes, reaction mixtures from	11.9	
dehydrohalogenation of 2-chloro-2-methylbutane	11.10	
2-Methyl-3-butyn-2-ol	8.2	8.1
3-Methyl-2-cyclohexenone		4.2
3-Methyl-3-hydroxy-2-butanone	8.4	8.3
4-Methyl-2-pentanol	6.3	6.2
2-Methyl-1-pentene	6.13	6.12
2-Methyl-2-pentene	6.7	6.6
4-Methyl-1-pentene	6.5	6.4
cis-4-Methyl-2-pentene	6.11	6.10
trans-4-Methyl-2-pentene	6.9	6.8
2-Methylpropanal (isobutyraldehyde)	14.4	14.3
Nitrobenzene		17.7
1-Nitropropane	4.5	
t-Pentyl alcohol	11.14	11.13
1-Phenylethanol	17.3	17.2
Polystyrene		17.5
Propylene oxide	12.6	12.5
cis-Stilbene	15.15	15.13
trans-Stilbene	15.16	15.14
Styrene		17.4
Sulfanilamide	17.12	17.11
Sulfathiazole		17.15
3-Sulfolene (2,5-dihydrothiophene-1,1-dioxide)	9.6	9.5
Triphenylmethane	5.17	5.16
Triphenylmethanol	13.6	

index

Abbe refractometer, 17
Absorbance, definition of, 71, 131
Accidents, chemical burns, 5
 cuts, 5
 fire, 5
Acenaphthene, *357**
Acetamide, *366*
p-Acetamidobenzenesulfonyl chloride, Ammonolysis of, 286
 ammonolysis of, discussion, 280
 ir spectrum of, 290
 preparation of, 279, 285
 sulfathiazole from, 282, 288
Acetanilide, *366*
 bromination of, 132, 133, 134, 293, 297
 chlorosulfonation of, 279, 285
 ir and nmr spectra of, 290
 preparation of, 278, 285
Acetic acid, *361*
Acetic anhydride, as dehydrating agent, 311
 in acetylation, 269, 309
acetone, *356*
Acetonitrile, *366*
Acetophenone, 244, 246, *356*
 by acetylation, 266, 269
 reduction of, 271
m-Acetotoluide, *366*
Acetylation, 269
Acetyl chloride, hydrolysis of, 259
 in acetylation, 310
Acetylferrocene, acetylation of, discussion, 308
 ir and nmr spectra of, 313
 preparation of, 309
Acetylsalicylic acid, bromination of, 134
Acid anhydrides (*see* Carboxylic acid
  anhydrides)
Acid halides (*see* Carboxylic acid halides)
Acidity, of hydrocarbons, 137, 305
 of carboxylic acids, 251
Acrolein, *355*
Acyl halides (*see* Carboxylic acid halides)
Addition (*see* Alkenes, Alkynes)
Adipic acid, ir spectrum of, 221
 preparation of, 218

Alcohols, classification tests for, 339
 conversion to alkyl halides, discussion, 172
 dehydration of, 99, 103, 271
 derivatives of, 340
 from esters, 207
 in esterification, 252, 258
 oxidation of, 214, 217, 218
 table of, and derivatives, 359
Aldehydes, 1, 4-addition to, discussion, 245
 aldol reactions of, 243
 Cannizzaro reactions of, 223
 classification tests for, 334
 conversion to alkenes by Wittig reaction, 233,
  234
 derivatives of (*see also Specific type*), 335
 haloform reaction with, discussion, 242
 hemiacetal formation in, 320
 oxidation of, 216, 223
 preparation of, 215, 217
 with ammonia derivatives, 157, 226
 table of, and derivatives, 355
Aldol condensation, 246
Aldol reactions, theory of, 243
Alkenes, addition reactions, concerted, 114,
  144
 addition reactions, theory of, 111
 by dehydration, 99, 103, 271
 by dehydrohalogenation, 161, 163
 by Wittig reaction, 233, 234
 classification tests for, 335
 geometric isomerization of, 113, 114
 hydrogenation of, 114
 relative stabilities of, 100, 155
Alkyl halides, classification tests for, 336
 elimination of, 163
 from alcohols, 172, 174–176
 with Lewis acids, 121
 with sodium iodide, 336
 with sodium nitrate, 336
Alkynes, 136
 classification tests for, 335
 hydration of, 138, 139
Amides (*see* Carboxylic acid amides)
Amines, acetylation of, 285
 classification tests for, 345

* Numbers in italics refer to entries in tables of Chapter 20.

379

Amines (*cont.*)
   derivatives of, 347
   diazotization of, 298
   from imines, 231, 232
   from nitro compounds, 284
   preparation of, discussion, 227
   tables of, and derivatives, 362, 363
2-Aminopyridine, *362*
2-Aminothiazole, ir and nmr spectra of, 291, 292
   preparation of, 281, 287
   sulfathiazole from, 282, 288
n-Amyl acetate, *365*
Analysis, qualitative, classification tests, *see class of compound*
   derivatives in, *see specific derivative*
   discussion of, 329
   elementary analysis in, 330
   solubility tests in, 331
   spectroscopy in, 353
Anhydrides (*see* Carboxylic acid anhydrides)
Aniline, 232, *362*
   acetylation of, 278, 285
   ir and nmr spectra of, 262
   preparation of, 276, 284
   with benzoyl chloride, 259
o-Anisaldehyde, *355*
p-Anisaldehyde, *355*
Anisic acid, *361*
o-Anisidine, *362*
p-Anisidine, *362*
Anisole, bromination of, 132
Anthracene, in Diels-Alder reaction, 198
   in Diels-Alder reaction, discussion, 197
   ir spectrum of, 200
Anthranilic acid, 197, 198, *362*
Apparatus, lubrication of, 4, 163
Arenes (*see* Hydrocarbons, aromatic)
Aromaticity, discussion of, 304
Aryl halides, classification test for, 337
   derivatives of, 338
   table of, and derivatives, 358
Arynes, 196
Azeotropic distillation, drying of liquids by, 33
   drying of solids by, 368
   drying of styrene by, 272
Azeotropic mixture, maximum boiling, 28
   minimum-boiling, 27
Azobenzene, separation of isomers, 61, 63
Azulene, 304

Benedict's reagent, 335
Benzalacetophenone, *356*
   ir and nmr spectra of, 248, 249
   preparation of, 246
   reactions of, discussion, 245
   with aniline, 247
   with bromine, 247
Benzaldehyde, 232, 234, *355*
   imines from, 232
   in Cannizzaro reaction, 223
   ir and nmr spectra of, 225
Benzamide, *366*
   benzonitrile from, 260
   ir and nmr spectra of, 261, 262
   preparation of, 259
Benzanilide, 259, *366*
   ir and nmr spectra of, 263
Benzene, *357*
   acetylation of, 269
   alkylation of, 124

Benzene (*cont.*)
   bromination of, 90, 132
   drying of, 33, 124
   nitration of, 284
   nitration of, discussion, 276
   separation from toluene, 33
   use in removal of water, 228, 231
Benzenediazonium-2-carboxylate hydrochloride, benzyne from, 198
   preparation of, 198
Benzenesulfonyl chloride, with amines, 346
Benzilic acid, *361*
Benzoic acid, *361*
   acid chloride from, 258
   from acetophenone, 246
   from benzaldehyde, 223
   from benzoyl chloride, 258
   from ethylbenzene, 257
   from phenyl magnesium bromide, 208
   ir and nmr spectra of, 209, 210
Benzoic anhydride, ir spectrum of, 263
   preparation of, 255, 260
Benzoin, *356, 359*
Benzonitrile, *366*
   ir spectrum of, 264
   preparation of, 260
Benzophenone, *356*
p-Benzoquinone, ir spectrum of, 150
   as dienophile, 146
Benzoyl chloride, alcoholysis of, 259
   ammonolysis of, 259, 347
   benzoic anhydride from, 260
   hydrolysis of, 259
   preparation of, 255, 258
Benzoyl peroxide, as initiator, 83
   precautions for, 84
Benzyl acetate, *365*
Benzyl alcohol, 223, *359*
   ir spectrum of, 225
Benzylamine, *262*
N-Benzylanilinium chloride, 232
Benzyl chloride, 233
   ir spectrum of, 239
Benzyl cinnamate, *365*
N-Benzylideneaniline, 232
N-Benzylidene-*m*-nitroaniline, 232
   ir spectrum of, 238
Benzyl methyl ketone, *356*
N-Benzyl-*m*-nitroaniline, 232
   ir and nmr spectra of, 239
Benzyne, Preparation of, discussion, 197
   with anthracene, 198
Betaines, 229
Biphenyl, 205, *357*
Blocking group (*see* Protective group)
Boiling points, definition of, 15
   determination of, 16
   effect of pressure on, 14
   theory of, 14
Bromination, of acetanilide, 297
   free radical substitution, 90
   test for unsaturation, 103
   test for unsaturation, discussion, 102
Bromination, electrophilic, alkenes, theory of, 111
Bromination, electrophilic aromatic, rate expressions for, 130
   substituent effects on rate of, 129
Bromine, precautions for, 91
   test for, 331
   treatment of burns from, 5
Bromine water, with phenols, 342

4-Bromoacetanilide, *366*
  chlorination of, *297*
  ir and nmr spectra of, 300
  preparation of, 293, 297
p-Bromoacetophenone, *356*
p-Bromoaniline, *262*
p-Bromobenzaldehyde, *355*
o-Bromobenzamide, *366*
m-Bromobenzanilide, *366*
o-Bromobenzanilide, *366*
Bromobenzene, *358*
  Grignard reagent from, 206
  ir spectrum of, 211
o-Bromobenzoic acid, *361*
1-Bromo-3-chloro-5-Iodobenzene, ir and nmr
    spectra of, 303
  preparation of, 296, 298
  synthesis of, chart, 293
o-Bromonitrobenzene, *364*
p-Bromonitrobenzene, *364*
o-Bromophenol, *360*
p-Bromophenol, *360*
  bromination of, 132
o-Bromotoluene, *358*
Buffers, in reactions of carbonyl compounds,
    156, 157
Burns, 5
1, 3-Butadiene, preparation of, 147
  with maleic anhydride, 147
Butanone, *356*
n-Butyl acetate, *365*
t-Butyl acetate, *365*
n-Butyl alcohol, *359*
  bromide from, 175
  ir and nmr spectra of, 177
sec-Butyl alcohol, *359*
t-Butyl alcohol, *359*
  chloride from, 176
n-Butylamine, *362*
sec-Butylamine, *362*
n-Butylbenzene, 124
  ir spectrum of, 123
sec-Butylbenzene, 124
  ir spectrum of, 123
t-Butylbenzene, *357*
  bromination of, 90
  ir and nmr spectra of, 93, 123
Butylbenzenes, spectra of mixtures of, 126, 127
n-Butyl benzoate, *365*
n-Butyl bromide, preparation of, 172, 175
t-Butyl bromide, solvolysis of, 185
n-Butyl chloride, alkylation of benzene with,
    124
sec-Butyl chloride, alkylation of benzene with,
    125
t-Butyl chloride, preparation of, 174, 176
  solvolysis of, 183
Butyl chlorides, spectra of mixtures from alky-
    lation of benzene with, 126, 127
p-t-Butylphenol, *360*
n-Butyraldehyde, *355*
n-Butyranilide, *366*

d,l-Camphor, *356*
Cannizzaro reaction, 223
Carbazole, *362*
Carbenes, 189
Carbon, activated, as catalyst support, 114
  as decolorizing agent, 37
Carbon dioxide, 208
Carbohydrates, derivatives of, 326
  determination of specific rotation, 326

Carbonium ions, as intermediates in elimina-
    tion reactions, 98
  as intermediates in nucleophilic substitution,
    170
  in dehydration of alcohols, 99
  rearrangement of, 101, 121
  stability of, 100
Carbonyl compounds (*see specific class of com-
    pound*)
Carboxylic acid amides, classification test for,
    352
  dehydration of, 260
  derivatives of, 353
  discussion of, 256
  from carboxylic acid halides, 259
  from carboxylic acids, 344
  hydrolysis of, 353
  table of, 366
Carboxylic acid anhydrides, discussion of, 254
  in acylation, 269, 309
Carboxylic acid esters, by oxidation of alcohols,
    216
  classification tests for, 349
  derivatives of, 351
  discussion of, 252
  hydrolysis of, 253
  saponification equivalent, procedure, 350
  table of, 365
  with Grignard reagents, 207
Carboxylic acid halides, discussion of, 254
  from carboxylic acids, 258
  hydrolysis of, 259
  in acetylation, 310
Carboxylic acids, acidity of, 258
  acidity of, discussion, 251
  classification test for, 344
  as derivatives, 339
  discussion of, 250
  derivatives of, 344
  equivalent weight, procedure, 344
  esterification, 258
  esterification of, discussion, 252
  from aldehydes, 216
  from arenes, 251, 257
  from carboxylic acid halides, 259
  from esters, 253
  from Grignard reagent, 208
  table of, and derivatives, 361
Ceric nitrate, with alcohols, 340
  with phenols, 343
Chemical shift, 74
  effect of ring current on, 307
  table, 75
Chloral, *355*
Chlorination, of 4-bromoacetanilide, 297
Chlorination, free radical, procedures for, 84
  theory of, 82
Chlorine, preparation of, 294, 297
  test for, 331
Chloroacetaldehyde, preparation of, 287
  with thiourea, 282, 287
p-Chloroacetanilide, *366*
Chloroacetic acid, *361*
Chloroacetone, *356*
Chloroacetonitrile, *366*
m-Chloroaniline, *362*
p-Chloroaniline, *362*
o-Chlorobenzaldehyde, *355*
p-Chlorobenzaldehyde, *355*
m-Chlorobenzamide, *366*
p-Chlorobenzanilide, *366*

Chlorobenzene, *358*
  bromination of, 132
m-Chlorobenzoic acid, *361*
o-Chlorobenzoic acid, *361*
p-Chlorobenzoic acid, *361*
m-Chlorobenzonitrile, *366*
o-Chlorobenzonitrile, *366*
p-Chlorobenzonitrile, *366*
2-Chloro-4-bromoacetanilide, iodination of, 298
  ir and nmr spectra of, 301
  preparation of, 294, 298
2-Chloro-4-bromo-6-iodoaniline, diazotization of, 296, 298
  ir and nmr spectra of, 302
1-Chlorobutane, chlorination of, 86
  ir and nmr spectra of, 88, 89
Chlorocyclohexane, ir spectrum of, 89
  preparation of, 84
2-Chloro-2-methylbutane, dehydrorehalogenation of, 162, 163
  ir and nmr spectra of, 167, 168
  preparation of, 176
2-Chloronaphthalene, *358*
  formation of, 192
  ir spectrum of, 196
o-Chloronitrobenzene, *364*
p-Chloronitrobenzene, *364*
o-Chlorophenol, *360*
p-Chlorophenol, *360*
α-Chloropropionic acid, *361*
Chlorosulfonation, 285
Chlorosulfonic acid, 285
m-Chlorotoluene, *358*
o-Chlorotoluene, *358*
p-Chlorotoluene, *358*
Cholesterol, *359*
Chromatography, gas-liquid (*see* Gas-liquid chromatography)
Chromatography, liquid-liquid (*see* Extraction *and* Paper chromatography)
Chromatography, liquid-solid (*see* Column chromatography *and* Thin-layer chromatography)
Chromic acid, as classification reagent, 222, 249
  as oxidant, discussion, 215
  as oxidant for side-chains, 339
  preparation of, 217, 222
  with isobutyl alcohol, 217
Cinnamaldehyde, *355*
  imines from, 231, 232
  in Wittig reaction, 234
  ir and nmr spectra of, 236
Cinnamanilide, *366*
Cinnamic acid, *361*
trans-Cinnamic acid, determination of phase diagram with urea, 14
Cinnamyl alcohol, *359*
N-Cinnamylaniline, 232
  ir and nmr spectra of, 238
N-Cinnamylideneaniline, 232
  ir spectrum of, 237
N-Cinnamylidene-*m*-nitroaniline, 231
  ir spectrum of, 236
Citronellal, *355*
Classification tests (*see* Analysis)
2,4,6-Collidine, *363*
Column chromatography, adsorbents for, 59
  application of, 61, 309
  methods of detection in, 58
  preparation of column, 61

Column chromatography (*cont.*)
  solvents for, 59
  theory of, 57
Coupling constants, 76
m-Cresol, *360*
o-Cresol, *360*
p-Cresol, *360*
m-Cresyl acetate, *365*
o-Cresyl acetate, *365*
p-Cresyl acetate, *365*
p-Cresyl benzoate, *365*
Crotonaldehyde, *355*
1,3,5-Cycloheptatriene, 311
  ir and nmr spectra of, 314
Cycloheptatrienyl cation (*see* Tropilium ion)
Cyclohexane, bromination of, 90
  chlorination of, 84
  ir spectrum of, 89
cis-Cyclohexane-1,2-dicarboxylic acid, ir spectrum of, 118
  preparation of, 114
Cyclohexanol, *359*
  Dehydration of, 103
  ir and nmr spectra of, 110
Cyclohexanone, 157, 246, *356*
  ir and nmr spectra of, 160
  oxidation of, 218
  preparation of, 218
Cyclohexanone semicarbazone, formation of, 157
Cyclohexene, ir and nmr spectra of, 110, 111
  preparation of, 103
3-Cyclohexenecarboxaldehyde, *355*
4-Cyclohexene-*cis*-1,2-dicarboxylic acid, hydrogenation of, 114
  ir spectrum of, 118
  preparation of, 147
4-Cyclohexene-*cis*-1,2-dicarboxylic anhydride, hydrolysis of, 147
  ir and nmr spectra of, 151
  preparation of, 147
Cyclohexyl acetate, *365*
Cyclohexylamine, *362*
Cyclohexylbenzene, *357*
Cyclohexylcarboxylic acid, *361*
Cyclopentadiene, acidity of, 305
  ir and nmr spectra of, 149
  preparation of, 145
  with *p*-benzoquinone, 146
  with maleic anhydride, 145
Cyclopentanol, *359*
Cyclopentanone, *356*

D, 319
Dalton's Law, 14, 22, 29
Dehydrohalogenation (*see* Elimination reactions)
Delta (δ), in nmr, 74
Density, determination of, 18
Derivatives (*see* specific type and Table of)
Desiccants (*see* Drying, agents for)
Desoxybenzoin, *356*
1,1'-Diacetylferrocene, detection of, 310
  structure of, 308
Diastereomers, definition of, 316
  of glucose, 320
Diazotization, 346
  discussion of, 296
  procedure for, 298
Dibenzyl, *357*
Dibenzylamine, *362*
Dibenzyl ketone, *356*

p-Dibromobenzene, *358*
Dichloroacetanilide, *366*
Dichloroacetic acid, *361*
2,5-Dichloroaniline, *362*
o-Dichlorobenzene, *358*
p-Dichlorobenzene, *358*
Dichlorocarbene, from sodium
    trichloroacetate, 192
3,5-Dichlorophenol, *360*
Dicyclopentadiene, cracking of, 145
Diels-Alder reaction, benzyne as dienophile,
    198
  benzyne as dienophile, discussion, 197
  theory of, 144
Dienes, in Diels-Alder reaction, 143
N,N-Diethylaniline, *363*
m-Diethylbenzene, *357*
2,3-Dihydroxyphenol, *360*
N,N-Dimethylaniline, *363*
2,3-Dimethylbutane, ir spectrum of, 70
Dimethyl fumarate, from dimethyl maleate,
    114
  ir and nmr spectra of, 117
2,6-Dimethyl-4-heptanol, *359*
2,6-Dimethyl-4-heptanone, *356*
Dimethyl maleate, ir spectrum of, 117
  isomerization of, 114
2,4-Dimethyl-3-pentanone, *356*
2,5-Dimethylphenol, *360*
3,5-Dimethylphenol, *360*
m-Dinitrobenzene, *364*
3,5-Dinitrobenzoyl chloride, with alcohols, 341
2,4-Dinitrochlorobenzene, *364*
2,4-Dinitrophenylhydrazine, as classification
    reagent, 249
2,4-Dinitrophenylhydrazones, as derivatives of
    aldehydes and ketones, 228
  preparation of, 233
Diphenylamine, *362*
1,4-Diphenyl-1,3-butadiene, preparation of,
    234
  ir and nmr spectra of, 241
Diphenyl ether, bromination of, 132, 133, 134
Diphenylmethane, *357*
Di-N-propylamine, *362*
Dissymmetry, 316
Distillation, precautions for, 3, 33
  see also Azeotropic distillation
Distillation, fractional, apparatus for, 25
  column packing, 25, 33
  procedure for, 33
  theory of, 22
Distillation, simple, apparatus for, 21
  procedure for, 33
  theory of, 20
Distillation, steam, apparatus for, 32
  application of, 192
  limitations of, 32
  theory of, 29
  utility of, 30
Distillation, vacuum, discussion, 28
  application of, 272
Dry Ice, 208
Drying, agents for, discussion, 367
  agents for, table, 369
  of solids, 39, 286
  see also Azeotropic distillation

Electrophilic aromatic substitution (*see* Substi-
    tution, electrophilic aromatic)
Elementary analysis (*see* Analysis)

Elimination reactions, in formation of arynes,
    196
  in formation of carbenes, 190
  competition with substitution, 99
  theory of, 98
Elimination reactions, base-induced, 163
  competition with nucleophilic substitution,
    162, 171
  theory of, 161
Emulsions, remedy for, 50
Enantiomers, definition of, 315
Equilibrium control, theory of, 152
Equivalent weight, 252
  procedure for, 344
Esters (*see* Carboxylic acid esters)
Ethanol, *359*
Ethyl acetate, *365*
N-Ethylaniline, *362*
o-Ethylaniline, *362*
p-Ethylaniline, *362*
Ethylbenzene, *357*
  bromination of, 90
  ir and nmr spectra of, 91, 92
  oxidation of, 257
Ethyl benzoate, *365*
Ethylene glycol, *359*
Ethyl m-nitrobenzoate, *365*
Ethyl p-nitrobenzoate, *365*
p-Ethylphenol, *360*
Ethyl phenylacetate, *365*
Ethyl salicylate, *365*
Extraction, continuous, liquid-liquid, 46
  continuous, solid-liquid, 47
  effect of pH on, 48
  selection of solvent for, 48
  technique of, 48
  theory of, 44

Fehling's reagent, 335
Ferric chloride, with esters, 350
  with phenols, 343
Ferrous hydroxide, with nitro compounds, 348
Ferrocene, acetylation of, 309
  ir and nmr spectra of, 312, 313
  structure and properties of, 307
Filter-aid, 38, 40, 219
Filtration, vacuum, apparatus for, 39
Fire, 3, 5
First-aid, 5
Fluoboric acid, 311
Fluorene, *357*
Fractional distillation (*see* Distillation,
    fractional)
Free radical reactions, 82
  selectivity in, 90
Friedel-Crafts acylation, 309
  discussion of, 266, 308
Friedel-Crafts alkylation, kinetic and
    equilibrium control in, 154
  procedure for, 124
  theory of, 120
Fructosazone, preparation of, 326
D-Fructose, determination of specific rotation,
    326
  in sucrose, 323
2-Furaldehyde, 157, 234
  ir and nmr spectra of, 159
2-Furaldehyde semicarbazone, formation of,
    157
Furan, 305
Furfural, *355*
Furfuryl alcohol, *359*

Gas-liquid chromatography, apparatus for, 52
  application of, 34, 85, 103, 125, 164, 224
  quantitative analysis in, 57
  preparative uses of, 57
  retention time, factors affecting, 53, 55
  solid supports in, table, 54
  stationary phases for, table, 54
  theory of, 52
Glassware, care and cleaning of, 4
  lubrication of, 4, 163
Glucosazone, preparation of, 326
  structure of, 324
α-D-Glucose, in sucrose, 323
  mutarotation of, 321
  mutarotation of, discussion, 319
  preparation of, 321
D-Glucose, determination of specific rotation, 326
Glycolic acid, *361*
Glyoxal, *355*
Grignard reagent, carbonation of, 208
  carbonation of, discussion, 205
  discussion of, 202
  preparation of, 206
  with esters, 207
  with esters, discussion, 204
  side reactions in preparation of, 203
  solvents for, 202

Haloform reaction, procedure for, 246
  theory of, 242
Halogens, test for, 331
Halogenation (*see* Chlorination, Bromination)
Hazards (*see* Safety)
Hemiacetal, glucose as, 320
Hemiketal, fructose as, 324
Heptanal, *355*
n-Heptane, chlorination of, 86
1-Heptanol, *359*
2-Heptanone, *356*
3-Heptanone, *356*
4-Heptanone, *356*
n-Heptyl acetate, *365*
1-Hexadecanol, *359*
Hexanal, *355*
1-Hexanol, *359*
2-Hexanone, *356*
n-Hexylamine, *362*
Hinsberg test, 346
Huckel's rule, 306, 307, 308
Hydrobromic acid, preparation of, 173, 175
Hydrocarbons, acidity of, 137
  aromatic, classification test for, 337
  aromatic, side-chain oxidation of, discussion, 251
  chlorination of, 82, 84
  derivatives of, 338
  see also Substitution, electrophilic aromatic
  table of, and derivatives, 357
Hydrocinnamic acid, *361*
Hydrogen, generation and use of, 115
Hydroxamic acids, from amides, 352
  from esters, 350
  from nitriles, 352
p-Hydroxybenzaldehyde, *360*
p-Hydroxybenzoic acid, *361*
p-Hydroxybenzophenone, *356*
Hydroxylamine hydrochloride, with esters, 350
  with nitriles, 352
  with amides, 352
4-Hydroxy-2-pentanone, *356*

m-Hydroxyphenol, *360*
o-Hydroxyphenol, *360*

Ideal solution, definition of, 22
Ignition test, 330
Imines, formation of, 231, 232
  formation of, discussion, 226
Indene, ir and nmr spectra of, 195, 196
  with dichlorocarbene, 191, 192
Infrared spectroscopy, analysis of unknowns by, 353
  application of, 125, 126, 140
  functional group frequencies, table, 69
  identification using, 70
  practical considerations in, 71
  theory of, 67
  with nmr spectroscopy, 77
Inverse addition, 206
Iodination, of 2-chloro-4-bromoaniline, 298
Iodine, test for, 331
Iodoform, 246
Iodoform test, application of, 310
  procedure for, 246
  theory of, 243
Isoamyl nitrite, 198
Isobutyl alcohol, *359*
Isobutylbenzene, 124
  ir spectrum of, 123
Isobutyl chloride, alkylation of benzene with, 125
Isobutyl alcohol, ir and nmr spectra of, 78, 220
  oxidation of, 217
Isobutyl isobutyrate, 217
Isobutyraldehyde, *355*
Isopropyl acetate, *365*
Isopropyl alcohol, 246, *359*
Isopropylamine, *362*
Isopropylbenzene, *357*
  bromination of, 90
  ir and nmr spectra of, 92
Isopropyl benzoate, *365*
Isoquinoline, *363*

Ketones, 1,4-addition to, 247
  1,4-addition to, discussion, 245
  aldol reactions of, 244
  classification tests for, 334
  conversion to alkenes by Wittig reaction, 229
  derivatives of (*see also specific type*), 335
  haloform reaction with, 242, 246
  hemiketal formation in, 324
  oxidation of, 216, 218, 222
  preparation of, discussion, 215
  reduction of, 271
  table of, and derivatives, 356
  with ammonia derivatives, 157, 226
Kinetic control, theory of, 153
Kinetics, first-order rate expression for, 320
  first-order rate expression for, 130
  for electrophilic aromatic substitution, 119, 133, 134
  for mutarotation, 320, 321
  for nucleophilic substitution, discussion, 178
  least-squares treatment in, 180
  for solvolysis, 181, 183

Lambert-Beer law, 132
Limiting reagent, considerations in synthesis, 282

Limiting reagent (*cont.*)
  definition of, 372, 374
Lubrication, 4, 163
Lucas test, 340
2,4-Lutidine, *363*

Maleic anhydride, as dienophile, 146, 147, 199
  ir spectrum of, 150
Markownikoff's rule, 111, 136, 138
Melting points, depression of, 9
  effect of heating rate on, 13
  effect of impurities on, 8
  mixture, 10
  phase diagram, 9
  theory of determination of, 7
Menthol, *359*
L-Menthone, *356*
Meso isomer, example of, 317
Methanol, *359*
p-Methoxyacetanilide, *366*
p-Methoxyacetophenone, *356*
o-Methoxybenzoic acid, *361*
p-Methoxybenzonitrile, *366*
2-Methoxyphenol, *360*
1-Methoxy-2-propanol, *359*
N-Methylacetanilide, *366*
Methyl acetate, *365*
p-Methylacetophenone, *356*
N-Methylaniline, *362*
Methyl benzoate, from benzoic acid, 258
  from benzoyl chloride, 259
  ir and nmr spectra of, 210
  with phenyl magnesium bromide, 207
2-Methylbutanal, *355*
2-Methylbutane, ir spectrum of, 70
3-Methyl-1-butanol, *359*
3-Methyl-2-butanone, *356*
2-Methyl-1-butene, by dehydrohalogenation, 163
  ir and nmr spectra of, 169
2-Methyl-2-butene, by dehydrohalogenation, 163
  ir and nmr spectra of, 168
2-Methyl-3-butyne-2-ol, acetylide from, 139
  hydration of, 139
  hydration of, discussion, 138
  ir and nmr spectra of, 141
Methyl *m*-chlorobenzoate, *365*
Methyl *p*-chlorobenzoate, *365*
Methyl cinnamate, *365*
Methylcyclohexane, bromination of, 91
3-Methyl-2-cyclohexenone, ir spectrum of, 68
N-Methylcyclohexylamine, *362*
2-Methyl-4-ethylpyridine, *363*
5-Methyl-3-heptanone, *356*
6-Methyl-3-heptanone, *356*
5-Methyl-2-hexanone, *356*
3-Methyl-3-hydroxy-2-butanone, ir and nmr spectra of, 142
  preparation of, 139
Methyl iodide, with amines, 348
Methyl *m*-nitrobenzoate, *365*
2-Methyl-1-pentanol, *359*
4-Methyl-2-pentanol, dehydration of, 103
  ir and nmr spectra of, 106
  products from dehydration of, 101
4-Methyl-2-pentanone, *356*
2-Methyl-1-pentene, ir and nmr spectra of, 109
2-Methyl-2-pentene, ir and nmr spectra of, 107
4-Methyl-1-pentene, ir and nmr spectra of, 106, 107

cis-4-Methyl-2-pentene, ir and nmr spectra of, 108, 109
trans-4-Methyl-2-pentene, ir and nmr spectra of, 108
4-Methyl-3-penten-2-one, *356*
Methylphenylcarbinol, *359*
2-Methylpropanal, ir and nmr spectra of, 220, 221
  preparation of, 217
Methyl *o*-toluate, *365*
Methyl *p*-toluate, *365*
Mixtures, separation of, in analysis, 328
Mole fraction, definition of, 22
Morpholine, *362*
Mutarotation, discussion of, 319
  kinetics of, 321

Naphthalene, *357*
α-Naphthol, *360*
  bromination of, 132
β-Naphthol, *360*
α-Naphthyl acetate, *365*
β-Naphthyl acetate, *365*
α-Naphthylamine, *362*
α-Naphthyl isocyanate, with alcohols, 341
  with phenols, 341
Neopentyl alcohol, *359*
Neutralization equivalent, discussion of, 252
  procedure for, 344
Nitration, in preparation of derivatives, 338
  procedure for, 284
  theory of, 276
Nitriles, classification test for, 351
  derivatives of, 352
  discussion of, 256
  from carboxylic acid amides, 260
  hydrolysis of, 352
  table of, 366
Nitro compounds, as derivatives, 338
  classification test for, 348
  derivatives of, 349
  table of, and derivatives, 364
m-Nitroacetanilide, *366*
o-Nitroacetanilide, *366*
m-Nitroaniline, 231, 232, *362*
o-Nitroaniline, *362*
p-Nitroaniline, *362*
o-Nitroanisole, *364*
p-Nitroanisole, *364*
p-Nitrobenzaldehyde, *355*
m-Nitrobenzamide, *366*
p-Nitrobenzamide, *366*
m-Nitrobenzanilide, *366*
Nitrobenzene, *364*
  bromination of, 132
  ir spectrum of, 289
  preparation of, 276, 284
  reduction of, 276, 284
m-Nitrobenzoic acid, *361*
p-Nitrobenzoic acid, *361*
m-Nitrobenzonitrile, *366*
o-Nitrobenzonitrile, *366*
p-Nitrobenzonitrile, *366*
4-Nitrobiphenyl, *364*
Nitrogen, test for, 331
p-Nitro-N-methylaniline, *362*
β-Nitronaphthalene, *364*
m-Nitrophenol, *360*
o-Nitrophenol, *360*
p-Nitrophenol, *360*
  bromination of, 133

1-Nitropropane, nmr spectrum of, 74
m-Nitrotoluene, *364*
o-Nitrotoluene, *364*
Nitrous acid, with amines, test, 346
Non-ideal solution, definition of, 26
endo-Norbornene-*cis*-5,6-dicarboxylic acid,
　　preparation of, 146
endo-norbornene-*cis*-5,6-dicarboxylic
　　anhydride, hydrolysis of, 146
　　preparation of, 145
Normal addition, 206
Notebook, format for, 372, 373
　　importance of, 2
Nuclear magnetic resonance spectroscopy,
　　analysis of unknowns by, 353
　　application of, 126, 140, 164
　　chemical shift in, 74
　　chemical shift, table, 75
　　coupling constants in, 76
　　criterion for aromaticity, 307
　　identification using, 77
　　integration in, 77
　　practical considerations in, 79
　　spin-spin splitting in, 76
　　theory of, 72
　　with ir spectroscopy, 77

1-Octadecanol, *359*
2-Octanol, *359*
2-Octanone, *356*
Optical activity, 317
Optical isomerism, 315
Optical rotation, factors affecting, 318
　　measurement of, 321
　　see also Specific rotation
Organolithium reagents, 211
Osazones, preparation of, 324, 326
Oxidation, aromatic side-chain, 257
　　of alcohols, 217, 218
　　of aldehydes, 221, 223
　　of alkenes, 102
　　of ketones, 216, 218, 222
　　of methyl ketones, 246
　　of side-chains, in preparation of derivatives,
　　339
Oximes, as derivatives of aldehydes and
　　ketones, 228
　　preparation of, 233

Paper chromatography, 65
Pentanal, *355*
2,4-Pentanedione, *356*
1-Pentanol, *359*
2-Pentanol, *359*
3-Pentanol, *359*
2-Pentanone, *356*
3-Pentanone, *356*
t-Pentyl alcohol, *359*
　　chloride from, 176
　　ir and nmr spectra of, 178
t-Pentyl chloride, preparation of, 176
　　see also 2-Chloro-2-methylbutane
　　solvolysis of, 183
Petroleum ether, 39
Phase diagram, determination of using urea
　　and *trans*-cinnamic acid, 14
　　liquid-vapor, azeotropic mixture, 27, 28
　　liquid-vapor, typical, 23
　　solid-liquid, examples of, 9
Phase distribution, principle of, 44
β-Phenethyl alcohol, *359*

Phenol, *360*
　　bromination of, 132
　　esterification with, 259
Phenols, bromine water with, 342
　　classification tests for, 342
　　derivatives of, 343
　　table of, and derivatives, 360
Phenylacetaldehyde, *355*
Phenyl acetate, *365*
Phenylacetic acid, *361*
Phenylacetonitrile, *366*
Phenyl benzoate, 259, *365*
1-Phenylethanol, dehydration of, 271
　　from acetophenone, 271, 367
　　ir and nmr spectra of, 273, 274
Phenylhydrazine, 326
Phenylhydrazones, as derivatives of aldehydes
　　and ketones, 228
Phenyl isocyanate, with alcohols, 341
　　with phenols, 341
Phenyl magnesium bromide, carbonation of,
　　208
　　carbonation of, discussion, 205
　　preparation of, 206
　　with methyl benzoate, 207
N-Phenyl-*α*-naphthylamine, *362*
N-Phenyl-*β*-naphthylamine, *362*
β-Phenylpropionic acid, *361*
Phenyl salicylate, *365*
Phthalic acid, *361*
　　recrystallization of, 40
2-Picoline, *363*
3-Picoline, *363*
4-Picoline, *363*
Picric acid, with amines, 348
Picryl chloride, *364*
Pinacolone, 246, *356*
Piperidine, *362*
Pivalic acid, *361*
pK_R^+, 307
Polarimeter, discussion of, 318
　　use of, 321, 326
Polarizability, 172
Polarized light, 317
Polymerization, 268, 272
Polystyrene, from styrene, 268, 272
　　ir spectrum of, 274
　　synthesis of, chart, 266
Potassium *t*-butoxide, 166
Potassium permanganate, as oxidant, 218, 257,
　　339
　　test for presence of, 219
　　test for unsaturation, 102, 103
Propionaldehyde, *355*
Propionamide, *366*
Propionic acid, *361*
Propionitrile, *366*
Propiophenone, *356*
N-*n*-Propylacetanilide, *366*
n-Propyl alcohol, *359*
n-Propylamine, *362*
n-Propyl benzoate, *365*
Propylene oxide, as scavenger for hydrochloric
　　acid, 198
　　ir and nmr spectra of, 200, 201
Protective group, use in synthesis, 281, 292,
　　293
Pyridine, *363*
　　in anhydride formation, 255
Pyrrole, 305

Quinoline, *363*

Racemic modification, 317
Raoult's law, deviations from, 27
  liquid-vapor equilibria, 22
  solid-liquid equilibria, 8
Rate constant, definition of, 178
  determination of by least-squares method,
    180
  graphical determination of, 133, 179
  quantitative determination of, 134, 321
  semiquantitative determination of, 133
Recrystallization, discussion of, 35
  mixed solvents in, 36
  oil formation in, 38
  procedure for, 36
  selection of solvent for, 36
  vacuum filtration in, 39
Reduction, of alkenes, 114
  of diazonium group, 296, 298
  of imines, 227, 231
  of ketones, 267, 271
  of nitro group, 276, 284, 349
Reflux ratio, 26
Refraction, index of, 17
R_f values, definition of, 62
  use of, 64, 65, 310
Ring current, 307

Safety, chemical precautions, 3
  explosion precautions, 3, 33
  fire precautions, 3
Salicylaldehyde, *355*
Salicylamide, *366*
Salicylic acid, *361*
Saponification equivalent, discussion of, 253
  procedure for, 350
Semicarbazide hydrochloride, 157
Semicarbazones, as derivatives of aldehydes
    and ketones, 228
  dependence of formation on pH, 156
  preparation of, 140, 157, 232
Shaw polarimeter (*see* Polarimeter)
Silver acetylide, formation of, 139
Silver nitrate, with alkyl halides, 336
Simple distillation (see Distillation, simple)
Sodium bisulfite addition product, 217
Sodium borohydride, as reductant, 227, 231,
    271
  in hydrogenation of alkenes, 115
Sodium ethoxide, preparation of, 166
Sodium fusion, 330
Sodium hypochlorite, 246
Sodium hypoiodite, 246
  as classification reagent, 249
Sodium iodide, with alkyl halides, 336
Sodium isopropoxide, preparation of, 166
Sodium methoxide, 233
  preparation of, 164
Sodium trichloroacetate, preparation of, 193
  thermal decomposition of, 192
Solubility, determination of, 39
  principles of, 36
Solubility tests (*see* Analysis)
Solution, ideal, definition of, 22
  non-ideal, definition of, 26
Solvolysis, discussion of, 180
  factors affecting rate of, 181
  kinetics of, 183
Soxhlet extraction apparatus, 47

Specific rotation, definition of, 318
  measurement of, 326
Spectroscopy (*see* Infrared spectroscopy *and*
    Nuclear magnetic resonance
    spectroscopy)
Spin-spin splitting, 76
Steam, generation of, 31
Steam distillation (*see* Distillation, steam)
Stereoisomerism, discussion of, 315
cis-Stilbene, ir and nmr spectra of, 240
trans-Stilbene, ir and nmr spectra of, 240, 241
  preparation of, 233
Styrene, ir spectrum of, 274
  from 1-phenylethanol, 271
  from 1-phenylethanol, discussion, 268
  polymerization of, 272
  polymerization of, discussion, 268
2-Styrylfuran, preparation of, 234
Sublimation, apparatus for, 42
  limitations of, 43
  theory of, 41
Substitution, electrophilic aromatic, acetyla-
    tion, 269, 309
  discussion of, 119, 276
  examples of, table, 120
  halogenation, 297
  kinetics of, 119
  relative rates of, discussion, 128
  relative rates of, table, 128
  substituent effects on, discussion, 129
  substituent effects on, table, 129
  sulfonation, 285
Substitution, nucleophilic, competition with
    elimination, 99, 162, 171
  factors affecting rate of, 181
  kinetics of, 178, 183
  side reactions of, 173
  theory of, 170
Sucrose, determination of specific rotation, 326
  hydrolysis of, 326
  hydrolysis of, discussion, 323
Sulfanilamide, ir and nmr spectra of, 291
  preparation of, 286
  preparation of, discussion, 280
Sulfathiazole, ir spectrum of, 292
  preparation of, 282, 288
  synthesis of, chart, 275
3-Sulfolene, as source of 1,3-butadiene, 147
  ir and nmr spectra of, 150, 151
Sulfur, test for, 331
Sulfuryl chloride, 84, 85

Table of derivatives, acids, 361
  alcohols, 359
  aldehydes, 355
  amides, 366
  amines, primary and secondary, 362
  amines, tertiary, 363
  aromatic hydrocarbons, 357
  aromatic nitro compounds, 364
  aryl halides, 358
  carboxylic acid esters, 365
  ketones, 356
  nitriles, 366
  phenols, 360
meso-Tartaric acid, *361*
Tau (τ), in nmr, 75
1-Tetradecanol, *359*
Tetrahydroisoquinoline, *362*
Tetralin, *357*
Tetramethylsilane, as nmr reference, 74

Theoretical yield, definition of, 372, 374
Thermodynamic control, theory of, 152
Thermometer, calibration of, 12
Thiele tube, 12, 16
Thin-layer chromatography, adsorbents for, 62
    application of, 63, 310
    methods of detection in, 63
    R_f values in, 62, 63
    theory of, 62
Thionyl chloride, 258, 260
    discussion of use, 255, 256
Thiophene, 305
Thiourea, with chloroacetaldehyde, 282, 287
Tollens' test, discussion, 334
    procedure for, 335
o-Tolualdehyde, *355*
p-Tolualdehyde, *355*
o-Toluamide, *366*
p-Toluamide, *366*
m-Toluanilide, *366*
o-Toluanilide, *366*
p-Toluanilide, *366*
Toluene, *357*
    bromination of, 90
    drying of, 33
    relative rates of substitution on, table, 128
    separation from benzene, 33
m-Toluic acid, *361*
o-Toluic acid, *361*
p-Toluic acid, *361*
m-Toluidine, *362*
o-Toluidine, *362*
p-Toluidine, *362*
m-Tolunitrile, *366*
o-Tolunitrile, *366*
p-Tolunitrile, *366*
Transmittance, percent, definition of, 68
Trap, cold, 164
    gas, 85, 286
    water from steam line, 32
Tri-*n*-butylamine, *363*
Trichloroacetanilide, *366*
Trichloroacetic acid, *361*
Triethylamine, *363*
Triethylphosphite, 230, 233
Trimethylacetaldehyde, *355*

3,5,5-Trimethylcyclohexene-3-one, *356*
2,4,6-Trimethylphenol, *360*
Triphenylmethane, *357*
    bromination of, 95
    ir and nmr spectra of, 96, 97
Triphenylmethanol, 311
    nmr spectrum of, 211
    preparation of, 204, 207
Triphenylmethyl bromide, preparation of, 95
Triphenylmethyl fluoborate, preparation of, 311
Triphenylmethyl peroxide, preparation of, 95
Tri-*n*-propylamine, *363*
Triptycene, 197, 198
Tropylium fluoborate, preparation of, 311
Tropylium iodide, preparation of, 311
Tropylium ion, 308

Ultraviolet spectroscopy, application of, 231
2-Undecanone, *356*
Unsaturation, qualitative tests for, 102, 103, 137
Urea, determination of phase diagram with *trans*-cinnamic acid, 14

Vacuum distillation (*see* Distillation, vacuum)
Vanillin, *355*
Vapor pressure, dependence on temperature, 20
    effect of non-volatile impurities on, 20
    of solids, 43
    relationship to boiling points, 14
    relationship to melting points, 8
Visible spectroscopy, 131
    application of, 134

Water, removal of by azeotropic distillation, 228, 231
Wittig reaction, phosphonate ester modification of, discussion, 229
    procedures for, 233, 234
    theory of, 229
m-Xylene, *357*
o-Xylene, *357*
p-Xylene, *357*

Ylides, 229